The classification of stars

The classification of stars

CARLOS JASCHEK & MERCEDES JASCHEK
Centre de Données Stellaires, Observatoire de Strasbourg, France

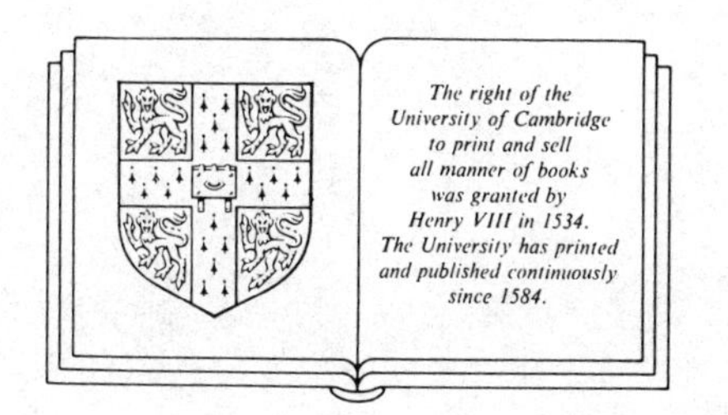

CAMBRIDGE UNIVERSITY PRESS
Cambridge
New York   Port Chester   Melbourne   Sydney

Published by the Press Syndicate of the University of Cambridge
The Pitt Building, Trumpington Street, Cambridge CB2 1RP
40 West 20th Street, New York, NY 10011, USA
10 Stamford Road, Oakleigh, Melbourne 3166, Australia

First published 1987
Reprinted with corrections 1990

Printed in Great Britain by the University Press, Cambridge

British Library cataloguing in publication data

Jaschek, C.
The classification of stars.
1. Stars - Classification
I. Title II. Jaschek, Mercedes
523.8 QB881

Library of Congress cataloguing in publication data

Jaschek, Carlos.
The classification of stars.

Includes index.
1. Stars – Classification. I. Jaschek, Mercedes,
1926– . II. Title.
QB881.J37 1987 523.8′012 86–24929

ISBN 0 521 26773 0

TM

Contents

Preface

The purpose of this book is to provide an outline of the methods and the results of classification. Classification is one of the fundamental branches of astronomy, in the same way as astrometry is. Classification is used whenever we need to order a large number of objects, or a variety of phenomena. Classification methods appeared in astronomy in the nineteenth century when a large number of stellar spectra suddenly became available, through two technical developments – the objective prism and the photographic plate.

Because the number of spectra was so large, it was impractical to study each of them in detail and later to generalize the similarities found between them. In fact, classification takes just the opposite approach: first groups of similar members are formed from the N individuals, and then typical members of each group are studied in detail. Since the members of a group are similar, the conclusions of the detailed study of typical objects can then be applied to all group members. By replacing a large number of individual stars by a much smaller number of groups and typical objects, a significant economy of work is achieved.

Classification methods use observational data, like objective prism spectra, with minimal interpretation. This stems from the fact that large amounts of data are used, let us say several thousand spectra; if the data were interpreted in a elaborate way, one would be performing analysis and not classification.

Classification methods thus precede further analysis; they are not intended to substitute it, but they constitute a necessary first step. On the other hand, since classification is closely based on observational data, its results are not likely to be changed by more elaborate interpretations; it provides the basic framework for further work. Interpretations of the observations change rapidly, however, as progress is made in the mathematical and physical tools used. To illustrate these points, let us quote a few examples.

Astronomers of the Harvard Observatory showed that the spectra of the 2×10^5 brighter stars ($V < 9$) can be put together into a few groups, and then arranged into a unidimensional list. This list incorporates 95% of the spectra in a sequence symbolized by the letters O, B, A, F, G, K and M. The sequence was established without any theoretical guidance, and it was only years later, through the work of Saha and followers, that the sequence was identified as a temperature sequence. It took the next fifty years to calibrate the temperatures through stellar atmosphere analysis, and many different effective temperatures were assigned over the years to stars labelled A1, for instance. On the other hand, the classifications of the stars remained the same – an A1 star is still called A1. Classification provided a conveniently stable reference frame for interpretation and explanation.

Another example is the analysis of the chemical composition of stars. This was started in the 1930s with the analysis of the sun, followed by the analysis of the first stars by Unsöld and his group. In view of the large amount of work involved in each analysis, it is clear that nobody would have analysed objects taken at random. All objects were therefore chosen from among the typical objects classifiers had previously selected.

Stellar atmosphere analysis kept in close touch with classification and was able to provide in the next decades a rapidly growing view of stellar constitution and evolution. In retrospect, we must be grateful to the classifiers who were able to provide such a vast array of groups of both normal and peculiar spectra, independently of any theory.

Another aspect of classification work appeared in the 1940s through the addition of luminosity classes to spectral types. Since luminosity classes are derived from observed line ratios, they maintain a close link with observation, but still allow freedom in their interpretation, explanation and calibration. It suffices to note that forty years after luminosity classes were introduced, they were still being calibrated. Through the introduction of luminosity classes Morgan was able to demonstrate the existence of the spiral structure of our galaxy, again independently of theoretical arguments.

When nuclear astrophysics appeared, theoreticians were able to make some predictions of what could happen in the course of evolution to a star. Very soon they found that many of their predictions (and also many unpredicted results) had already been discovered by classifiers – stars with no hydrogen, stars with strong helium, stars with CNO excesses, stars with varying degrees of line strength, stars with enhancements of *s*-type elements. All these groups were there – sorted out by classifiers without any theoretical guidance.

These four examples illustrate some of the uses and achievements of classification methods; let us add simply that classification methods have become so 'basic' that many astrophysicists tend to consider them old-fashioned and useless. This is perhaps especially because many people associate classification methods with visual classification of photographic spectra on blue sensitive plates.

From what we have said, it should be clear that such an association is simply a reflection of the situation at the end of the nineteenth century. In fact, the power of classification methods consists in that they can be applied both to a variety of objects (galaxies, nebulae, stars, meteorites) and to a variety of techniques.

In recent decades detectors have been vastly improved, and large stretches of hitherto untouched wavelength regions are now open to us, presenting us with a clear invitation to re-examine methods and to apply them to new data. Whenever a new spectral region is opened, classification must be done first. This is an important point, which many astronomers have not realized. Radio-astronomers would certainly not use flux predictions based upon extrapolations from visual observations, because they know that objects may behave differently in the radio region. But up to now astronomers studying the ultraviolet $\lambda\lambda$1000–3000 region are happily using extrapolated MK standards, i.e. types classified in a different spectral region. This is incorrect both in principle and in practice, and sometimes produces big surprises, since a star normal in one region is not necessarily normal in a different wavelength region. Such a situation shows rather convincingly that classification is needed as much today as it was a century ago.

Classification methods are also needed when a new technique opens up another unexplored spectral region, as happened recently with IRAS, the infrared satellite. More than 2×10^5 point sources were observed, of which 1.3×10^5 could be linked with stars, on the basis of positional coincidence. Such links can be checked easily if spectral classifications of the stars are available, because all late type stars, for instance, are infrared emitters. Classification, the 'old-fashioned' technique, is again called on to help in the interpretation of some very modern observations.

Coming now to the book itself, it is divided in two parts.

The first part is devoted to the methods and techniques of classification, both spectroscopic and photometric (chapters 1 to 5). A comparison of both methods is given in chapter 6.

In the second part (chapters 7 to 15) we apply the methods explained in part 1. Chapter 7 describes the content of chapters 8 to 15; each chapter deals with one family of stars. We consider first the normal stars of the family in section 0, and the objects with peculiar spectra in the sections that follow.

We have included in general only stable stars, i.e. non-variable objects, although some exceptions are made. The reader shall thus find no discussion of Cepheids, long-period variables or novae, nor of extended objects such as planetary nebulae, nor of Herbig Haro objects. We also exclude binaries with strong interactions, such as β Lyr or W UMa stars. Such a selection emphasizes the more common, stable stars, although the limits between variables and non-variables are becoming fuzzy with progress in observation and interpretation.

The reader is warned that this is a book on classification and not on the physics of the stars; it is definitely not a series of monographs on each type of star. There are in fact many excellent publications of that kind, and these are quoted in the reference section at the end of each chapter.

The book relies heavily on some projects which were carried out in the last few years at Strasbourg and which are discussed in chapter 7. The first is the *Catalog of stellar groups*, by M. Jaschek and D. Egret, which assigns about thirty thousand stars to some fifty spectroscopically defined groups with various spectral peculiarities. The second is the *Atlas of spectrophotometric tracings of stars,* based upon spectra taken at the Haute Provence Observatory. The atlas is the result of collaboration with G. Goy of the Geneva Observatory. The majority of the illustrations of spectra in this present book are taken from the atlas. Thirdly, many of the data quoted in the book come from the SIMBAD database of the Centre de Données Stellaires (CDS) at Strasbourg, which is the world's largest database.

The spectroscopic part of this book is the result of a long-standing involvement of the authors with spectral classification, starting at the La Plata Observatory and continuing at Lick, Mt Wilson, Leuschner, Yerkes, Ohio State, Geneva, Lausanne and Strasbourg. Over the years we had the privilege of meeting many of the colleagues who produced the results assembled in this book, and of working with them.

The photometric part is the outcome of many stimulating and enriching discussions with Dr M. Golay at Geneva, to whom we express our gratitude.

We would also like to thank W. Bidelman, P. Keenan, E. Mendoza, D. Philip and A. Slettebak, who read and kindly criticized parts of this book. We especially thank M. Holtzer for his careful reading of the whole book.

We thank Dr A. Florsch, director of the Strasbourg Observatory, and staff of the observatory for their help, especially Mr J. Marcout who prepared most of the illustrations. Thanks are also due to Miss C. Bruneau, A. Bielecka and I. Strukeli for typing the manuscript.

Bibliographic references

In the bibliographic references at the end of each chapter we have used a number of abbreviations for the titles of Journals, which are as follows:

AA *Astronomy and Astrophysics*
AA Suppl. *Astronomy and Astrophysics Supplement Series*
AdA *Annales d'Astrophysique*
AJ *Astronomical Journal*
AN *Astronomische Nachrichten*
Ann. Rev. *Annual Review of Astronomy and Astrophysics*
Ap. J. *Astrophysical Journal*
BICDS *Bulletin d'Information du Centre de Données Stellaires*
Mem. RAS *Memoirs of the Royal Astronomical Society*
MNRAS *Monthly Notices of the Royal Astronomical Society*
PASP *Publications of the Astronomical Society of the Pacific*
QJRAS *Quarterly Journal of the Royal Astronomical Society*
Sp. Sc. Rev. *Space Science Reviews*
Z.f. Astroph. *Zeitschrift für Astrophysik*

Acknowledgements

We would like to thank the following for permission to reproduce the figures listed below.

The authors and the *Astrophysical Journal* published by the University of Chicago Press © 1986, The American Astronomical Society; for figures 5.1, 8.3, 8.8, 8.10, 8.11, 9.5, 9.24, 9.25, 10.4, 10.8, 10.15, 10.17, 11.10, 11.14, 11.15, 11.18, 11.21, 12.21, 12.22, 13.11, 13.18, 14.11, 14.13, 14.14, 15.2, 15.5, 15.9, 15.10

D. Reidel Publishing Company for figure 9.11 from *IAU Colloquia 'Stellar rotation'* (ed. A. Slettebak) and figure 13.20 from *IAU Coll.* **70**

The International Astronomical Union for the figures from *IAU Symposia* **50** (12.8, 12.9, 12.10), **70** (9.10, 9.12), **80** (8.16, 10.18, 10.19, 11.19) and **102** (11.11 and 11.12)

The Editor of *IAU Coll.* **32**, for figures 9.22 and 9.27

The Editor of *IAU Coll.* **76**, for figure 14.9

The Editor of the *Dudley Observatory Report* for figures 5.13 and 5.14

The Editor of the *Dominion Astrophysical Observatory Publications* for figures 2.4, 9.3, 13.12, 13.13 and 13.14

The *Revista Mexicana de Astronomia y Astrofisica* for figures 12.5 and 12.6

The Editor of the *Astronomical Journal* for figures 9.13, 9.20, 9.21, 11.6, 12.7, 12.15, 15.6, 15.7 and 15.8

The Director of the Vatican Observatory for table 3.5 and figure 14.8

The Department of Astronomy of University of Michigan, for figure 3.1 and table 3.4

Subsidiary Rights Department, Pergamon Press, Oxford, for figure 14.2

The *Publications of the Astronomical Society of the Pacific*, for figures 5.8, 11.23, 12.19, 12.20, 13.15, 14.5 and 14.12

The authors and the Royal Astronomical Society for figures 5.18, 9.14, 9.29, 9.30, 11.16, 11.17, 12.12, 12.13 and 12.14

The Board of Directors of *Astronomy and Astrophysics. A European Journal* for the following figures: 4.2, 8.12, 8.14, 8.15, 9.23, 10.11 and 10.14 from the main journal; 5.10, 5.11, 5.12, 8.9, 9.8, 9.9 and 10.23 from the *Supplement Series*; 4.4 from *Zeitschrift für Astrophysik* and 4.5 from *Annales d'Astrophysique*

The Lick Observatory, for figures 13.8 and 13.9 (Lick Observatory Photograph)

The authors and the Editor of the *Colloque International d'Astrophysique de Liège*, for figures 9.17 and 9.18

The Kitt Peak National Observatory, for figures 13.5 and 13.7

The Joint IRAS Science Working Group, for figure 14.6

Part 1

1

Stellar taxonomy

Taxonomy, according to the dictionary, is the classification of objects into ordered categories. All natural sciences – botany, zoology, geology – classify the objects of their study: in a certain sense, taxonomy is the backbone of everything else.

The classification can be done on the basis of a single observable property of the objects, or a combination of properties. For instance, Linnaeus used plants' flowers for his botanical classification system, but modern authors use a variety of plant characteristics. In astronomy, the main difficulty consists in that the objects are far away and we can study certain properties only for a rather small number of objects. Whereas for a plant, an animal or a rock we can search leisurely for those parameters which are best for classification purposes, and then measure the parameters in all plants of a certain type, such an operation is impossible in astronomy. In fact the astronomer must use those parameters which are available for a large number of stars, even if they might turn out not to be the best for classification. Stellar parameters that can be estimated or measured are temperature, color, spectral type, proper motion, radial velocity, radius, magnetic field, rotational velocity, chemical composition and so on. It is obvious that one essential condition for classification is that the parameter used should be known for a large number of objects. Table 1.1 summarizes for how many stars these parameters are known; the numbers are taken from compilations existing at the Centre de Données Stellaires (CDS).

For comparative purposes let us recall that the number of stars brighter than visual magnitude $V = 9$ is about 5.5×10^5. Obviously, only the first three parameters in the table are potentially useful for stellar taxonomy, because all the others are known for comparatively few stars. We then have to drop proper motion from our list because the information it provides – change of position with regard to a reference frame – is not directly related to the nature of a star.

It is thus clear that stellar taxonomy can be based either upon stellar spectra or upon photoelectric colors, and we shall have a spectral taxonomy (also called spectral classification) and a photometric taxonomy. We will now describe briefly how each technique works and what it provides. We shall return later to the details (chapters 2 to 5).

1.1 Spectral classification

For this, the spectra of a large number of stars obtained with the same instruments are examined visually, ensuring that all features are looked at. If N spectra are available, they are compared by pairs; after each comparison, the conclusion is made whether the spectra look different or similar. If they are similar, both spectra are assigned to the same group. The process is continued with the next pair, until the smallest possible number of groups is established.

Spectra belonging to the same group are regarded as being identical, a fact which is only true as long as the spectra are photographed at the same dispersion. If stars belonging to the same group are photographed at a higher dispersion, they may not look identical. This means essentially that the groupings established are bound to a certain dispersion, and resolution (and technique? See chapter 2.) Of course, the term 'stellar spectrum' is a gross exaggeration, because whereas stars radiate at all wavelengths, a 'spectrum' conventionally covers only a short interval of wavelengths. We shall use the

Table 1.1. *Number of stars with known parameters.*

Parameter	Number
Spectral type (HD)	5×10^5
Photoelectric color (all systems)	1×10^5
Proper motion	3×10^5
Radial velocity	3×10^4
Rotational velocity	6×10^3
Radius	4×10^3
Magnetic field	1×10^3
Chemical composition	2×10^3

term 'classical region' to refer to the interval $\lambda\lambda$3500–4800. (In this book, all wavelengths are expressed in angstroms. 1 Å = 0.1 nm = 10^{-10} m.)

In addition, if spectra in two different wavelength regions are compared, there is no logical requirement that stars identical in the 'classical region' say, should also be identical in the 'red region' or 'ultraviolet region'.

With this in mind let us assume that we have arranged the spectra of N stars, taken with the same dispersion and at the same wavelength region, in m groups, each group consisting of a number n_m of similar looking spectra. We can pick out any star of a given group and call it a 'typical object of the group' or, more succinctly, a 'standard'. The process of classification has thus substituted the descriptions of only m standard for the separate descriptions of N spectra. This operation thus facilitates further studies, since it greatly reduces the number of objects to study. For instance, it was found that about 2×10^5 stellar spectra classified by Harvard astronomers could be put into fewer than 10^2 groups.

It always happens that a few spectra out of the N are not identical to any standard. These we termed 'abnormal spectra' and are put to one side. Since by definition normality is what the majority are, 'abnormal' stars should always be in a minority; the success of any classification system is to be able to assign places to as many spectra as possible.

The next step in classification is to try to relate the different groups; in biology this corresponds to the grouping of species into genera. If we start with the idea that nature proceeds by continuous transition, we realize that stellar spectra should show only small differences from one group to the next. In stellar spectra, this is indicated by the strength of the different features visible. If the same features are present in the spectra of two standards, with only slightly different intensities, we would consider both groups (or standards) to be contiguous or alike. The practical question of how to establish this 'kinship' will be considered in detail in chapter 2, so for the moment we shall assume that we have found that two groups are neighboring ones. How can we describe this relationship?

The similar problem in biology is resolved by attributing a name to particular kind of animal – *Canis domesticus* for dog, for instance. The first part of the name refers to the genus, the second part to the species: so *Canis lupus* is an animal of the same genus, but of a different species (i.e. wolf). We notice in passing that this system is somewhat cumbersome, if the number of species (groups in our case) becomes large.

Initially, classifiers of stellar spectra called each group by the name of a bright star it contained, for instance 'alpha-CMa'. This notation is colorful but does not include the idea of relationship with other groups; in fact it was replaced very soon by abbreviations like 'A1'. Here the groups are expressed in a double notation, where 'A' stands for the larger grouping and '1' for the smaller grouping. Then A1 is related to A0 and A2. This very practical method enables us to order the groups into an (unidimensional) sequence. Two- or three-dimensional classifications can be introduced later, and can be reflected in the nomenclature by using more symbols. More details are given in chapter 2.

You will probably ask if this assignment of 'likeness' is done in a purely empirical way, or if the process is guided by our knowledge of the physics of stars. Historically, classification schemes were worked out before physical theory was capable of handling stellar physics – and when the rudimentary theory then available was applied it was usually wrong. Harvard astronomers started by using groups labelled A, B, C, D,...; the order O – B – A – F – G – K – M used today shows that lengthy process of simplification and reaccommodation of groups has taken place. That the Harvard classification has survived suggests that there is something sound in it, independent of the physical theories which explain the classification scheme. However, this idea should not be over-emphasized, simply because classification schemes are worked out by professional astronomers and so some physics enters inevitably, even if subconsciously.

Before leaving this section, we should mention briefly the question of the number of spectra needed to set up a system of classification. The first precaution we have to take is to ask why we do classification. If we wish, in the example quoted above, to see if the group A1 can be further subdivided, we would need to consider several dozen spectra – let us say between 10 and 10^2. It is quite a different matter if we want to set up a general system for all kinds of stars, because in this case we need to use many more spectra – let us say between 10^3 and 10^4.

As an example, let us consider the spectra obtained by satellites, which opened the ultraviolet region ($\lambda\lambda$1000–3000), inaccessible from the ground. If the number of stars for which UV spectra became available were large, we could set up an independent classification system, which may, or may not, coincide with the classifications of spectra obtained by ground-based observatories. Let us start by regarding the number of stars observed by different satellites, assembled in table 1.2. It is clear that the number of spectra is still insufficient to produce entirely new classification schemes. The number is, however, large enough to study smaller parts of the whole area, or to check the consistency of existing schemes.

These problems were discussed in a conference on *Ultraviolet stellar classification* (1982), and in the workshop, *The MK process and stellar classification* (1985); the reader can find more details in these references. The workshop reference in particular contains discussion of the philosophy of classification in astronomy. To avoid getting the wrong impression that classification is something specifically astronomic, the reader should consult *The growth of biological thought* (Mayr 1982), where the development of classification is described in detail from Aristotle (384 to 322 BC) onwards.

1.2 Photometric classification

The observational basis of photometric classification is provided by the measurement of stellar radiation at certain pre-specified wavelengths (λ_i) for a large number (N) of stars. The number of specified wavelengths is generally small, i.e. $i \leqslant 10$.

The specification of the wavelengths, plus the description of the instruments used and a list of reference stars observed with these procedures establishes what is called a 'photometric system'. (See chapter 3 for more details.)

Stellar radiation is measured as illumination (flux/surface), but astronomers prefer a logarithmic scale of fluxes, called magnitude. A magnitude of a star shall be written as

$$m(\lambda_i) = C - D \log E(\lambda_i)$$

In this formula C and D are constants and $E(\lambda_i)$ is the illumination at λ_i; 'log' is decimal logarithm.

A star may have n different magnitudes, as many as the different $E(\lambda_i)$ that were measured. The difference between two $m(\lambda_i)$ of the same star

$$m(\lambda_i) - m(\lambda_k) = c(i, k)$$

Table 1.2. *Number of stars observed with satellites for which spectra were obtained.*

Experiment	Number of stars	Source
Skylab S-109	500	Henize *et al.* (1979)
TD1	1800	Jamar *et al.* (1976) Macau-Hercot *et al.* (1978)
IUE	~2000	Merged log (1985)
OAO	~300	Meade and Code (1980)

is called the 'color index' or simply 'color'. The difference is always calculated with $\lambda_i < \lambda_k$.

A photometric system using n specified passbands λ_i (with $i = 1, 2, \ldots, n$) allows us to define

$$C_2^n = \frac{n(n-1)}{2}$$

colors, but only $n-1$ are independent. We shall therefore speak interchangeably of a system with n passbands or with $n-1$ colors.

The practical set-up of a photometric system requires a description of the instruments used, and the listed measurements of the magnitudes of a number of reference stars. The set of reference stars guarantees the repeatability of the system and permits others to check the observational procedures. Each measurement is affected by an observational error, ε_i, whose size depends upon the system; for repeatability, new measurements must agree with previous ones within the errors.

The use of a photometric system produces ultimately a series of n colors for N stars, which constitutes the basic material for classification. We can imagine this data collection as a set of (N) points in an n-dimensional photometric space. Because of observational errors, the position of a given point is uncertain within the limits of precision of the measured color indices. So instead of a 'hard' point with coordinates $(c_1, c_2, \ldots, c_n)$, we should think of a 'soft' point defined by

$$\left.\begin{array}{c} c_1 - \varepsilon_1 \leqslant c_1 \leqslant c_1 + \varepsilon_1 \\ c_2 - \varepsilon_2 \leqslant c_2 \leqslant c_2 + \varepsilon_2 \\ \vdots \\ c_n - \varepsilon_n \leqslant c_n \leqslant c_n + \varepsilon_n \end{array}\right\} \tag{1.1}$$

A new measurement of the star should produce a point falling within the intervals specified above around the first measure.

It is clear that two points falling in the intervals specified by (1.1) are indistinguishable. The set of conditions (1.1) defines an (n-dimensional!) box in the photometric space, and points within one box are indistinguishable, i.e. they are photometrically identical.

The notion of a photometric box, first introduced by Golay (1974), can be generalized in several ways, which we shall examine in chapter 3. The easiest generalization is to change the definition by introducing a (small) positive number a:

$$c_1 - a\varepsilon_1 \leqslant c_1 \leqslant c_1 + a\varepsilon_1$$

and similarly for the other coordinates. If $a > 2$, obviously the boxes become 'contaminated'. Such photometric boxes are strictly analogous to spectroscopic groups, except that spectroscopic groups are discontinuous, whereas photometric boxes are not.

Having thus defined photometric boxes, we can next look for larger structures. Such structures become immediately obvious if we represent photometric systems with $n = 2$ graphically. (Obviously, we could do the same for $n = 3$ but the representation becomes more difficult, and for $n > 3$ only partial projections are possible.) By plotting indices (see for instance figure 5.1) we become aware that photometric space is not uniformly populated, but rather that points tend to accumulate around certain curves (or sequences). These sequences are equivalent to the 'spectroscopic similarity' we mentioned before. If we consider a point in a two-color diagram which lies on such a sequence, then starting from the given point in different directions there are certain chances of 'hitting' a neighbor. The probability of hitting a neighbor is a maximum in the direction of the sequence; the sequences might thus be considered as 'maximum neighborhood' tracers.

References

Golay M. (1974) *Introduction to astronomical photometry*, D. Reidel Publ. Co.

Henize K.G., Wray J.D., Parsons S.B. and Benedict G.F. (1979) *NASA Reference Publ.* 1031

Jamar C., Macau-Hercot D., Monfils A., Thompson G.I., Houziaux L. and Wilson R. (1976) *ESA SR*-27

Macau-Hercot D., Jamar C., Monfils A., Thompson T.I., Houziaux L. and Wilson R. (1978) *ESA SR*-24

Mayr E. (1982) *The growth of biological thought*, Belknap Press of the Harvard University Press

Meade N. and Code A. (1980) *A.J. Suppl.* **42**, 283

'Merged log of IUE observations' (1985) *IUE NASA newsletter* **23**, 104

The MK process and stellar classification (1985), R.F. Garrison (ed.), Toronto

Ultraviolet stellar classification (1982) *ESA SP*-182

2

Spectral classification

This chapter describes spectral classification in general, without referring to a particular classification system.

We start by discussing certain specifications of the spectroscopic material. Then we describe the features of the spectra and the factors which may affect them. Next we consider the process of classification itself and finally we deal with certain aids to classification.

2.1 The material of spectral classification

A *spectrum* is the display of stellar radiation as a function of wavelength. The names of the different wavelength regions are given in table 2.1. Astronomers also talk of the 'ultraviolet spectrum' of a star; this is a short way of speaking of the 'ultraviolet region of the spectrum' of the star.

Once a spectrum has been obtained, certain details about it must be provided. In first place, what and how it was observed must be specified. Among the most important specifications is the *wavelength range* covered. Note that one should specify what the (usable) spectrum region is, rather than state that '103aO plates used' because in a few years this will not be understandable. Next the *plate factor* (D) should be specified, which is the number of angstroms of the spectrum entering 1 mm length of receiver. Much confusion originates here because astronomers call D the 'dispersion', whereas opticians call it (correctly) the 'reciprocal dispersion'. A small plate factor D implies highly dispersed spectra (i.e. large scale), and a large value of D, low dispersion spectra. Authors differ somewhat as to the designations given to the different plate factors; however, table 2.2 summarizes current opinion.

If the spectrum is taken on a photographic plate with a grating dispersion device, the relation between abscissae measured on the plate and wavelength is (to a first approximation) given by a formula of the type:

$$x = A + B\lambda \quad (A, B \text{ constants})$$

therefore

$$D = \frac{\mathrm{d}\lambda}{\mathrm{d}x} = \frac{1}{B}$$

For prismatic dispersion devices the relation is given by a formula of the type:

$$x - C = \frac{E}{(\lambda - F)^n}$$

where C, E, F, n are constants. Thus

$$D = \frac{\mathrm{d}\lambda}{\mathrm{d}x} = f(\lambda).$$

In the classical wavelength range, the plate factor D is always specified by convention by its value at Hγ ($\lambda = 4340$ Å).

Table 2.1. *Wavelength regions.*

Name	Wavelength interval	Unit
Extreme ultraviolet	< 1000	Å
Ultraviolet	1000–3000	Å
Classical	3000–4900	Å
Visual	4900–7000	Å
Near infrared	7000–10^5	Å
Far infrared	1–10^3	μm
Radio	0.1–10^4	cm

Table 2.2. *Plate factors.*

Name	Plate factor
Very high	< 5 Å/mm
High	5–20 Å/mm
Intermediate	20–80 Å/mm
Classification	80–120 Å/mm
Low	120–400 Å/mm
Very low	400–2000 Å/mm

Next to the plate factor, the *resolving power* (R) of the detector should be quoted; this is the minimum distance for two lines to be seen as separate features. In other words, all details of the spectrum falling within R are obliterated or smeared out. Thus if a detector on a satellite has $R = 7$ Å, two lines separated by 5 Å form a *single* feature. In photographic material, the resolving power depends upon the type of emulsion and, to some extent, upon the plate handling techniques (sensitization, development); usually it lies between 10 and 20 μm.

If a *slit spectrograph* is used, the *projected slit width* (d) should also be given; this is the width on the plate of the entrance slit of the spectrograph. Since the entrance slit is fixed by the astronomer, $d \leqslant R$ is usually chosen; if $d > R$, each spectral line is wider than the instrument can resolve. It should be remembered that the image of the star on the slit has a diameter (δ) which depends strongly upon atmospheric turbulence. Usually $\delta \sim 1''$–$2''$ or more. Thus if the slit width of the spectrograph has to be small to have $d \sim R$, a large value of δ implies a large loss of light. If the spectrograph is slitless, it is clear that we will have on the plate not d, but the equivalent of δ. This means that in general we will have a line which is larger than R, and thus the lines are blurred. A slitless device like an objective prism therefore produces spectra which depend critically upon the observing conditions (turbulence); at the same dispersion, objective prism spectra are worse than spectra taken with slit spectrographs.

The *width* of a spectrum is the height of the spectrum perpendicular to the direction of the dispersion. It is usually produced by trailing the star image along the slit. Widening improves considerably the visibility of faint lines. The minimum width is 0.15 mm = 150 μm; it should be used only if absolutely necessary. If plates are to be used for classification work, a width of 500–1000 μm is preferable. However, since observing time increases in a way which is practically proportional to height, 600 μm or so often has to be used. Beyond 1 mm it seems that no real gain is achieved in the contrast for faint lines.

2.2 Description of the material

The outcome of the observations is either a photographic plate or a recording (or tracing); the recording can be made either in density, or in intensity. The conversion from density (or pseudo-density) to intensity is a laboratory technique for which the reader should consult any of the many books on astronomical techniques, and we shall not deal with it here.

The final result is a continuous band – the continuum – interrupted by indentations called *lines*. The spectra are always examined with longer wavelengths to the right (*red* to *right*), with absorption lines pointing downward. Note that because astronomers usually work with photographic negatives, any absorption line is transparent. If there is doubt whether a plate is a negative or a positive, look for the lines of the comparison spectrum; these are always emission lines.

In theory the continuous spectrum should be a smooth curve, but in practice it is always observed through a mask of background receiver noise (plate grain). The noise can be diminished by various techniques like widening the spectrum or superimposing several spectra. It is important to make sure that no minor faults (dark spots on the emulsions, for instance) are present, because in any averaging process they reappear with a different shape. When different spectra are superimposed, they should be inspected visually to check that no individual differences exist from plate to plate, since these differences will be obliterated in averaging. Care should also be paid to real features that have nothing to do with the object observed, such as emission lines due to airglow or to street illumination, or absorption lines due to atmospheric water-vapor, or calibration marks on satellite spectra, or instrumentally produced lines.

Let us now turn our attention to the *spectral lines*. In an isolated line there is a central part or *line core* (also called simply 'core'), accompanied by more or less extended parts leading into the continuum. These are the *line wings* (or 'wings'). Figure 2.1 shows schematically lines as seen on a microphotometric tracing. Wings can be almost absent, as for instance in interstellar lines (Figure 2.1(c)).

If the registration is made in terms of intensity, the lines can be characterized numerically. The *line depth* (l), the *central line depth* (l_c) or the *central residual intensity* ($1 - l_c$) can be used for this. (See figure 2.2.)

The line continuum in these definitions is always taken as unity.

To characterize the line width – which is sometimes rather difficult – the

Figure 2.1. Line structure.

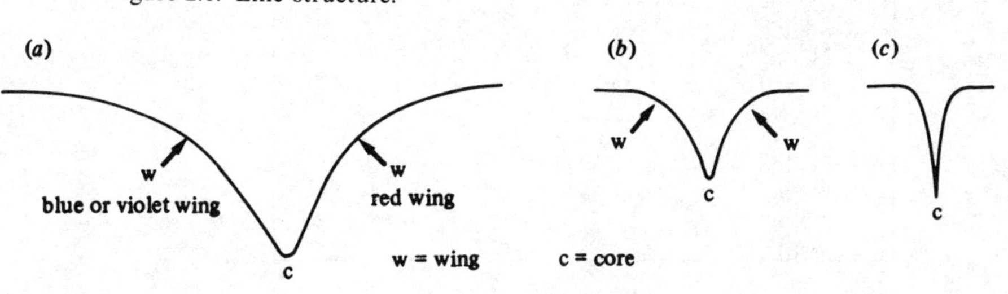

line width ($AC = \Delta$) at half central depth is used. There is some confusion over the name: some call Δ the half width; others call AB the half width. (See figure 2.3.)

To characterize the total line strength, the '*equivalent width*', W_λ, is used. This is defined as

$$W_\lambda = \int_{-\infty}^{\infty} \frac{I_c - I}{I_c} \mathrm{d}\lambda$$

Figure 2.2. Parameters characterizing a spectral line. l = line depth; l_c = central line depth; $1 - l_c$ = central residual intensity. The continuum is defined as $I_c = 1$.

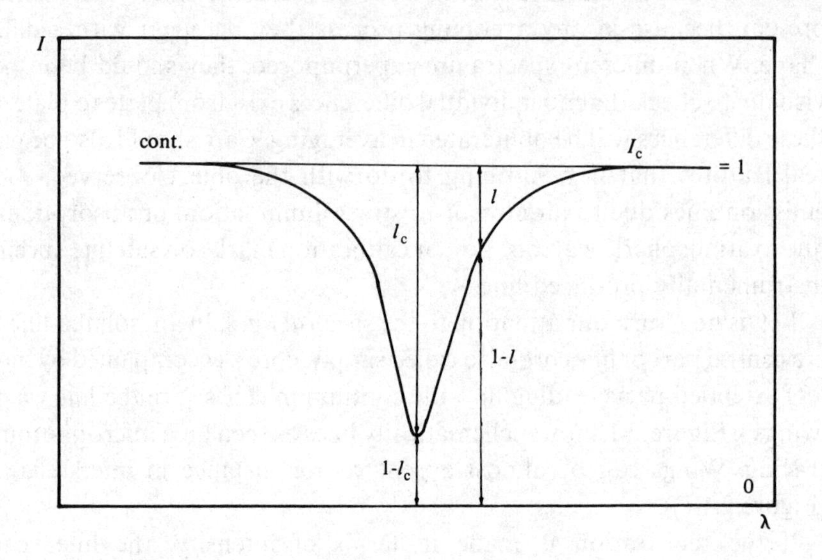

Figure 2.3. Line width and equivalent line width. $\Delta = AC$ = line width at half central depth; W_λ = equivalent width (for definition, see text).

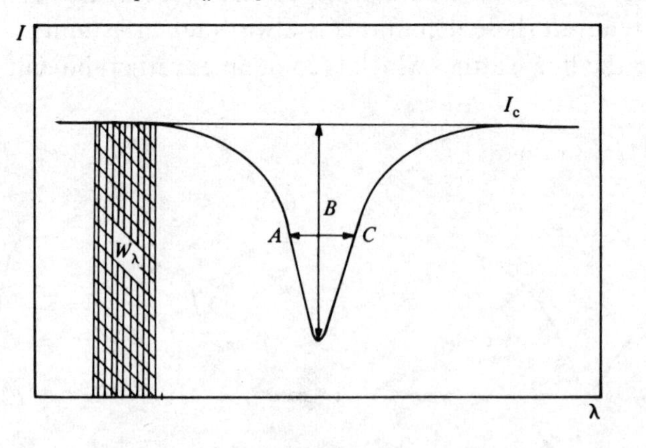

(Usually the integration need only be done over a quite short interval.) The equivalent width is measured in angstroms or in milliangstroms. The name is due to the fact that W_λ can be envisaged as the area of a rectangle of unit height and of length W_λ; it is therefore the width of the rectangle whose area is equal to the integral of the energy removed by the line from the emitted radiation.

In strong lines, because of the influence of the wings of the line upon the calculation of the integral, W_λ is very sensitive to the exact tracing of the continuum. However, determination of the continuum is not usually easy. In the classical region, the continuum for the early type stars can be drawn without too much trouble except below λ3900 where the confluence of the Balmer lines lowers the true continuum considerably. In the regions which are free, the continuum is drawn as an average line, in order to smooth out the random fluctuations of plate grain. (See figure 2.4.)

When considering later stars, the regions without lines become shorter and we cannot be sure that these regions are really free from lines. This happens at 2 Å/mm for G-type stars and at 10 Å/mm for F-type. In these cases the continuum is drawn through the highest points, the justification for this being that it takes care of the lines which are so faint as to simulate plate grain.

It is clear that the determination of the continuum is the crucial step; most difficulties connected with the interpretation of tracings start at this point.

Since the equivalent width characterizes line strength, it is clear that the former is related to what is visible on a plate. As a general rule, Seitter (private communication, 1981) found that the equivalent width of the faintest visible feature in m Å is approximately equal to the plate factor in Å/mm. Not taking the faintest features but rather the easily visible ones, we find

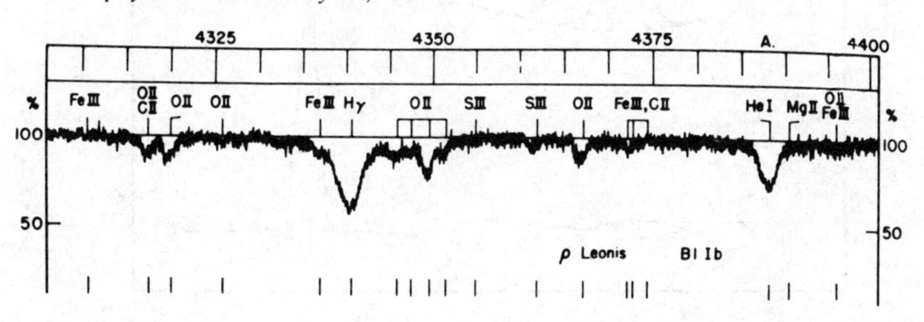

Figure 2.4. Tracing of the continuum. From *Publications of the Dominion Astrophysical Observatory* **12**, 198.

that

$$\text{at } 240\,\text{Å/mm we see lines with } W_\lambda \simeq 0.5$$
$$\text{at } 100\,\text{Å/mm we see lines with } W_\lambda \simeq 0.2$$
$$\text{at } 40\,\text{Å/mm we see lines with } W_\lambda \simeq 0.1$$

The profile of emission lines is usually of the type given in figure 2.5. The width of an emission line can be specified by the half width AB, expressed in km/s (in figure 2.5, $c\Delta\lambda/\lambda$ where $\Delta\lambda$ is AB measured in angstrom). The base level from which I is measured must be specified carefully.

Figure 2.5. Profile of an emission line. The half width (AB) is the line width at half central intensity (see text) expressed in km/s.

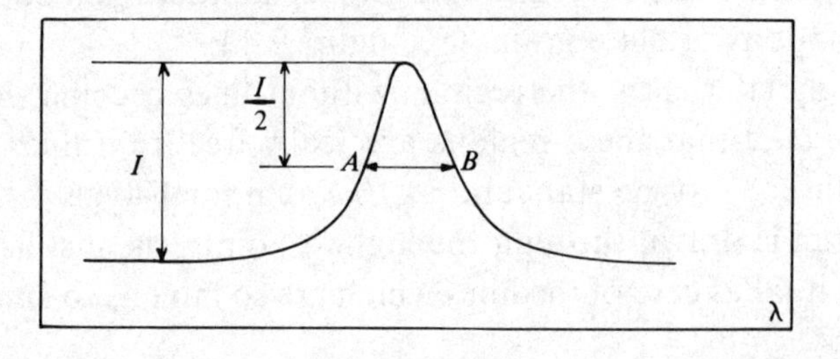

Figure 2.6. Five different types of emission lines. **1**: central emission; **2**: self absorption; **3**: fill in; **4**: P Cygni profile; **5**: inverse P Cygni profile.

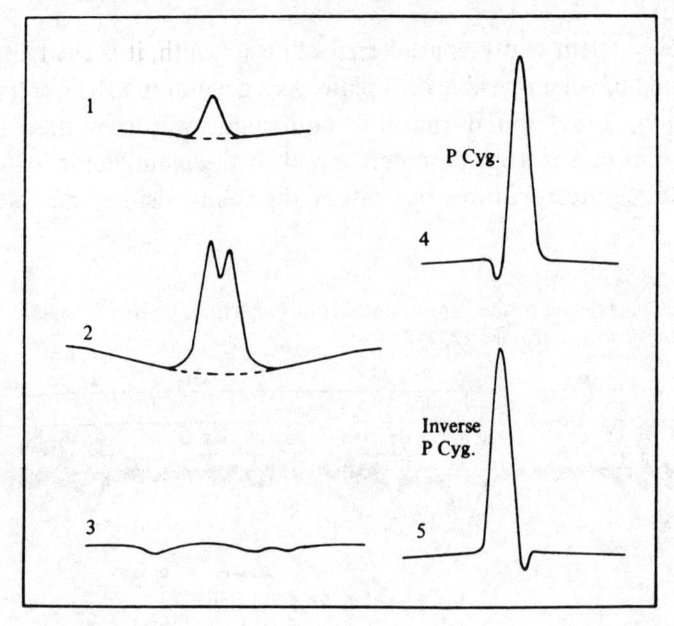

If an absorption line is accompanying the emission, several cases are possible (figure 2.6). Case 1 is called a *central emission*, 2 a central absorption core or *self-absorption*, 3 is usually called a '*fill in*' (the line is almost invisible), 4 is a '*P Cygni profile*' (emission on the red border) and 5 is an '*inverse P Cygni profile*' (emission on the violet border).

It often arises that several lines fall close together, and become indistinguishable; we then speak of a '*blend*' of lines. Blends always depend on the dispersion used and can be overcome at higher dispersion. If the blends come from the overlapping of molecular lines, we speak of 'bands' (figure 2.7), degrading to the violet (case b) or to the red (case a).

When referring to lines, blends and bands in general, we speak of spectral 'features'. In late type stars, the continuum ceases to exist, because of the overlapping of many faint lines. The top of the lines then produces a 'pseudo-continuum', which can lie considerably below the true continuum. A large break in the continuum is called a 'discontinuity', like for instance the Balmer discontinuity.

Special difficulties appear when faint features are examined; the first question is usually if they are real or not. Often this can be decided with the help of additional plates, provided that the star is non-variable. In any case it is always preferable to go back to the plate, because there the feature appears in two dimensions, whereas a tracing provides only one dimension. A white spot may appear on the tracing as a line, but it is easily seen on a plate to be distinct from a line.

2.3 Factors which influence the line spectrum

Spectral lines reflect the physical conditions of that part of the stellar atmosphere where they were formed. This implies that the line shape (or profile) depends upon a number of factors such as temperature, electron pressure, magnetic field and so on. It is certainly not a matter for this book to discuss line formation, but we will deal simply with some factors affecting all lines, independent of their origin. These factors are stellar rotation and binarity, and we shall also say something about the magnetic field.

Figure 2.7.(a) Bands degraded to the red; (b) bands degraded to the violet.

2.3.1 *Stellar rotation*

Abney (1877) made the initial discovery that some stars do rotate, but it was Shajn and Struve (1929) who in the twenties showed that stellar rotation is very common phenomenon. Under the influence of rotation, spectral lines become blurred, and the blurring is greater the higher the speed of rotation. At very high rotation speeds, the profiles are 'dish-shaped' and their visibility diminishes, because of the loss of contrast with the nearby continuum (see figure 2.8). In general, the rotational velocity is characterized by $V \sin i$, where V is the equatorial rotational velocity of the star and i the angle of inclination between the visual and the rotation axis. Theoreticians obtain from the study of the profile of a spectral line a value of '$V \sin i$'. Note that all lines of a stellar spectrum are affected by rotation and should lead to similar values of $V \sin i$, unless the places from where the lines originate greatly differ (for instance, if the lines come from the upper and the lower parts of the stellar atmosphere). In practice $V \sin i$ is usually determined from the profiles of one or two lines, like He I λ4026 or Mg II λ4481. These lines are chosen because they vary slowly with spectral type and can thus be used over a large range of spectral types. They are lines of intermediate intensity and are thus usable even on low dispersion

Figure 2.8. The influence of stellar rotation upon line shape. The same line is shown as seen for different rotational velocities, indicated at each profile (0, 100, 200 and 400 km/s).

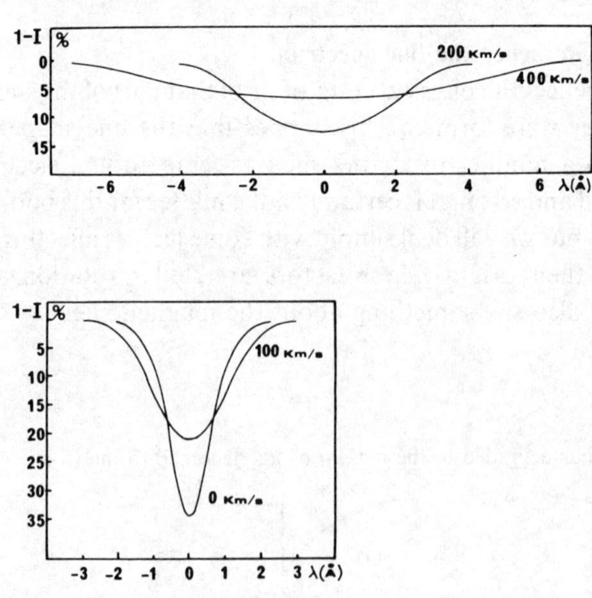

spectra. Hydrogen lines on the other hand cannot be used because their profile depends very much on Stark broadening.

The point at which rotation starts blurring the lines depends on the plate dispersion D. As a rule, rotation becomes imperceptible when

$$V \sin i \leqslant D$$

The reason for this is as follows. Suppose a star rotates around an axis contained in the plane of this page (i.e. perpendicular to the line of sight). The extremes of the disc rotate therefore with radial velocities $V_0 + V_{ec}$ (right border) and $V_0 - V_{ec}$ (left border), where V_0 is the radial velocity of the barycenter of the star. If we imagine that each point of the stellar disc produces a separate spectral line, the lines produced by both borders would differ by $2V_{ec}$ in velocity. If the stellar axis is inclined by an angle i (angle between visual and rotation axis) we can see that the same reasoning leads to $2V_{ec} \sin i$. The complete line will be produced by all the profiles arising from different points of the stellar disc, i.e. it is the 'envelope' of the individual lines. But since the region near the border has the largest velocity but comparatively the smallest area, we see effectively a line which is not $2\,V_{ec} \sin i$ broad, but somewhat less, say $1.5\,V_{ec} \sin i$. If we now reason that this width is equivalent to a certain wavelength interval given by the Doppler formula $\Delta\lambda = \lambda v/c$, we find

$$\Delta\lambda = \frac{\lambda \times 1.5\,V_{ec} \sin i}{c}$$

If the plate dispersion is D Å/mm, this $\Delta\lambda$ corresponds to

$$\frac{\lambda \times 1.5\,V_{ec} \sin i}{c} \times \frac{1000}{D} \quad \text{in microns}$$

Finally this has to be larger than the resolving power of the plate (20 μm), so that

$$\frac{\lambda \times 1.5\,V_{ec} \sin i}{c} \times \frac{1000}{D} \geqslant 20$$

implying

$$V \sin i \geqslant \frac{20cD}{\lambda \times 1.5 \times 1000}$$

With $\lambda \simeq 4000$ Å, this becomes $V \sin i \geqslant D$.

This means that at $D = 20$ Å/mm, if $V \sin i = 20$ km/s, the lines are still sharp; at $D = 100$ Å/mm, $V \sin i = 100$ km/s does not affect the line profiles very much. If the lines are blurred, we attach an 'n' (for nebulous) – or 'nn' (very nebulous) in extreme cases – to the spectral classification. If the lines are very sharp, we call the stars 's' (sharp). These letters only make sense when the plate dispersion (D) is specified, because 's' lines can be blurred when a higher dispersion is used. These letters provide an estimate of $V \sin i$ to a certain order of magnitude, in the sense that 'n' usually implies $V \sin i > D$ and 'nn' $V \sin i \gg D$, whereas 's' means $V \sin i \leqslant D$.

One word of caution must be added here. Although it is true that rotation 'blurs' spectral lines, other physical causes may produce the same effect (for instance, turbulence or large magnetic fields). In O stars, for instance, rotation may not be the main cause of the line widening, because macroturbulence certainly contributes too.

High rotation has some undesirable consequences. The lines become less visible because of diminishing contrast, and they also overlap more easily. Line visibility can be improved by diminishing the background noise. A smaller dispersion can also be used; by compressing the spectrum into a lesser linear extent its visibility is very much enhanced. The white dwarfs provide a good example of this: whereas at 10 Å/mm the broad hydrogen lines are barely visible, at 100 Å/mm they are well seen.

The overlapping of lines because of blurring is more difficult to deal with and cannot be avoided. Since fewer lines implies a larger uncertainty in stellar atmosphere studies, theoreticians have usually avoided analyzing rapidly rotating stars. It is, however, an open question whether an undesirable bias is not introduced in this way.

We have already observed that if a star rotates, *all* spectral lines should be affected by rotation. Now in some spectra are observed an additional set of very sharp lines, which are produced by the interstellar medium. To qualify as such, the lines must be sharp and have a radial velocity which differs from that of the star (this latter condition is because they originate from different places). When examined at very high dispersion, these sharp lines usually split up into even sharper components, each one originating in a different interstellar cloud.

For an interstellar line to become visible, there must be no stellar line at the same wavelength. Thus interstellar Ca II is only visible in spectra which usually do not have Ca II lines of their own, i.e. from stars earlier than B8. Table 2.3 (Münch 1968) lists the most common interstellar lines visible at intermediate dispersion in the classic range.

The lines at the bottom of table 2.3 are wide and shallow. They are also called 'interstellar bands' and their origin is uncertain. In most cases the equivalent widths of interstellar lines are of the order of a few tenths of an angstrom, thus being well visible only at higher dispersion.

Besides the case of a few sharp interstellar lines seen in a spectrum with broad lines, there are other cases in which two types of line profile appear in the spectrum. So for instance, some He I lines may appear systematically broader than lines of Mg II and F II. This is incompatible with a normal star; this peculiar spectrum is called a 'shell spectrum'.

2.3.2 *Binaries*

Another factor that may influence line width is whether the star is a spectroscopic binary. The problem arises when two components of a line should be visible but are not because of inadequate resolving power. The condition for this is that the difference in radial velocity between the components should be less than or equal to the resolving power of the plate (R). If $R = 20\,\mu$m and $\lambda = 4000$ Å, then $\Delta v < \frac{3}{2}D$.

In addition to this purely geometrical condition, stellar rotation also intervenes. Only if $v \gg R$ are the lines of both components separated. It may

Table 2.3. *Interstellar lines.*

Atomic lines		Molecular lines		Diffuse lines
Na I	λ3302, λ3303	CH	λ3137	λ4430*
	λ5890, λ5896		λ3143	λ5780
			λ3146	λ6284
K I	λ7665		λ3878	
	λ7699		λ3886	
			λ3890	
Ca I	λ4226		λ4300*	
Ca II	λ3933			
	λ3968	CN	λ3874	
			λ3875	
Ti II	λ3073			
	λ3229	CH^+	λ3745	
	λ3242		λ3957	
	λ3284		λ4232*	
Fe I	λ3720			
	λ3859			

*The most important feature

happen that in a stellar spectrum the line profiles become blurred at certain epochs when $\Delta v \sim R$; in such cases the star may be a binary.

2.3.3 *Magnetic field*

A third factor which may influence line sharpness is the magnetic field. Because of the Zeeman splitting of spectral lines, a magnetic field introduces a separation of the line components. The degree of splitting depends on the spectroscopic terms involved and differs therefore from line to line, even within a given element. The number of components of a line, their spacing, polarization and relative intensity are thus different from line to line, but the scale of the pattern is proportional to the intensity of the field. It can be shown that the width of the pattern is

$$\Delta\lambda = 2 \times 4.67 \times 10^{-3} \lambda^2 H$$

where $\Delta\lambda$ and λ are given in angstrom and H in gauss. For $H = 10^3$ G at $\lambda =$ 4000 Å, this gives $\Delta\lambda \sim 1.5 \times 10^{-2}$ Å only. Thus except for very high fields observed at high dispersion, the Zeeman splitting is only a minor broadening agent as far as classification is concerned.

The matters change considerably if the fields are very strong, let us say more than 10^6 G. In this case the simple formula given is no longer valid and we must consider the quadratic terms in the theory of the Zeeman effect. Such fields are found in degenerates. There the Balmer lines are split up into several components whose separation increases rapidly. Figure 15.10 illustrates three spectra of degenerates (Angel 1977). We see clearly that very large magnetic fields show up at even low dispersion.

2.4 **Spectral feature anomalies**

We have said several times that spectral classification assigns each star to a place in a classification scheme, and that once this has been done there usually remains a residuum of stars which cannot be fitted into the scheme. Take the case of classification of 1700 A-type stars by Cowley *et al.* (1969). Here about 1400 stars were ordered into the MK scheme. The residual of 300 stars did not enter into the scheme and thus had to be left aside: these were the 'peculiar' looking objects. Further subdivision into families of peculiar stars resulted in the following breakdown: 140 Am stars, 182 Ap, 8 δDel, 5 A-shell and 14 miscellaneous cases. Expressed in percentages, Am stars constituted 11%, Ap 10%, and the others 1.6%. Thus in general peculiar stars are only a fraction of normal stars. This is so, of course, because we defined normality to be the condition of the majority.

When it comes to the definition of the peculiarity groups, these are given in terms of spectral features, whether they be the presence of enhanced features or their weakness or absence, with regard to other features whose behavior is similar to that of normal stars. In other words, among a number (n) of features (lines, blends) in the peculiar looking star, we can assign a certain number (p) to a 'normal' spectrum, and $n - p$ lines are either stronger or weaker than the corresponding lines in the 'normal' star. The number of lines behaving differently depends on the spectral type, the wavelength range and the dispersion. For instance at 60 Å/mm, in the wavelength range $\lambda\lambda$3900–4800, there exists among the late B dwarfs a group of stars characterized by the strong enhancement of the Si II lines at $\lambda\lambda$4128–30. These stars are called 'Si stars' and, because of their relation to the peculiar stars of type A, 'Ap stars of the Si-type'. In other cases the situation is more complex since $n - p > p$, as happens in K-type giants where a number of molecular bands of carbon molecules (C_2, CN, CH) are sometimes enhanced. These stars are called 'C stars'. Since these bands dominate the spectrum, it requires some analysis to find out that the (faint) atomic lines behave as in a normal K-type giant.

As a third example we have the early B-type stars whose spectra are normal, except that the He I lines are all very strong. It is tempting to call these stars 'He stars', following the example of the 'Si stars'. Here however a difficulty appears in that other stars – middle B-type dwarfs – have weak He I lines, when compared with normal (middle B-type) stars. We are thus obliged to distinguish these two groups by calling them 'He rich' and 'He poor'. These two designations are really abbreviations for the phrases 'stars with stronger than normal helium lines' and 'stars with weaker than normal helium lines', but it must always be remembered that they are just abbreviations. They do *not* mean that in these stars the number of helium atoms is larger or smaller than in normal stars, because this is something for the theoretician to find out from the analysis of the spectrum. What the classifier sees is an enhancement, a weakening or the disappearance of features; what this means in physical terms is something else.

In some cases theoreticians have been able to show that there is a real enhancement in the number of atoms of certain species of element (or elements!), whereas in other cases more complex situations are involved, like a sorting out of certain elements from the stellar atmosphere by gravitation or by radiation pressure.

If we consider two stars, one with 'normal' composition and the other with an overabundance of – say – helium, we would expect to see an anomaly in the line strength of the second star. This is true *only if* the star is hot enough to excite He I lines – if it were a cool star, the abundance anomaly would be impossible to discover, because no He I lines would be seen in its spectrum.

We have emphasized these points in order to dissociate in the reader's mind the observed facts (abnormal line intensity) from the interpretation. Please remember through the remainder of the book that terms like 'strong' and 'weak' refer to observed facts, not to interpretation.

In the analysis of what is seen in the spectrum, we must still consider some further complications. Up to now we have mentioned mostly cases in which just one element is enhanced or weakened, like He or Si, but a simultaneous enhancement (or weakening) of a group of elements may also occur.

Since terminology has not always been clear, we summarize briefly the most frequent groupings of elements.

Iron group elements or 'iron peak elements'. In this group are usually placed Cr (24), Mn (25), Fe (26), Co (27) and Ni (28), but some authors add Sc (21) and V (23).

Rare earths or lanthanides. This group comprises elements with atomic numbers between La (57) and Lu (71).

Heavy elements are usually elements between Sr (38) and Xe (54); however, usually only Sr, Y and Zr can be seen distinctly in the spectra.

'*Metals*' (a term which is used very equivocally). Stellar interior specialists use 'metals' to designate any element other than hydrogen and helium, and in consequence 'metal abundance' implies all elements other than the first two. For spectroscopists this is very misleading, because they use the word in the chemical sense. On the other hand photometrists, who observe the combined effects of all lines (i.e. without distinguishing the different elements), often use the term 'metal abundance', which may also include the effect of the hydrogen lines. It is important to make sure in each particular case what the author really meant.

2.5 The technique of classification

Spectrograms are examined visually with the help of a magnifying device giving an amplification of about ten times. Higher magnification usually enhances the plate grain too, so that in the end little is gained. Spectrograms should be examined against a uniformly illuminated background, whose intensity can be regulated. The examination is best carried out with microscope-type binocular viewers, rather than magnifying glasses, because then both eyes are used and hands are free.

Since classification is based upon the comparison of spectra, it is vital to be able to look simultaneously at two plates. This can be achieved by superimposing the two plates (emulsion against emulsion), but the plates must be handled carefully to avoid scratching them. Using a spectrocomparator, in which both plates are brought into the same focal plane at slightly different positions in the field of view, solves this problem. Both plates need to be set up carefully, but they are less likely to be scratched.

We have said that two plates have to be compared; clearly, this can be carried out only if they were photographed at the same dispersion. This is an absolute requirement: it is impossible to compare spectra taken at different dispersions. This means that for classification all standard stars must be observed with the same instrument as the program stars, because classification cannot be carried out by comparing spectra with the standard stars of a printed atlas. Photographic atlases can never substitute for a direct plate because of the different aspects of the spectral features and of the plate grain, the differences in density and illumination, and the different scales and degrees of widening.

If plates taken with the same instrument are compared, one important factor is the density of the spectra. It is essential that both spectra are as similar in density as possible; nowadays this is easy to achieve by using exposure meters. If no exposure meter is available, three exposures of each object need to be taken, the first with the correct exposure, the second with one-half and the third with double the exposure time. This is, of course, very wasteful of exposure time, but ensures that a plate with a convenient density to match with a standard can always be found. Experience shows that a spectrum's aspect changes very much with density and therefore all precautions should be taken to compare only similar plate material. Such precautions are essential in objective prism work, where the best exposure cannot always be chosen; here plates with different exposure times are essential.

Assuming then that we have two similar spectra side by side, the next question is what to look for. The answer to this is contained in the *spectral atlases*, which illustrate the spectra of different types of stars and point out the essential features. These features should be sought in the actual spectra and compared. Usually a little experience helps us to determine approximately the spectral type of the object. Let us say that the spectrum examined is of early A-type. The second step is to compare it successively with the standards of type A0, A1, A2 and A3 to discover which it most resembles. Having decided that it most resembles, say, an A2 star, we consult the atlas to obtain the criteria for classifying an A2 star, which are then applied.

Most classification criteria use the ratio of line intensities, for example He I $\lambda 4471$/Mg II $\lambda 4481$. This means that we compare visually the two lines and decide which is stronger. We notice for instance that both are equal in strength and write $\lambda 4471 = \lambda 4481$, which is a criterion for a (main sequence) B8 star. Usually this operation strikes most people as being rather inaccurate, but in fact with a little experience these estimates can be remarkably consistent, in the sense that comparisons of line strength are repeatable and give similar results even when done by different people.

If a bidimensional scheme is used, which specifies spectral type and luminosity class, the procedure is to fix an approximate spectral type first, then to fix the luminosity class and finally to determine the accurate spectral type. It is clear that we should do these operations leisurely, and without knowledge of the classifications assigned by other people, or of the photometrically suggested type.

A word of warning about using the classification criteria: these are only to be used as a guide, and the whole spectrum must be looked at. The purpose of classification is to ensure that a spectrum is identical to that of the corresponding standard star. If the criteria recommended suggest a type B8 V, but closer examination of the spectrum reveals an unexpected line (unexpected because it is not present in the standard), this fact is important and needs to be pointed out in a note. In short, criteria should never be applied blindly.

You find out by experience that classification is an occupation which requires a careful examination of the material, as often or as many times as you need to convince yourself that you have seen everything present on the plate. This should be done under the best working conditions, when your eyes are not tired and when you still remember what was done the day before. It is rather like playing a musical instrument: once learned you never forget it entirely, but it is best to keep practicing before performing.

By examining the plate material several times, you start to see things not perceived before. The experience acquired in classifying is what makes a classifier.

Because it takes some time to build up experience, and also because it is certainly time-consuming to classify a large number of stars, several schemes for automatic classification have been put forward. We shall deal with this in the next section.

2.6 Automated classification

We shall cover here all methods of obtaining a spectral classification without using the visual intervention of a classifier.

In principle, the problem can be solved by having a library of standard stars and by comparing a given star with the stars of this library. Statistical tests may then be set up to decide if objects are or are not identical; the spectral classification then depends on the outcome of these tests.

In practice, however, this scheme is difficult to apply. The difficulties arise between the origin of the spectrum (the atmosphere of the star) and its final plot. First of all, the electromagnetic radiation must traverse interstellar space and the earth's atmosphere before it arrives at the telescope. Then the radiation may be altered in the telescope before it is recorded by an instrument. Finally, a plot of intensity (or a quantity related to it) versus wavelength appears. Clearly, our automated spectral classification can only be applied to the final plot; the complications arise from the distorting influence of the several intermediate factors. Among these the biggest difficulty is the conversion of photographic density into density; in fact, many attempts at automated spectral classification have exhausted most of their energy on this step. To visualize what could ideally be done, let us therefore assume that this step is taken care of and that we have arrived at a plot of $I(\lambda)$, which can be imagined as a set of n intensity points at prefixed values of λ_i $(i = 1,2,\ldots, n)$, or as one point in an n-dimensional space.

The next step is to compare this point with the points representing standard stars, and to judge if it is sufficiently close – within the observational errors – as to be identifiable with one of the standards. If it *is* close, the star can be identified with the standard, and the process of classification is thus ended.

It is easy to think of other ways to carry out classification; for instance, not to use all n points, but only a few. This is equivalent to applying classification criteria alone, disregarding all other information.

The interested reader may find a detailed explanation of the principles involved in the summary paper by Kurtz (1984) and in a recent paper by Heck *et al.* (1986).

Up to now automated methods have not produced large amounts of new data, so their capabilities have not yet been fully tested.

2.7 Auxiliaries

Classification work needs some auxiliaries, which we shall group under four headings, namely line identification tables, atlases, lists of standards and catalogs.

2.7.1 *Line identification tables*

These tables provide the spectral lines characterizing a chemical element. Usually wavelengths, line intensities, identification of the corresponding ionization stage of the element and the spectroscopic designation of the transition are given. A vast literature exists concerning spectroscopic work, which is surveyed regularly in the *Reports on astronomy*, published by the IAU for every General Assembly, chapter Comm. 14 'Atomic and molecular data' (Martin 1983). Most classifiers, however, prefer general tables where all elements are combined, for example *A multiplet table of astrophysical interest* by C. Moore (1945). This work is now outdated. More recent material is:

- *An ultraviolet multiplet table* (Moore 1950, 1952, 1962)
- *Atomic and ionic emission lines below 2000 angstroms* (Kelly and Palumbo 1973)
- *Tables of spectral lines of atoms and ions* (Striganov and Odintsova 1982)
- *Line spectra of the elements* (Reader and Corliss 1982)

Updated bibliographies of atomic line identification lists are published by Adelman (1978, 1983), and by the National Bureau of Standards (1972).

2.7.2 *Atlases*

Spectral atlases are graphic documents which illustrate spectral classification systems. Therefore in principle each system has its own atlas. Since nowadays the most widely used system is the MK system, we quote its atlases:

- *Revised MK spectral atlas for stars earlier than the sun* (Morgan, Abt and Tapscott 1978)
- *An atlas of spectra of the cooler stars. Types G, K, M, S and C* (Keenan and McNeil 1976)

At somewhat higher plate factors we have:

- *An atlas of representative stellar spectra* (Yamashita, Nariai and Norimoto 1977; 73 Å/mm)
- *An atlas of grating spectra at intermediate dispersion* (Landi–Dessy, Jaschek and Jaschek 1977; 42 Å/mm)

For objective prism spectra we mention:

- *An atlas of objective prism spectra* (Houk, Irvine and Rosenbush 1974; 108 Å/mm)

Outside the strict MK domain, but linked to it we have:

- *An atlas for objective prism spectra* (Seitter, vol. I, 1970, 240 Å/mm; vol. II, 1975, 645 and 1280 Å/mm)

The preceding atlases all provide reproductions of photographic plates. An atlas of density tracings of photographic plates has been produced by Goy, Jaschek and Jaschek (1986).

Intensity tracings of spectra are provided in the *Library of stellar spectra* (Jacoby, Hunter and Christian 1984), but the stars reproduced are not always MK standards.

In addition to the optical region we have atlases covering the ultraviolet spectral region. Here all atlases provide tracings, no bidimensional images. We have here:

- *An atlas of ultraviolet stellar spectra* (Code and Meade 1979, 1980; both references cover the region $\lambda\lambda$1200–3600 Å.)

At higher plate factors (resolution 37 Å):

- *An atlas of ultraviolet stellar spectra* (Cucchiaro, Jaschek and Jaschek 1978, $\lambda\lambda$1350–2740)

The region $\lambda\lambda$1200–3200 is covered at a resolution of 2 Å by:

- *The IUE low dispersion reference atlas* (Heck, Egret, Jaschek and Jaschek 1984)

Besides these general atlases, we shall mention in the different chapters of this book the more specific atlases refering to small subsets of the stellar population.

2.7.3 *Lists of standard stars*

As we said before, each classification system is defined by its own standards. This implies that for any classification work in a well-defined system we must observe all or a large fraction of its standards.

Standards of the MK system were collected by Philip and Egret (1985). The recommended lists are Morgan and Abt (1972), Morgan and Keenan (1973), Morgan, Abt and Tapscott (1978) and Keenan and Yorka (1985). Some of the stars have been reclassified slightly over the years, so the user is recommended to take the latest classification of a star.

2.7.4 *Catalogs*

Catalogs are collections of data of a given type. We can distinguish roughly between *observational catalogs* (OC) or lists of an author's own observations, *bibliographic compilation catalogs* (BCC) or lists of observations from many different sources, and *critical compilation catalogs* (CCC) or lists of observations from many different sources with a critical treatment. For more details the reader is referred to Jaschek (1984b). Since the number of catalogs is very large, we shall concentrate on classification catalogs. Among the most widely used spectral classification catalogs we have:

- *Henry Draper catalog* (OC) by Cannon and Pickering (1924 on), which provides unidimensional classifications for 225 000 stars
- *Catalogue of stellar spectra classified in the Morgan–Keenan system* (1964) (BCC) by Jaschek, Conde and de Sierra (1964), which provides classifications for about 21 000 stars. The update of this work is the MK extension catalog (ed. 1982) (BCC) by Kennedy (1983). It exists only on tape.

In recent years Houk has been reclassifying in the MK system all stars contained in the HD. This is done by zones of declination, and up to 1986 volumes I, II and III have been published, which cover the southern hemisphere from $-90°$ to $-26°$. This constitutes the *Michigan catalog of two-dimensional spectral types for the HD stars* (OC) by Houk and Cowley (1975), Houk (1978, 1982).

A critical compilation catalog (CCC) is the *Catalog of selected spectral types* by M. Jaschek (1978), providing a 'best' spectral type for about 30 000 stars.

Besides these general catalogs, there are a number of more specialized ones, referring either to a selection of stars or to a restricted region. For a general survey of what is available, see Jaschek (1985). Updated lists of catalogs available are published in the *Bulletin d'Information du Centre de Données Stellaires* (Strasbourg, France) and the *Astronomical Data Centre Bulletin* (NASA Space Science Data Centre, Greenbelt, USA).

References

Abney W. de W. (1877) *MN* **37**, 278
Adelman S.J. (1978) *PASP* **90**, 766
Adelman S.J. (1983) *PASP* **95**, 897
Angel J.R.P. (1977) *Ap. J.* **216**, 1
Cannon A.J. and Pickering E.C. (1924 on) *Harvard Obs. Annals*, vols. 91 to 100
Code A.D. and Meade M.R. (1979) *Ap. J. Suppl.* **39**, 195
Code A.D. and Meade M.R. (1980) *Ap. J. Suppl.* **42**, 283
Cowley A., Cowley C., Jaschek M. and Jaschek C. (1969) *A. J.* **74**, 375
Cucchiaro A., Jaschek M. and Jaschek C. (1978) Liège and Strasbourg Observatories

Goy G., Jaschek M. and Jaschek C. (1986) (to be published)
Heck A., Egret D., Jaschek M. and Jaschek C. (1984) ESA SP-1052
Heck A., Egret D., Nobelis P. and Turlot J.C. (1986) *Astroph. Sp. Sc.* **120**, 223
Houk N. (1978) *Michigan catalog*, vol. II, University of Michigan
Houk N. (1982) *Michigan catalog*, vol. III, University of Michigan
Houk N. and Cowley A.P. (1975) *Michigan catalog*, vol. I, University of Michigan
Houk N., Irvine N.J. and Rosenbush D. (1974) *An atlas of objective-prism spectra*, University of Michigan
Jacoby G.H., Hunter D.A. and Christian C.A. (1984) *Ap. J. Suppl.* **56**, 257
Jaschek C. (1984a) in '*The MK process and stellar classification*', Garrison R. (ed.), Toronto, p. 94
Jaschek C. (1984b) *QJRAS* **25**, 259
Jaschek C. (1985) *IAU Symp.* **111**, 331
Jaschek C., Conde H. and de Sierra A.C. (1964) *Publ. La Plata, Ser. Astr.* **27**, (2)
Jaschek M. (1978) CDS Catalog 3042 and microfiche, Strasbourg
Keenan P.C. and McNeil R.C. (1976) *An atlas of spectra of the cooler stars*, Ohio State Univ. Press
Keenan P.C. and Yorka S. (1985) *BICDS* 29
Kelly R.L. and Palumbo L.J. (1973) NRL Report N. 7599
Kennedy P. (1983) CDS Catalog 3078, Strasbourg
Kurtz M.J. (1984) in *The MK process and stellar classification*, Garrison R. (ed.), Toronto
Landi-Dessy J., Jaschek M. and Jaschek C. (1977) Observatorio Astronomico de Cordoba–Argentine
Martin W.C. (1983) *Highlights of astronomy* **6**, 775, West R.M. (ed.)
Moore C.E. (1945) *Princeton Obs. Contr.* N. 20
Moore C.E. (1950) *An ultraviolet multiplet table*, NBS Circ 488, section I
Moore C.E. (1952) *An ultraviolet multiplet table*, NBS Circ. 488, section II
Moore C.E. (1962) *An ultraviolet multiplet table*, NBS Circ. 488, section III-IV
Morgan W.W. (1979) *Spec. Vaticana* **9**, 59
Morgan W.W. and Abt H.A. (1972) *A.J.* **77**, 35
Morgan W.W. and Keenan P.C. (1973) *Ann. Rev. AA* **11**, 29 (Palo Alto. Ann. Rev. Inc.)
Morgan W.W., Abt H.A. and Tapscott J.W. (1978) *Revised MK special atlas for stars earlier than the sun*, Yerkes Observatory and Kitt Peak National Observatory
Münch G. (1968) *Stars and stellar systems*, vol. VII, p. 365, Chicago Univ. Press
Ochsenbein F. (1981) Thesis, Strasbourg
Philip D. and Egret D. (1985) *IAU Symp.* **111**, 353
Reader J. and Corliss C.H. (1982) in *CRC Handbook of Chemistry and* Physics, 63rd edition, Weast R.C. (ed.) CRC Press, Boca Raton FL
Rufener F. (1981) *AA Suppl.* **45**, 207
Seitter C. (1968) *Mitt. Astr. Gesellschaft* **25**, 105
Seitter W.C. (1970) *Atlas für Objektiv-Prismen Spektren*, vol. I, F. Dümmler Verlag, Bonn

Seitter W.C. (1975) *Atlas für Objektiv-Prismen Spektren*, vol. II, F. Dümmler Verlag, Bonn
Shajn G. and Struve O. (1929) *MN* **89**, 222
Striganov A.R. and Odintsova G.A. (1982) *Tables of spectral lines of atoms and ions*, Energoizdat, Moscow, USSR
van Altena W. (1983) in *IAU Coll.* **76**, 29
Yamashita Y., Nariai K. and Norimoto Y. (1977) *An atlas of representative stellar spectra*, University of Tokyo Press

3

Spectral classification systems

Spectral classification started in the second half of the nineteenth century and has had a long and interesting development, which is detailed by Curtiss (1932), Fehrenbach (1958) and Seitter (1968). Instead of following the historical developments in their complicated interactions, we shall study in detail only two classification systems, because they synthetize the developments. These two systems are the Harvard and Yerkes systems. In the last section of this chapter we shall consider briefly classifications at higher plate factors.

3.1 The Harvard system

The material for this system consists of objective prism spectra of a large number of stars. Regrettably not all spectra were obtained with the same cameras and/or objective prisms, and this produced some undesirable consequences. In particular, brighter stars were observed with lower plate factors than fainter ones, the latter having a total length of 2.2 mm between Hβ and Hε. For stars of intermediate brightness this length was 5.6 mm and for the brightest stars up to 80 mm. However, all spectra were widened considerably to more than 2 mm.

The Harvard scheme divided the spectra into classes symbolized by capital letters, arranged in alphabetic order. Class A was described as 'only the hydrogen series and generally K are visible'. In class B the lines of class A were supplemented with λ4026 and λ4471 of He I. Class C was the class of stars of types A and B but with Hδ and Hγ observed as double lines. Successive revisions rearranged the order of the groups (for example, types A and B exchanged places) and others (like type C) were suppressed. The use of letters, introduced by Pickering, was complemented by Cannon with subdivisions denoted by numbers from zero to nine, thus providing symbols at decimal intervals of a class. The order finally retained was:

O – B – A – F – G – K – M

Stars of classes O and M were subdivided by affixing the letters a, b, c, etc. after the type, to denote subdivisions not necessarily in a sequence. Not all decimal subdivisions were used for classes B to K, so that finally only the ones given in table 3.1 were used.

Miss Cannon (1901, 1912) produced a short description of each type, adding the name of one or two standard stars. These descriptions can be found in the preface of each of the volumes constituting the *Henry Draper catalog* (Cannon and Pickering, 1918); except for late N stars the descriptions are identical in each volume.

We do not wish to reproduce here the description of all the types, but will select rather arbitrarily the sequence of types A0 to F0.

Class A0. Typical star, α Canis Majoris. The hydrogen lines are at their maximum intensity, and line K is 0.1 as intense as Hδ, or less. On plates having sufficient dispersion, the calcium line H, at λ3968.6, is separated from Hε, λ3970.3, and is nearly as intense as line K. Line λ4481.3 is the strongest after the hydrogen lines and line K. On a photograph taken with the 13-inch Boyden telescope, with the dispersion of three prisms, 93 solar lines were measured.

Class A2. Typical stars, δ Ursae Majoris and ι Centauri. Line K is 0.3 or 0.5 as intense as Hδ. Solar lines are well marked, especially lines λ4481.3, λ4226.9 and λ4233.8. The latter two form a nearly equal pair. No helium lines are seen in this or any of the following classes.

Table 3.1. *Spectral types used in the Harvard system.*

O	a, b, c, d, e
B	1, 2, 3, 5, 8, 9
A	0, 2, 3, 5
F	0, 2, 5, 8
G	0, 5
K	0, 2, 5
M	a, b, c, d

Class A3. Typical stars, α Piscis Austrini, and τ^3 Eridani. Line K is more than 0.5 as intense as the compound line H and Hε, and is 0.8 as intense as Hδ. The metallic lines are more numerous and more intense than in class A2, while the hydrogen lines are slightly fainter.

Class A5. Typical stars, β Trianguli and α Pictoris. Line K is 0.9 as intense as the compound line H and Hε, and about 3.0 as intense as Hδ. Lines λ4305.6, λ4308.0, λ4309.5, and other lines which form the absorption band called G by Fraunhofer, are faint and inconspicuous.

Class F0. Typical stars, δ Geminorum and α Carinae. The hydrogen lines are about 0.5 as intense as those of α Canis Majoris. The K line is as strong as the compound line H and Hε, and about 3.0 as intense as Hδ. Lines λ4305.6, λ4308.0, λ4309.5, and other lines which form the absorption band called G by Fraunhofer, are faint and inconspicuous.

As can be seen from these descriptions, Cannon does not use one single criterion to subdivide stars, but three, namely:

- line ratios (for instance, K with regard to H + Hε)
- absence of lines (He I lines absent)
- strength of lines

and no preference is given to any particular one. It can be noted that the use of 'absence' or 'strength' of the lines is somewhat dangerous, especially in the case of objective prism plates, where the spectrum depends very much on the stellar image quality. To complicate things further, the use of different dispersions does not help either. It was found, for instance, that many B-type stars near the plate limit were misclassified because the He I lines were too weak to be recorded. The spectrum was thus called B9 or A0 because only the hydrogen lines remained visible, whereas correctly it should have been called B2 or B3.

The spectral sequence defined in the Harvard system is illustrated in the frontispiece of *Harvard Observatory Annals* **91** with a few typical spectra. However, these spectra were observed with a slit spectrograph, whereas the material for the *Henry Draper catalog* consists, as we have seen, of objective prism plates.

Besides the spectral classes from O to M and class S which are still used today, Cannon also used the following types, which have since been dropped.

Class P, corresponding to spectra of gaseous nebulae, subdivided into a, b, . . . , f.

Classes R and N, corresponding to carbon stars. They were subdivided into R0 – 3 – 5 – 8 – Na – Nb and (in part of the work)Nc.

Class 'Pec': all spectra which could not be assigned to any known class, considering their principal characteristics. This includes the spectra of novae, a few variables, very red stars and some others.

Class 'Con': spectra which are apparently continuous. This includes spectra of nebulae without bright lines or of clusters which resemble such nebulae with the dispersion employed.

The Harvard system was used to classify all stars in a large survey (the above-mentioned *Henry Draper catalog*) which covers the whole sky and includes 225 000 stars. The work was carried out by Cannon, and is printed in *Harvard Observatory Annals* **91–100**. The survey has no fixed limiting magnitude; in the northern sky it is complete to about $m_v = 8^m25$, whereas in the south it goes to about 8^m75. The success of the HD system was that a simple scheme, consisting of no more than thirty types, was able to accommodate 99% of the objects observed (i.e. 2.2×10^5 stars). In the second place, no other classification scheme was ever applied to such a large number of spectra – it gave a sound basis for further research. In the third place, it was based upon easily obtained material, namely objective prism plates; each plate contains the images of many stars.

Miss Cannon also provided a wealth of remarks on peculiar looking spectra in her notes; a precious source of information for further research.

3.2 Later developments of the Harvard system

Subsequently, the Harvard system was officially recommended by the International Astronomical Union (IAU 1922), with some modifications. These fall into three categories:

(a) General underlying principles and methods
(b) Additions and redefinitions of classes or divisions, and
(c) Additions of new symbols of general significance.

For point (a) (general principles) we may quote two modifications, namely:

(1) The classification should describe the spectra, not the stars; that is, it

should be based solely on what can be seen in the spectrum of a given star, suitably observed with appropriate instruments.

(2) The Harvard system, which had already been adopted internationally, should be the basis on which further extensions are built.

The first principle stresses the very important point that a classification should only be based upon what we really see in the spectrum, not upon what we know besides the spectrum. This can be exemplified by the 'Ap' classification. 'Ap' denotes a peculiar spectrum, not a peculiar star, because the word 'peculiar' has no meaning outside a classification system since nothing is peculiar by itself.

The second principle is one of continuity. Clearly we should not leave to one side what has been done before, but rather we should use it for further developments. With more than 2×10^5 stars classified in the HD system, it would be impractical to set up a completely different system abandoning such a large number of classifications.

With regard to recommendations of type (b) (new classes), the class S was introduced, following a previous proposal from Cannon.

The most important proposals are of type (c) additions of new symbols denoting peculiarities which greatly extend the list of the few symbols used in the *Henry Draper catalog*. These symbols are often used even today and so we will describe them in some detail. For convenience we group them.

Width of lines. Spectra showing all lines wide or diffuse on good plates are denoted 'n' (nebulous), whereas 's' is used to qualify spectra in which the lines are sharp, as A5s. When extreme sharpness of lines is coupled with anomalous intensities of some lines, we speak of 'c' characteristics. 'c' is used as a prefix: cA5. Such stars were later found to be supergiants.

Interstellar lines. When these lines are observed in a spectrum, one uses the suffix 'k' (for interstellar 'K' lines). So, for instance, δ Ori would be designated B0nk.

Emission lines. The presence of emission lines is indicated by the suffix 'e'. However, in class Oe (also Pe and Qe, no longer used) the 'e' designates a subgroup, not an emission line.

Peculiarities. The letter 'p' denotes miscellaneous peculiarities, not sufficiently frequent or important to justify individual designations. The suffix qualifies the symbol immediately preceding; thus B2pe denotes a star of class B2, with peculiarities in the absorption spectrum and emission lines of the normal type. B2ep denotes an emission line star with peculiar emission lines.

Other symbols used are '!' to indicate a very marked degree of a phenomenon, for example 'e!', exceptionally strong emissions, and 'q' for P Cygni type lines (as in 'eq').

Luminosity effects. It was also recommended that prefixes should be used to denote the luminosity of the stars. Very luminous stars were to be prefixed with 'c', giants with 'g', and dwarfs with 'd'.

The introduction of the luminosity prefixes is largely the result of the pioneering work of the Mt Wilson astronomers, which was started by Adams and Kohlschütter (1914) and continued in the twenties and thirties by several other groups. The reader can find detailed reviews of the developments in these years in the papers by Curtiss (1932) and Fehrenbach (1958).

We must add that the luminosity prefixes were not universally used in the years following the IAU recommendation; for example, the Harvard astronomers did not use them.

3.3 The Yerkes system

The Yerkes system is the work of two astronomers, W.W. Morgan and P.C. Keenan, and their associates. Abbreviations such as MKK (for Morgan, Keenan and Kellman) and MK (for Morgan and Keenan) have been used to describe parts of the system, but it seems preferable to use the designation 'Yerkes system' in general.

The Yerkes system is based on slit spectrograms obtained with a prismatic spectrograph, covering the region $\lambda\lambda$3930–4860, at a dispersion of 115 Å/mm (at Hγ). The spectra were widened to about 1 mm. The advantage of slit spectrograms over objective prism spectrograms is that the definition of the spectra is independent of the quality of the images, so that the resulting material is of much higher quality, both in resolution and contrast.

With regard to the material used, the Yerkes system has tended over the years to move toward higher dispersions ($\sim$ 60 Å/mm), longer wavelength ranges ($\lambda\lambda$3500–4900), the use of grating spectrographs (therefore a uniform dispersion over all wavelengths), and the use of even greater widenings of the spectra.

The Yerkes system constitutes a development of the Harvard system. Its main innovation is the introduction of a second parameter, related to the luminosity of the star. As already said, Adams and Kohlschütter (1914) had shown that differences can be recognized in the spectra of giant and of dwarf stars when observed at about 40 Å/mm. They used pairs of lines which were not far from each other in the spectrum and were of practically the same intensity and character. By careful comparison they found that the ratios of certain lines are strongly dependent on luminosity. The line ratios were expressed on an arbitrary scale from 1 to 12, 1 being the smallest difference in intensity which could be detected. Average line ratios were then plotted against the absolute magnitudes of the stars and a relation derived.

More investigations of this kind showed the soundness of the approach, so that the use of luminosity sensitive line ratios became general. The Mt Wilson observers proposed therefore the introduction of symbols denoting stars of different kinds, namely the prefixes 'd' for dwarfs and 'g' for giants. A star is thus denoted a dK5 or gK5. For the very bright stars, the symbol 'c' introduced by Maury (1897) was retained.

Although these developments gave a great impetus to stellar astronomy, we should notice that the notation is ambiguous. When a star is called a 'giant', it is not clear if this conclusion comes from a determination of its absolute magnitude, or from a determination of its radius, or from the line ratios observed in the spectrum. To keep things clear, it is preferable to use notation referring only to phenomena observed in the spectrum.

This step was accomplished in the Yerkes system by introducing the luminosity classes I to V. The former class 'c' corresponds to luminosity class I, the former 'g' to III and the former 'd' to V. In this notation the luminosity class is an observable spectral characteristic; once established it can be correlated with absolute magnitude, radius or any other parameter. The introduction of luminosity classes adds a second dimension to the unidimensional spectral sequence of the Harvard system.

The procedure of classifying a star in the Yerkes system needs the use of a number of standard stars which define the system, both in spectral type and in luminosity class. Clearly the list of standards has evolved over the years. List of standards are given in Morgan, Keenan and Kellman (1943), Morgan and Roman (1950), Johnson and Morgan (1953), Morgan and Abt (1972), Morgan and Keenan (1973), Keenan and McNeil (1976), Morgan, Abt and Tapscott (1978), Keenan and Pitts (1980), Keenan (1983), and Keenan and Yorka (1985). A summary of them is provided by Philip and Egret (1985). A certain number of spectra are illustrated in the atlases (see section 2.7) which also describe the criteria to be used for classification. Atlases are didactic devices and are never to be used as a substitute for real spectrograms; for classification work, all stars, including the classification standards, need to be photographed with the same spectrograph and under the same conditions.

Later work in the Yerkes system refined the luminosity classes. Table 3.2 illustrates this.

For late type stars, the classification from the best spectrograms is precise enough to allow the use of more subdivisions in the range from II to V, as is illustrated on the right-hand-side of the table. When the use of the subclasses is not justified, the letters are simply omitted.

With regard to spectral class, there has also been progress over the Harvard system. First of all, the Yerkes system uses the classes given in table 3.3, although occasionally intermediate classes like A6, A8 and G7 are also used.

Table 3.2. *Luminosity classes in the Yerkes system.*

Symbol	Name	Subdivision
I	Supergiant	I
	(or hypergiant)	I a
		I ab
		I b
II	Bright giant	
III	Giant	II–III
		III a
		III ab
		III b
		III–IV
IV	Subgiant	
V	Dwarf	

Table 3.3. *Spectral subtypes used in the Yerkes system. From Keenan (1985b).*

O						4	5	6	7	8	9	9.5
B	0	0.5	1	2	3		5		7	8		9.5
A	0			2	3		5		7			
F	0			2	3		5		7	8	9	
G	0			2			5			8		
K	0			2	3	4	5					
M	0		1	2	3	4			7	8		

Classes O3 and B0.3 (halfway between B0 and B0.5) have also been proposed, but are not generally used (Morgan and Keenan 1973).

We refrain from listing the line ratios used for spectral types and luminosity classes, since we shall return to this in the chapters devoted to each spectral type. Here we shall confine ourselves to some general remarks. The first is that criteria or line ratios should not be used blindly, in the sense that of all the lines and features visible in the spectrum the classifier regards only those implied in the line ratios which the atlas quotes as being characteristic for a given type of objects. Quite to the contrary, the classifier should compare *all* features, and use the line ratios only once he or she is satisfied that the spectrum to be classified exhibits nothing abnormal, like for instance one (or several) additional lines, or the lack of a feature, or differences in the appearance of the lines. This consideration and weighing of all features is what characterizes the human spectroscopist, in contrast to a machine-made classification which handles only a small number of criteria, fixed beforehand.

In the second place, the classifier needs a large collection of standard stars. Since the classification process is essentially recognizing the identity of a spectrum as one of a given list of standards, ideally the classifier should observe all the Yerkes standard stars, to ensure a dense network of stars for comparison. This is crucial when the spectroscopist is working at plate factors which differ from those of the Yerkes system. In such cases the classification criteria cannot be translated, and the only way of guaranteeing a classification close to the Yerkes system is to have as many standards as possible.

Precision

We have seen in tables 3.2 and 3.3 that spectral types and luminosity classes are subdivided into tenths. This suggests that the precision of good classification should be slightly better than the smallest subdivisions used. The practical problem of assessing the precision can be solved by comparing classifications attributed to stars by different classifiers. Such an approach was followed by Jaschek and Jaschek (1965), who studied several samples of stars and concluded that $\sigma(S) = \pm 0.6$ ($S =$ spectral class in tenths of a spectral type, $\sigma =$ dispersion). Another approach is to use a parameter which is closely correlated with spectral type – color, for example – and to judge, for a given color, the intrinsic dispersion of the spectral types. (Obviously the colors must be freed from interstellar reddening effects.) Such an approach was followed by Jaschek and Jaschek (1973) who confirmed that $\sigma(S) =$

± 0.6. This result naturally represents an average over the different spectral types and is subject to variations from one type to another.

For luminosity classes this kind of procedure is more difficult to apply. Houk (private communication) concluded that $\sigma(L) \sim 0.8$ from the analysis of a large sample.

Both results taken together imply that errors are of the order of the finest subdivisions used in classification work, as expected.

3.4 Later developments of the Yerkes system

Over the years since its first formulation, the Yerkes system has won a wide acceptance, first of all because of its usefulness in providing a convenient reference frame. The fact that the system is defined in terms of spectral features and not in terms of physical parameters leaves it untouched by changes in the physical interpretation of stellar spectra. τ Sco remains a B0 V star although the values of its temperature and absolute magnitude have been changed several times in the past. Similarly a spectral peculiarity in the Yerkes system is independent of the exact physical explanation involved.

The value of such a reference system can be seen, for instance, in photometry. A color index can be correlated with temperature, but usually it is preferable to calibrate it on spectral type – simply because in this way changes in the temperature scale are circumvented. Of course, conversion from color index to temperature is needed if we want to know something about the physics of the star, but as long as we are interested only in statistics, this is not essential.

The second advantage of the Yerkes classification is that a luminosity class can be assigned to a star. The luminosity class can then be calibrated in terms of absolute magnitude (M_v), and through the relation

$$M_v = m_v + 5 - 5\log r - A_v$$

we can obtain the distance r of the object if its apparent magnitude (m_v) and the interstellar extinction (A_v) are known. According to van Altena (1983) there exist 7400 stars for which we have a measured trigonometrical parallax, but only 600 have standard errors not exceeding $\Delta M = \pm 0.3$. On the other hand we know luminosity classes for more than 10^5 stars, which can be translated into distances through the standard calibration procedure. This shows very convincingly that the spectroscopic method for obtaining stellar distances has no current rival. (In fairness it should be noted that stars in clusters should be included in the group of stars for which good distances can be determined from trigonometric parallaxes. So our remark properly refers to field stars.)

Unfortunately, if a system is successful it tends to be misused, either by omitting some of its basic requirements, or by using it outside its proper range of application.

The omission of some of the basic requirements corresponds mainly to the use of higher plate factors. It is clear that with 220 Å/mm plates one is not on the same ground as with the Yerkes material. Therefore the Yerkes notation for luminosity classes should *not* be used to describe the results. The same is true to a certain extent for objective prism material; by experience discordant results sometimes arise because of poor image quality. Similar warnings can be made with regard to resolving power and to plate widening.

Failing to bear in mind the range of application constitutes another kind of misuse. The Yerkes system is defined by a certain technique: visual inspection of photographic plates in the $\lambda\lambda$3600–4800 region. Therefore if spectra obtained in another region, say the satellite ultraviolet, are classified in the Yerkes system we should not expect the results to be compatible in both wavelength ranges (Jaschek and Jaschek, 1982, 1985); in fact disagreements do exist.

The same remark applies if we want to apply the Yerkes spectral classification to spectrophotometric or multicolor photometric data. Here again we should *never* use the same nomenclature as in the Yerkes system because we are not classifying spectrograms.

We shall refrain from providing more examples of misuse, because it is common sense that a system should be used only in the way it was defined. All extrapolations must be carefully justified.

3.5 Possible extensions of the Yerkes system

The Yerkes system was set up on a sample of bright stars, and in fact few stars are classified in it beyond the ninth or tenth magnitude (Ochsenbein 1981). This fact introduces a bias against low luminosity stars of all kinds since few of them are near enough to figure within $m_v < 10$. For dwarfs, this limit is 63 pc for K0, 25 pc for K7, 10 pc for M2 and 4 pc for M5. The situation is worse for subdwarfs and degenerates, which are fainter than main sequence stars. As we shall see later, the mode of the frequency curve of dwarfs of different spectral types lies beyond M2; therefore the Yerkes system refers only to intrinsically bright stars, which belong mostly to population I. A logical step would be to extend the system to stars of population II (Morgan 1979). An extension of the system to very late stars could also be possible, which should be done on material using the yellow and red regions ($\lambda\lambda$5000–8500) (Keenan 1985a, b).

Another relatively new chapter is the extension to stars in other galaxies. We know (see, for instance, Fehrenbach *et al.* 1977; Dubois, Jaschek and Jaschek 1977; Humphreys 1985) that the stars we can observe in each external galaxy differ by varying amounts from MK standards, but we have as yet no convincing evidence that a new classification must be used – perhaps a slight extension of the MK system will do.

3.6 Classification at higher plate factors

It is not always possible to observe stellar spectra under the conditions required for Yerkes classification. For instance, to reach fainter stars a higher plate factor may be used, or a different wavelength range, or both. It is clear from what was said before that we will then not be able to classify in the Yerkes system. On the other hand we would like to stay as much as possible within the general framework of the system. An illustration of what happens, for instance, at plate factors of 240 Å/mm is given by Seitter (1970) who reproduces spectra of MK standards at this dispersion. It is clear that at 240 Å/mm no differentiation between B6–A0V is possible, and the same is true for luminosity classes V, IV and III. This can be envisaged as a fusion of various MK boxes – the ones for B6 V, B7 V, B8 V, ... – into a single

Figure 3.1. Natural groups seen at 230 Å/mm. From Morgan (1951).

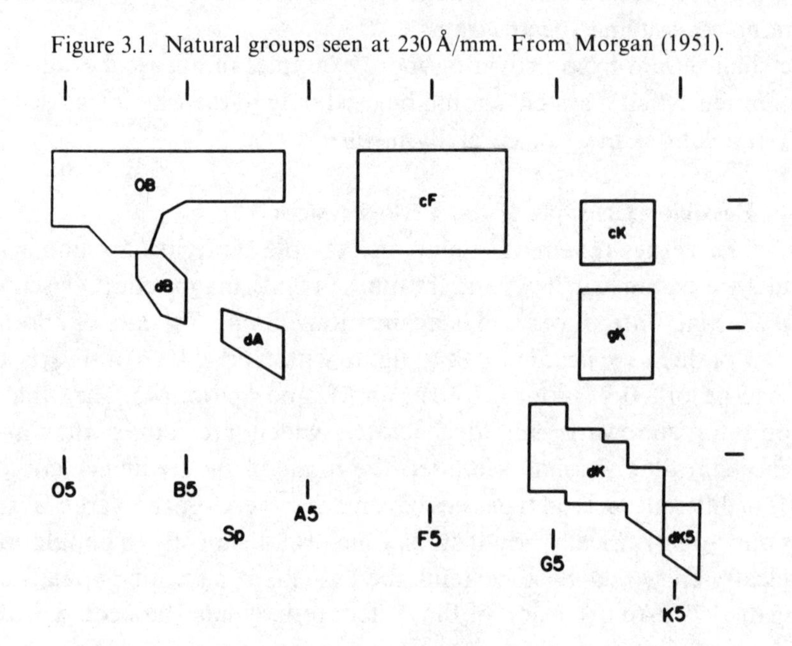

Table 3.4. *Data on natural groups for series* (*c*).

Natural group	LC-S classes	Group characteristics	Remarks
OB	O5–B0, B1 I–B5 I, B8 Ia, B1 II, B1 III, B2 II, B3	H weak or absent; all other lines weak or absent	Group is not unique, because of complication of main sequence B stars showing strong emission lines
dB	B2 V, B3 V	He I at maximum	Unique; compact group in luminosity and spectral type
dA	B9 V, A0 V, A2 V	H at maximum	Compact group in luminosity and spectral type; sensibly unique in low galactic latitudes
cF	F2 Ia–G0 Ia F2 Ib–G0 Ib	Ca II K ⩾ Ca II H; Hγ strongest absorption feature between Hβ and G band; blend near λ4170 outstanding	Unique; four subgroups in spectral type can be separated by ratio of G band to Hγ

Table 3.4. (*contd.*)

Natural group	LC-S classes	Group characteristics	Remarks
dK	G5 V–K2 V	CN absent; spectral type from growth of low-temperature features	Four subgroups in spectral type can be separated by change in ratio of G band to arc lines
cK	G8 Ib–K2 Ib	CN strong; Hδ strong	Several subgroups in spectral type can be separated from ratio of G band to Ca I λ4226
gK	G8 II–K2 II G8 III–K2 III	CN strong; Hδ weak	
dK5	K5 V–K6 V	Ca I λ4226 outstanding; no TiO	

one, which is 'the' corresponding box at 240 Å/mm. If we think in terms of a Hertzsprung–Russell (HR) diagram, the 'box' at 240 Å/mm appears as a rather large region, encompassing stars between absolute magnitude + 1.0 (A1 V) and − 1.8 (B6 III). Although this is rather coarse as far as absolute magnitude is concerned, it can be of interest to specific studies. When the size of the box in absolute magnitude is convenient for a certain purpose, the box is called a 'natural group', a designation introduced by Morgan (1951). As an illustration let us see which natural groups can be seen at 230 Å/min (Morgan 1951). Figure 3.1 illustrates the HR diagram and table 3.4 summarizes the description. Note the gaps in figure 3.1. For instance nothing is seen between 'dA' and 'dK'. This does not imply that these spectra are not seen – it implies only that they are all grouped in one box, which cannot be further subdivided. In this box we find stars between late A and middle G. From the point of view of the absolute magnitudes, this implies $0.5 < M < 5$, which is definitely too large to be useful and the dA–dG5 box is therefore *not* a 'natural group'. The same is true for the stars between B5 and B8. Notice also that what Morgan gives in table 3.4 is slightly different, as far as B6 V–A1 V stars are concerned, from the example we quoted above derived from Seitter (1970). The difference is due to the instrumentation used in both cases, and shows that natural groups depend very strongly upon the particular instrumentation. This is a fundamental fact.

McCarthy (1979) has listed the spectral features which are useful in the different spectral regions; the list is reproduced in table 3.5.

Table 3.5. *Spectral features at low dispersion.*

Near ultraviolet ($\lambda\lambda$3400–4000)	
Emission	: O II, Balmer
Absorp.	: O II, Balmer, Fe (λ3609, λ3635), CN Ca II (H and K)
Blue–yellow ($\lambda\lambda$4000–6000)	
Emission	: Balmer
Absorp.	: Balmer, G bd (λ4300), Ca I (λ4227), CN (λ4215), Na (D), Mg (G), TiO, VO
Red ($\lambda\lambda$6000–7000)	
Emission	: Hα
Absorp.	: Hα, TiO, VO
Near infrared ($\lambda\lambda$7000–9000)	
Emission	: Ca
Absorp.	: TiO, VO, LaO, A bd, Ca, C, CN

It can be seen that there are very few luminosity criteria – except the CN molecule and the D feature of Na – and few indicators of peculiar groups, except for emission line stars and CN anomalies. Illustrations of the different features can be found in Seitter (1975) who illustrates objective prism spectra taken at plate factors of 645 and 1280 Å/mm covering the region $\lambda\lambda$3500–8000 Å.

Practically all features listed in table 3.5 have been used in applications to specific problems, and it would take much space to consider all cases. We refer readers to two excellent summaries by McCarthy (1979, 1985) which cover the domain and provide a bibliography.

One common characteristic of these applications is that they were made for very specific problems – like the discovery of certain types of objects (C- and S-type stars), or the distribution of another (M-type dwarfs), or the discrimination of stars for further study (OB stars). In general no attempt was made to study *all* types of stars observed. From the point of view of classification, these methods are thus partial approaches, a fact which does not diminish their usefulness. When working at higher plate factors than those of the Yerkes system, we have to sacrifice the coverage of *all* types of stars, and limit it to a few specific regions – the 'natural groups'.

References

Adams W.S. and Kohlschütter A. (1914) *Ap. J.* **40**, 385

Cannon A. (1901) *Annals Harvard Obs.* **28**, 10

Cannon A. (1912) *Annals Harvard Obs.* **56**, 225

Cannon A. and Pickering E. (1918) Henry Draper catalog, *Annals Harvard Obs.* **91**

Curtiss R.H. (1932) in *Handbuch der Astrophysik*, vol. v, pt. 1, p.1, Springer Verlag

Dubois P., Jaschek M. and Jaschek C. (1977) *AA* **60**, 205

Fehrenbach C. (1958) *Handbuch der Physik*, ed. Flügge S. vol. 50, p. 1, Springer Verlag

Fehrenbach C., Jaschek M. and Jaschek C. (1977) *AA* **54**, 367

Humphreys R.M. (1985) in *The MK process and stellar classification*, Garrison R. (ed.), Toronto, p. 29

IAU (1922) *Transactions Intern. Astron. Union*, vol. I, p. 95

Jaschek C. and Jaschek M. (1965) in *IAU Symp.* **24**, 6, Loden K, Loden L.O. and Sinnerstad U. (ed.)

Jaschek C. and Jaschek M. (1973) in *IAU Symp.* **50**, 43, Fehrenbach C. and Westerlund B.E. (ed.)

Jaschek M. and Jaschek C. (1982) *ESA SP* **182**, 9

Jaschek C. and Jaschek M. (1985) in *The MK process and stellar classification*, Garrison R. (ed.), Toronto, p. 290

Johnson A.L. and Morgan W.W. (1953) *Ap. J.* **117**, 313

Keenan P.C. (1983) *BICDS* **24**, 19

Keenan P.C. (1985a) in *The MK process and stellar classification*, Garrison R. (ed.), Toronto, p. 29

Keenan P.C. (1985b) *IAU Symp.* **111**, 123

Keenan P.C. and McNeil R.C. (1976) *An atlas of the spectra of cooler stars*, Columbus, Ohio State Univ. Press

Keenan P.C. and Pitts R.E. (1980) *Ap. J. Suppl.* **42**, 541

Keenan P.C. and Yorka S.B. (1985) *BICDS* **29**, 25

McCarthy M.F. (1979) *Spec. Vaticana* **9**, 103

McCarthy M.F. (1985) in *The MK process and stellar classification*, Garrison R. (ed.), Toronto, p. 55

Maury C. (1897) *Annals Harvard Obs.* **28**, 13

Morgan W.W. (1951) *Publ. Michigan Obs.* **10**, 33

Morgan W.W. (1979) *Spec. Vaticana* **9**, 59

Morgan W.W. and Abt H.A. (1972) *A.J.* **77**, 35

Morgan W.W. and Keenan P.C. (1973) *Annual Rev. AA* **11**, 29, Palo Alto, Ann. Rev. Inc

Morgan W.W. and Roman N. (1950) *Ap. J.* **112**, 362

Morgan W.W., Abt H.A. and Tapscott J.W. (1978) *Revised MK spectral atlas*, Yerkes Obs. and Kitt Peak National Obs.

Morgan W.W., Keenan P.C. and Kellman E. (1943) *An atlas of stellar spectra*, Chicago Univ. Press

Ochsenbein F. (1981) Thesis, Strasbourg

Philip D. and Egret D. (1985) *IAU Symp.* **111**, 353

Seitter W.C. (1968) *Mitt. Astr. Gesellschaft* **25**, 105

Seitter W.C. (1970) *Atlas für Objektiv-Prismen Spektren*, vol. I, F. Dümmler Verlag, Bonn.

Seitter W.C. (1975) *Atlas für Objektiv-Prismen Spektren*, vol. II, F Dümmler Verlag, Bonn.

Van Altena W. (1983) in *IAU Coll. N.* **76**, Philip A.G.D. and Upgren A.R. (ed.)

4

Photometric classification

We start this chapter with a very general introduction to stellar photometry. The interested reader is advised to consult Golay (1974) for further details, whose book *Introduction to astronomical photometry* constitutes the standard reference. Readers familiar with the Russian language can also use the book *Multicolor stellar photometry* by Straizys (1977).

4.1 Stellar photometry

Stellar photometry is concerned with the measurement of radiative flux from stars.

Let $E(\lambda)$ be the illumination produced by the star's radiation at the top of the earth's atmosphere. If we measure all radiation in the wavelength interval between λ_b and λ_a $(\lambda_b > \lambda_a)$ we can write

$$E''_{ba} = \int_{\lambda_a}^{\lambda_b} E(\lambda)\,\mathrm{d}\lambda \tag{4.1}$$

If the radiation traverses the atmosphere, it is subject to a certain filtering which can be described by the function $T_{\mathrm{a}}(\lambda)$ giving the fraction of the original light which arrives at the observer. Therefore $0 \leqslant T_{\mathrm{a}}(\lambda) \leqslant 1$. Roughly speaking this function is zero for $0 < \lambda < 3100$ Å, about 1 for $3100 \leqslant \lambda \leqslant 10\,000$ Å and fluctuates strongly for $\lambda\lambda > 10\,000$ Å. It becomes practically zero for $\lambda > 17\,000$ Å, except for some narrow regions, and 1 for $1\ \mathrm{cm} < \lambda < 150\ \mathrm{m}$.

Therefore what actually reaches the observer is

$$E'_{ba} = \int_{\lambda_a}^{\lambda_b} T_{\mathrm{a}}(\lambda) E(\lambda)\,\mathrm{d}\lambda$$

The radiation then enters the telescope, traverses a filter and falls upon a receiver. We introduce similar filtering functions:

$T_t(\lambda)$ characterizing the transmission of the telescope,

$T_f(\lambda)$ characterizing the transmission of the filter,

$R(\lambda)$ characterizing the response of the receiver.

These functions vary between 0 and 1. Taking these factors in account, we measure not $E(\lambda)$ but

$$I = \int_{\lambda_a}^{\lambda_b} E(\lambda)\, T_a(\lambda)\, T_t(\lambda)\, T_f(\lambda) R(\lambda)\, d\lambda \tag{4.2}$$

The quantity I represents the measurement of the absolute flux, i.e. a number of photons. It is in general easier to make relative measurements between two sources; let us call them 1 and 2. Using the definition of magnitude, we write

$$m_2 - m_1 = -2.5 \log \frac{I_2}{I_1} \tag{4.3}$$

We have

$$m_2 - m_1 = -2.5 \log \frac{\int_{\lambda_a}^{\lambda_b} E_2(\lambda) T_a(\lambda) T_t(\lambda) T_f(\lambda) R(\lambda)\, d\lambda}{\int_{\lambda_a}^{\lambda_b} E_1(\lambda) T_a(\lambda) T_t(\lambda) T_f(\lambda) R(\lambda)\, d\lambda} \tag{4.4}$$

Please note that this cannot be transformed to

$$m_2 - m_1 = -2.5 \log \int_{\lambda_a}^{\lambda_b} \frac{E_2(\lambda)}{E_1(\lambda)}\, d\lambda$$

The aim of photometry is to find $E(\lambda)$, and this implies solving either equation (4.2) or equation (4.4). Mathematically this means solving an integral equation. Its solution presupposes the knowledge of T_a, T_t, T_f and R. Of these four, T_a is out of the control of the earth-bound astronomer, and can only be suppressed by observing from a satellite. The other three are selected by the astronomer and represent essentially today's technology. The easiest to handle is T_f, because it represents a filter which may be placed anywhere in the sensitivity range of R. With regard to filters, we may define three types of stellar photometry, namely

$\lambda_b - \lambda_a < 90$ Å narrow band photometry

$90 < \lambda_b - \lambda_a < 300$ Å intermediate band photometry

$\lambda_b - \lambda_a > 300$ Å wide band photometry

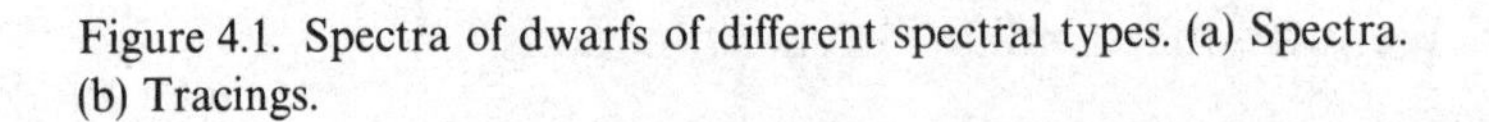

Figure 4.1. Spectra of dwarfs of different spectral types. (a) Spectra. (b) Tracings.

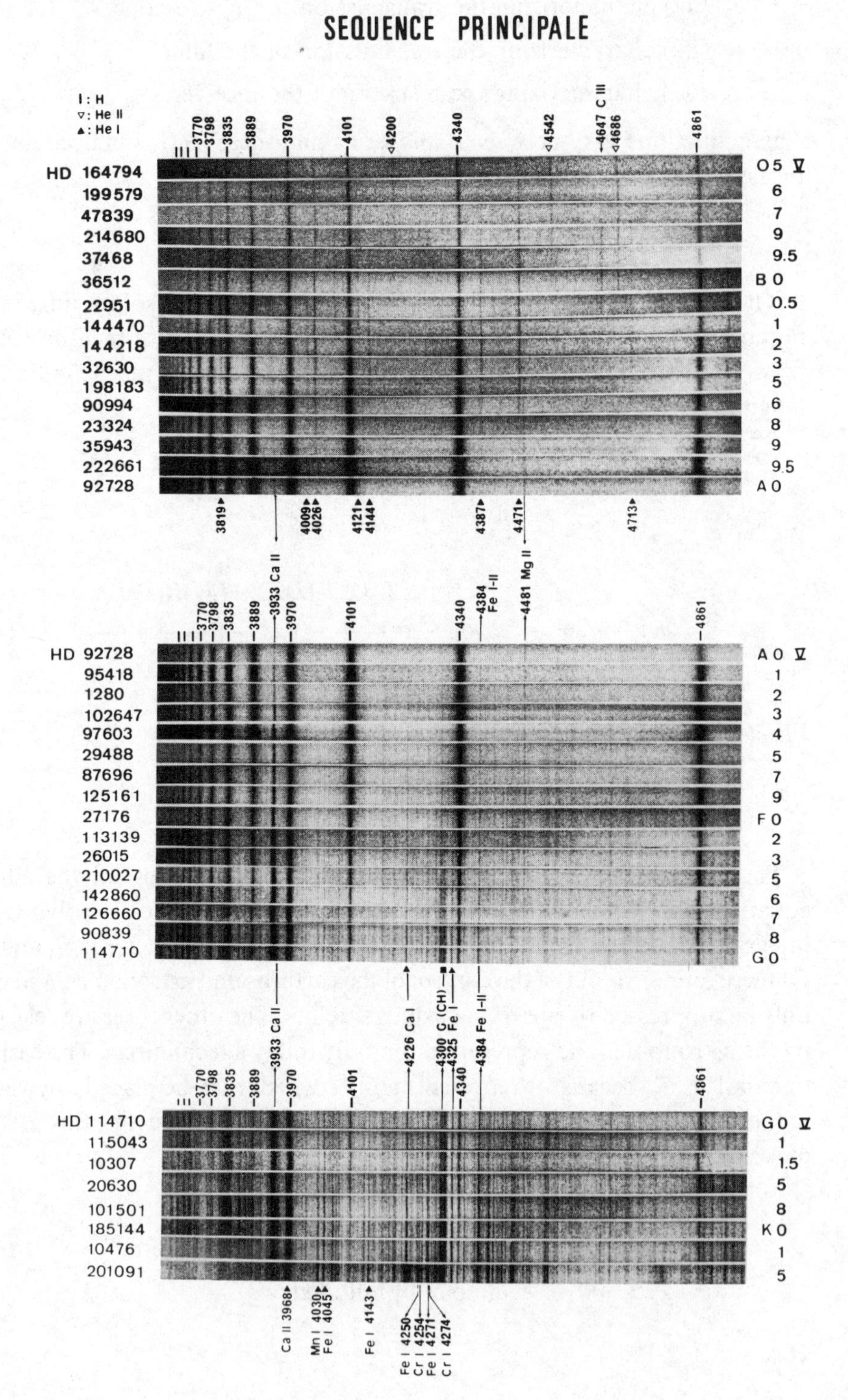

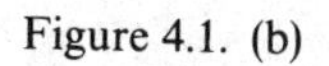
Figure 4.1. (b)

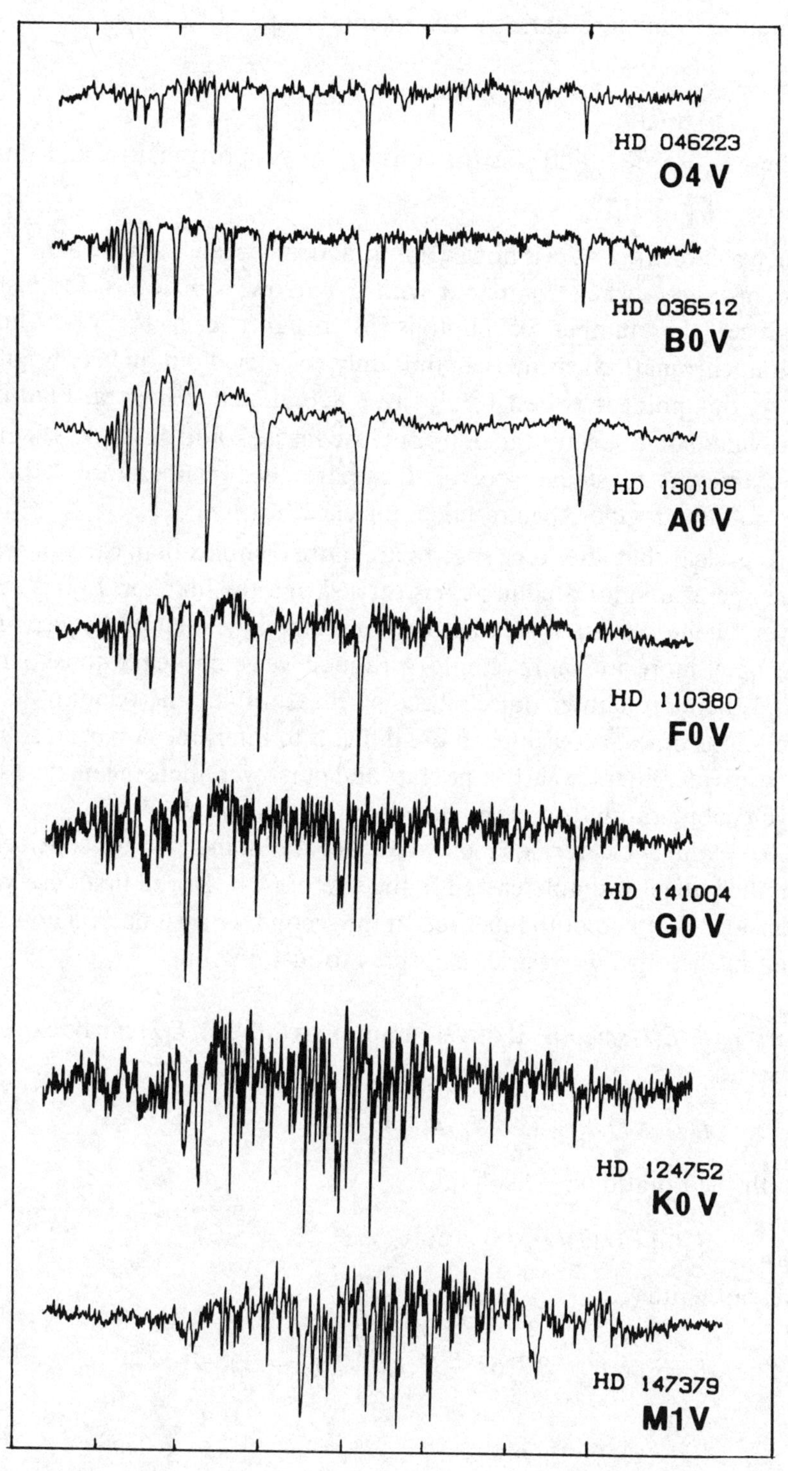
HD 046223
O4 V
HD 036512
B0 V
HD 130109
A0 V
HD 110380
F0 V
HD 141004
G0 V
HD 124752
K0 V
HD 147379
M1V

On the one hand it would be convenient to let $\lambda_b - \lambda_a \to 0$ because then we obtain a monochromatic measurement; in such a case

$$\lim_{\lambda_b \to \lambda_a} \int_{\lambda_a}^{\lambda_b} E(\lambda) T_\mathrm{a} T_\mathrm{t} T_\mathrm{f} R \, \mathrm{d}\lambda = E(\lambda')\beta$$

where $\lambda_b > \lambda' > \lambda_a$ and β is the constant of proportionality, and thus

$$I = E(\lambda)\beta$$

Such a choice ($\lambda_b \to \lambda_a$) is however impractical, because a wide band – 500 Å, for instance – when compared with a narrow band – 5 Å, for instance – reduces the number of photons by roughly a factor of a hundred. Monochromatic measures can thus only be carried out on very bright stars.

At this point it is best to illustrate actual stellar spectra. Figure 4.1(b) provides some spectra taken from Goy, Jaschek and Jaschek (1986). These are tracings of stellar spectra. Compare them with figure 4.1(a) which reproduces regular spectra taken for classification.

It is clear that later type spectra are more complex than early ones. When the operation indicated in (4.2) is carried out, the function $E(\lambda)$ is averaged over the interval (λ_b, λ_a); if this interval is large, the relation between E and I becomes more and more remote. Although we get more photons to measure (and therefore fainter objects become measurable), the relation between I and E becomes increasingly more difficult to interpret. A partial solution is to use wider filters in early type stars and narrower filters in late type stars, to avoid obliterating too many details of the spectrum.

To gain a clearer idea of what we are doing, let us consider three mathematically simple cases for the use of (4.4). In the first case we shall consider $E(\lambda)$ a smooth function. In the second we assume $E(\lambda)$ non-smooth and in the third we restrict measures to just one line.

First case. $E(\lambda)$ smooth. If this assumption is fulfilled, $E(\lambda)$ can be expanded in a series:

$$E(\lambda) = E(\lambda_0) + (\lambda - \lambda_0) E(\lambda_0) + \cdots$$

With the notation

$$T_\mathrm{a}(\lambda) T_\mathrm{t}(\lambda) T_\mathrm{f}(\lambda) R(\lambda) = S(\lambda)$$

we can write

$$I = E(\lambda_0) \int_{\lambda_a}^{\lambda_b} S(\lambda) \, \mathrm{d}\lambda + E'(\lambda_0) \int_{\lambda_a}^{\lambda_b} (\lambda - \lambda_0) S(\lambda) \, \mathrm{d}\lambda + \cdots$$

and the second term disappears if we choose

$$\lambda_0 = \frac{\int_{\lambda_a}^{\lambda_b} \lambda S(\lambda)\,\mathrm{d}\lambda}{\int_{\lambda_a}^{\lambda_b} S(\lambda)\,\mathrm{d}\lambda} \tag{4.5}$$

We have then

$$I = E(\lambda_0)c \tag{4.6}$$

where c is a constant. Thus the measure (I) is related to the energy radiated (E) at a certain specified wavelength λ_0 (with $\lambda_b > \lambda_0 > \lambda_a$). λ_0 is called the 'mean wavelength'; it depends upon $S(\lambda)$, i.e. the transparency of the earth's atmosphere, the telescope, the filter and the receiver, but not upon $E(\lambda)$. In consequence

$$m_2 - m_1 = -2.5 \log \frac{E_1(\lambda_0)}{E_2(\lambda_0)}$$

which is a very compact result. Its use is justified if second-order terms in the development of $E(\lambda)$ are unimportant. This excludes discontinuities and sharp changes in the spectrum, as for instance in the wavelength range $\lambda\lambda$3700–3900 where the accumulation of Balmer lines (the Balmer discontinuity) makes $E(\lambda)$ extremely ragged.

If we want to include second-order terms in the development of $E(\lambda)$, considerable complications arise. The reader is again referred for more details to Golay (1974); here we mention only the rather obvious fact that the second-order terms become important only if the interval $\lambda_b - \lambda_a$ becomes large, i.e. for wide band photometry.

From the above-mentioned consideration it is clear that if formula (4.6) is used, the precise knowledge of $E(\lambda)$ is linked to precise measurements of both I and $S(\lambda)$. If $E(\lambda)$ is to be known to an accuracy of 1% both I and $S(\lambda)$ have to be measured to 1% accuracy.

This implies a very good knowledge of the atmospheric conditions (extinction), of the condition of the telescope, the response of the receiver and the transmission of the filter, all of which must be known to a precision of 1%. Such severe requirements have often not been fulfilled in the past – for many measurements before 1950 even the value of λ_0 was not known to better than $\pm$ 5%. For a discussion of these factors see, for instance, Rufener (1982).

Second case. $E(\lambda)$ not smooth. In this case we introduce two functions, namely $E^*(\lambda)$, the continuous spectrum which is supposed to be a smooth function, and $e(\lambda)$, the real spectrum minus the energy taken away by the absorption lines. Of course $E^*(\lambda)$ is unobservable; it can only be extrapolated between the points where no absorption lines are found. Photometrists cannot study isolated lines (unless such lines are very strong) and therefore they are more interested in the integrated effects of the absorption lines on the continuous spectrum. We define thus a function $\eta(\lambda)$

$$\eta(\lambda) = \frac{\int_{\lambda-\varepsilon}^{\lambda+\varepsilon} [E^*(\lambda) - e(\lambda)]\,\mathrm{d}\lambda}{\int_{\lambda-\varepsilon}^{\lambda+\varepsilon} E^*(\lambda)\,\mathrm{d}\lambda} \tag{4.7}$$

which represents the fraction of the flux substracted by all lines from the continuum interval $(\lambda+\varepsilon, \lambda-\varepsilon)$. Usually ε is chosen to be between 5 and 50 Å.

$$\gamma(\lambda) = 1 - \eta(\lambda) = \frac{\int_{\lambda-\varepsilon}^{\lambda+\varepsilon} e(\lambda)\,\mathrm{d}\lambda}{\int_{\lambda-\varepsilon}^{\lambda+\varepsilon} E^*(\lambda)\,\mathrm{d}\lambda} \tag{4.8}$$

is called the blocking coefficient. $\gamma(\lambda)$ can be obtained from numerical integration of spectrograms. Its value has been tabulated by Ardeberg and Virdefors (1975) for different types of stars; the values are represented in figure 4.2. The integration step used was $\varepsilon = 25$ Å between $\lambda\lambda 3300$ and 5000 Å; for 5000–10 000 Å they took $\varepsilon = 50$ Å.

Inspecting figure 4.2, we see that blocking increases rapidly towards later spectral types. There exists also a clear dependence on luminosity class in the sense that supergiants do have more blocking than dwarfs. For a given star, blocking decreases sharply with increasing wavelength; in particular beyond $\lambda 6000$, the blocking is always small.

Third case. A single line. When we measure photometrically an isolated strong line, like Hβ for instance, the considerations made for the second case are simplified a little. Now $e(\lambda)$ represents the line profile and $\eta(\lambda)$ is then equal to the equivalent width $W(\lambda)$.

$$W(\lambda) = \frac{\int_{\lambda_0-\varepsilon}^{\lambda_0+\varepsilon} [E^*(\lambda) - e(\lambda)]\,\mathrm{d}\lambda}{\int_{\lambda_0-\varepsilon}^{\lambda_0+\varepsilon} E^*(\lambda)\,\mathrm{d}\lambda} \tag{4.9}$$

which is nothing but $\eta(\lambda)$. The physical measure is carried out through a filter centered on λ_0 and 2ε wide (Figure 4.3), which gives

$$\int_{\lambda_0-\varepsilon}^{\lambda_0+\varepsilon} e(\lambda)\,\mathrm{d}\lambda = A$$

The measure of $\int_{\lambda_0-\varepsilon}^{\lambda_0+\varepsilon} E(\lambda)\,\mathrm{d}\lambda = B$ would be the measure of the flux radiated if the line were absent. This can be simulated by measuring the fluxes in the vicinity of the lines at λ_1 and λ_2, chosen in such a way that

$$\lambda_0 = \frac{\lambda_1 + \lambda_2}{2}$$

and that $\lambda_2 - \lambda_1$ is not too large. Call these two measures C and D. We have then

$$\frac{C+D}{2} = B \qquad W = \frac{B-A}{B}$$

Instead of measuring the fluxes at λ_1 and λ_2 and averaging the result, we could place a broad band filter at λ_0 (covering both λ_1 and λ_2). It can be shown that we can obtain $W(\lambda)$ this way also.

Figure 4.2. Blocking coefficient at different wavelength regions as a function of spectral type and luminosity classes. The blocking coefficient is expressed in magnitudes. From Ardeberg and Virdefors (1975).

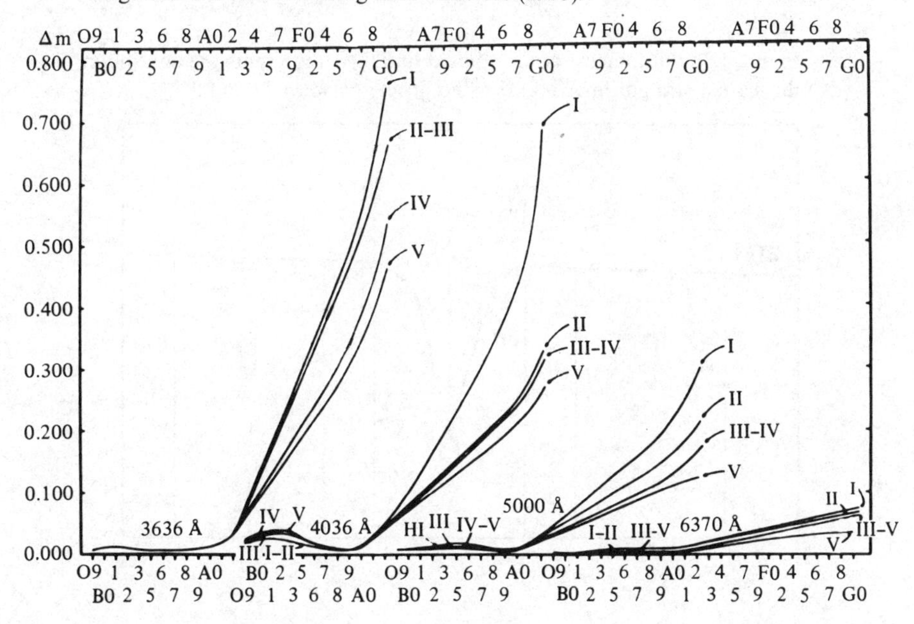

To use this technique we need strong lines. If we use a passband of 20 Å for instance, and the line has $W \sim 1$ Å, the line produces a flux decrease of only 5%. Since measuring accuracy is of the order of say $0^m\!.005$, we are getting uncomfortably close to large relative measuring errors. We could try to improve the situation by reducing the width of the passband, but then we must be careful not to cut off the extended wings that strong lines usually produce.

The preceding reasoning applies to strong lines which are isolated, i.e. not accompanied by weaker nearby lines which fall in the interval $(\lambda_0 + \varepsilon, \lambda_0 - \varepsilon)$. This happens only in early type stars, or at long wavelengths. Otherwise all strong lines are accompanied by a host of weak lines, and this is especially true for later type stars. Where this happens, the interpretation of the measures is obviously no longer simple.

A list of strong lines measured in this way is given in table 4.1.

4.2 Accuracy of measures

Quite obviously the accuracy of the measures depends upon the technique used. The general situation was studied by Young (1984); we summarize the situation simply by saying that photoelectric measures can be more precise than photographic ones, and the latter can be more precise than visual measures. Typical errors are given in table 4.2.

The errors are commonly expressed in magnitude. Because of

Figure 4.3. Photometry of an isolated line. $E(\lambda)$ represents the continuum (in this case a straight line); $e(\lambda)$ the line profile between $\lambda_{0-\varepsilon}$ and $\lambda_{0+\varepsilon}$.

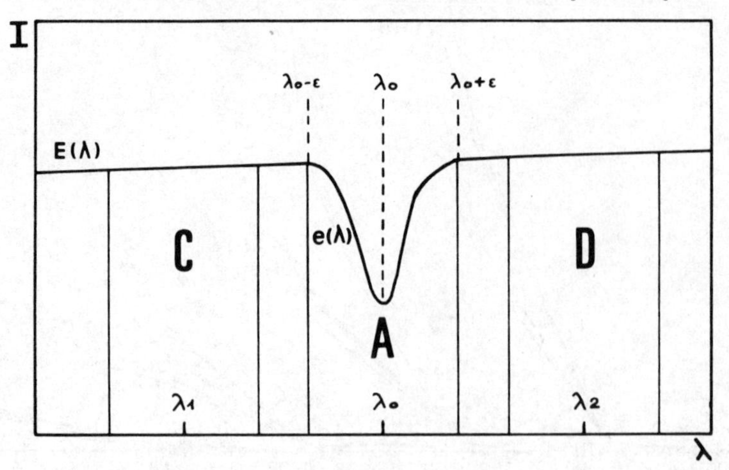

equation (4.3),

$$\frac{dI}{I} = 1.08\, dm$$

implying that an error of 0^m01 is equivalent to a relative error of 1% in the flux.

4.3 Color indices

Each stellar flux measurement ($I(\lambda_i)$) carried out at wavelength λ_i provides a magnitude $m(\lambda_i)$ through relation (4.3):

$$m(\lambda_i) = \text{const.} - 2.5 \log I(\lambda_i)$$

The difference between two magnitudes gives a color index or, shortened, a 'color', which corresponds to a flux ratio. The longer wavelength measurement is always substracted from the shorter, thus

$$m(3600) - m(4200)$$

A negative color index implies that at shorter wavelengths the star emits *more* than at longer wavelengths, whereas the opposite is true for positive color indices.

Because the definition of $m(\lambda_i)$ contains an arbitrary constant, to be fixed

Table 4.1. *Some strong lines used for narrow band photometry.*

Line	Reference	Line	Reference
Hα	Peat (1964)	Mg λ5170	Deeming (1960)
Hβ	Crawford and Mander (1966)	λ4300 G band	Kraft (1960)
Ca II K	Lockwood (1968)	Na I 5885	Price (1966)
He I λ4026	Nissen (1974)		

Table 4.2. *Typical errors of photometric measures.*

Visual (Uranometria Argentina)		$\pm 0^m15$
Photographic		$\pm 0^m07$–10
Photoelectric:	broad band UBV	$\pm 0^m01$–2
	narrow band ±	$\pm 0^m007$

Source: Data from Ochsenbein (1974).

by convention, the scale of *each* $m(\lambda_i)$ measurement also contains an arbitrary constant and the same should happen to the scale of color indices. Astronomers have solved this in two steps. First of all they fixed by convention the origin of the *visual* magnitudes, by stating that the star α U Mi has a magnitude of $2^{m}87$. Then they stated that an average A0 star should have all $m(\lambda_i)$ identical, i.e. all color indices are zero. Since this applies to a fictitious average A0-type star, any given A0 star may have indices differing slightly from zero. Such a convention implies that we measure relative to an A0 star, whose $E(\lambda)$ curve is taken as constant, i.e. $E(\lambda) = C$.

4.4 Photometric systems

Once an astronomer chooses a certain technique to measure stellar fluxes in several well-defined wavelength regions – i.e. $m(\lambda_i)$ – he or she establishes a photometric system. The specification of a photometric system must include:

(a) a description of the instrument used (for instance, aluminum-coated reflector),
(b) a description of the receiver used (a photon counting device, for instance),
(c) a description of the number, nature and characteristics of the filters used (for instance, by providing $T_f(\lambda)$ for each filter),
(d) a list of stars for which the magnitudes have been measured carefully within the system.

The purpose of points (a), (b) and (c) is to make possible a determination of $S(\lambda)$ and thus of λ_0. The purpose of point (d) is more complex. On the one hand it ensures the homogeneity of measures and permits others to measure 'new' stars against the basis provided by the standards. In practice the standards define the arbitrary constant mentioned in section 4.3 and permit a careful treatment of corrections for atmospheric extinction.

Note that here the term 'standard stars' is used in a slightly different sense to that used in chapter 1. Although both in spectral classification and in photometry they define a system, in photometry they also check the reduction and instrumental procedures – in short, the internal consistency of the systems. More details on the standard stars and their role may be found in *IAU Symp.* **111** 'Calibration of fundamental stellar quantities' in the review papers by Batten and Garrison.

Each observer measuring in a given system must observe some standards of the system each night; the more standards that are included, the more reliable the linkage to the system. Therefore it is desirable to have available as many standards as possible; they should be well distributed over all the sky, should cover a large range in magnitudes and in spectral types and should also include stars of different interstellar reddening.

Evidently these requirements are difficult to satisfy in practice, the more so since many of the systems in use were developed in the northern hemisphere. Usually some standard stars are selected along the celestial equator, to enable their use in both hemispheres, but for early type stars which follow the galactic plane closely, this results in rather severe restrictions. Very bright stars should not be included in the list of standards, because of well-known fatigue effects in receivers when illuminated brightly. Usually stars around 7^m are chosen, but for late type dwarfs this restricts the measure to very few objects. So the final list of standards is forced to be a compromise between many different requirements.

The set of standards finally chosen fixes the system, and in this sense the word 'standard' is used in the same way as in spectral classification (see chapter 1).

The set of standards and of the values of its measured color indices replaces in a way the exact description of $S(\lambda)$ (points (a), (b) and (c)) if we know in addition the $E(\lambda)$s for the stars of the list. This is because knowledge of the $m(\lambda_i)$ and $E(\lambda)$ values for many different stars enables us to solve

$$m(\lambda_i) = -2.5 \log \int_{\lambda_a}^{\lambda_b} S(\lambda) E(\lambda) \, d\lambda$$

for $S(\lambda)$. (It is, of course, assumed also that $E(\lambda)$ is constant in time.)

4.5 Choice of passbands

We come now to the selection of the passbands of the system; formerly we had simply assumed that some choice had been made. Essentially the choice is dictated by the nature of the problem we want to solve, and by the technology available.

As an example let us consider the problem of stellar temperatures; what we want to find out is where we should best place our filters. Since it is obviously preferable to place our filters where there are plenty of photons, we first enquire about the wavelength of maximum emission. Under the crude assumption that stars radiate as black bodies, this wavelength (λ_M) is given by the Wien displacement formula

$$\lambda_M T = C \tag{4.10}$$

Table 4.3 provides the values λ_M for different types of stars.

The table illustrates the fact that to measure temperatures of all kinds of stars, our measures have to cover an extended wavelength range. Now besides this, we also have to consider the transparency of the atmosphere, which is opaque to wavelengths below 3100 Å. If we observe from the ground, the radiation maxima of all stars earlier than about A5 fall in the invisible region. Since for these stars we should always use the most violet band usable, we find that for O – B – A – F and early G stars we can use the same type of photometry covering the range from 3100 to 6000 Å. For later type stars the violet measures become less useful because the radiation maximum is sliding into the red and infrared.

A second question concerns how many passbands are to be used in this wavelength region. In fact we have here two opposite requirements, one being the desire to choose a large number of filters in order to derive $E(\lambda)$ at as many points as possible, whereas for practical reasons the number should be a minimum, because each additional passband requires more observing time. Thus for our temperature problem we could perhaps settle on *one* measurement. The inconvenience with this is that stellar magnitudes measure illumination, and this depends both on the temperature of the star and the inverse of the square of the distance of the object. But if we make *two* measurements of a given star at different wavelengths, the ratio of the two will be independent of the distance. (A ratio of fluxes is a difference of magnitude, i.e. a color index.) We therefore have to use two passbands as a strict minimum for solving our problem.

If our problem were more complicated – involving, for instance, two physical parameters instead of one – we should measure at least *two* indices (i.e. three passbands). This result can be generalized to n parameters, for which at least n colors are needed, i.e. $n + 1$ passbands.

A further complication comes from the fact that receivers are usually only

Table 4.3 *Maximum emission wavelengths λ_m as a function of spectral type (Sp.t.).*

Sp.t.	λ_m(Å)	Sp.t.	λ_m(Å)
B0	1000	K0	5500
A0	2000	M0	8000
F0	4000	M5	10 000
G0	4500		

sensitive over a restricted wavelength range into which astronomers have to cram their $n+1$ passbands. Ideally each passband should be independent from its neighbors, but sometimes in practice this cannot be achieved, so that, strictly speaking, the indices may not be (completely) independent.

We thus see that the choice of the passbands and, in a more general way, of a photometric system is dictated by a number of different factors which may be difficult to reconcile. No 'best' system exists in abstract, but only the best solution for a given physical problem and a given technology. This is the natural explanation for the fact that many photometric systems have been and are still being developed.

Of the many systems used for specific purposes, a few have been used to measure large numbers of stars. The most 'popular' systems used are listed in table 4.4.

Before considering the interpretation of photometric measures we shall briefly discuss interstellar effects.

4.6 Interstellar effects

We know that in general part of the radiation emitted by a star is scattered or absorbed by interstellar particles before arriving at earth. The combined effect is called interstellar extinction (A) and is specified in magnitudes per kiloparsec. The extinction depends in general on wavelength, thus $A = A(\lambda)$. Furthermore the amount of extinction suffered on a particular path depends upon the location of the path in the galaxy. It is largest in the plane of the Milky Way; as an average value we can adopt $A(5500) \simeq 1^m/\text{kpc}$.

Outside the Milky Way (galactic plane) the extinction diminishes rapidly. We can describe statistically the amount of extinction by means of a formula

Table 4.4. *Photometric systems.*

System	Number of stars ($\times 10^3$)	System	Number of stars ($\times 10^3$)
UBV	69.0	UBVRI... N	4.5
uvby	19.0	Stebbins	1.3
Geneva	15.0	Walraven	2.7
TDI	30.0	Vilnius	1.9
		DDO	3.0

Source: Data from Hauck (1982).

of the type

$$A(r,b) = A_0 \sec b\left(1 - \exp\left(-\frac{r\sin|b|}{\beta}\right)\right)$$

where A_0 = extinction constant, b = galactic latitude, β = thickness of the extinction layer.

This formula should *not* be applied to individual stars, because of very large fluctuations in the interstellar medium. As an illustration we reproduce part of a map of the interstellar extinction produced by Neckel (1966) which shows the variations of $A(V)$ $(=A_v)$ as a function of galactic longitude (l) and galactic latitude (b) (see figure 4.4). For more recent data see Lucke (1978) and Ducati (1985).

The correction of photometric measures for interstellar extinction is a difficult problem. Progress has been made by comparing identical stars (i.e. stars with identical line spectra) situated at different distances and/or locations from us. Figure 4.5 (Divan 1954) illustrates what happens: the star which is further away emits less radiation at shorter wavelengths. If we plot the magnitude difference as a function of wavelength, we obtain

$$m(a) - m(b) = \frac{\alpha}{\lambda} + \gamma$$

where α and γ are two constants.
Detailed studies have shown that in our galaxy we can write

$$A = A_1(\lambda)A_2(r,b)$$

so that $A_1(\lambda)$ is a 'universal' function. Usually we write

$$A_1(\lambda) = F(\lambda)A(5500)$$

where $A(5500)$ is the absorption at $\lambda5500$. The function $F(\lambda)$ is given in table 4.5, taken from Savage and Mathis (1979) where more details can be found.

If it is assumed that this general curve is strictly obeyed, then we can obtain the extinction for any wavelength, provided we know the extinction at *one* wavelength.

For instance, we may compute A(B) and A(V), where B and V designate two colors at $\lambda4400$ and $\lambda5500$ respectively. We interpolate from table 4.5 and obtain

$$A(\mathrm{B}) = A(5500) \times 1.323 \text{ and } A(\mathrm{V}) = A(5500) \times 1.00$$

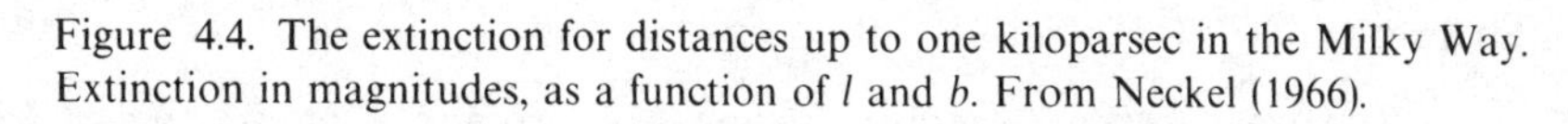

Figure 4.4. The extinction for distances up to one kiloparsec in the Milky Way. Extinction in magnitudes, as a function of l and b. From Neckel (1966).

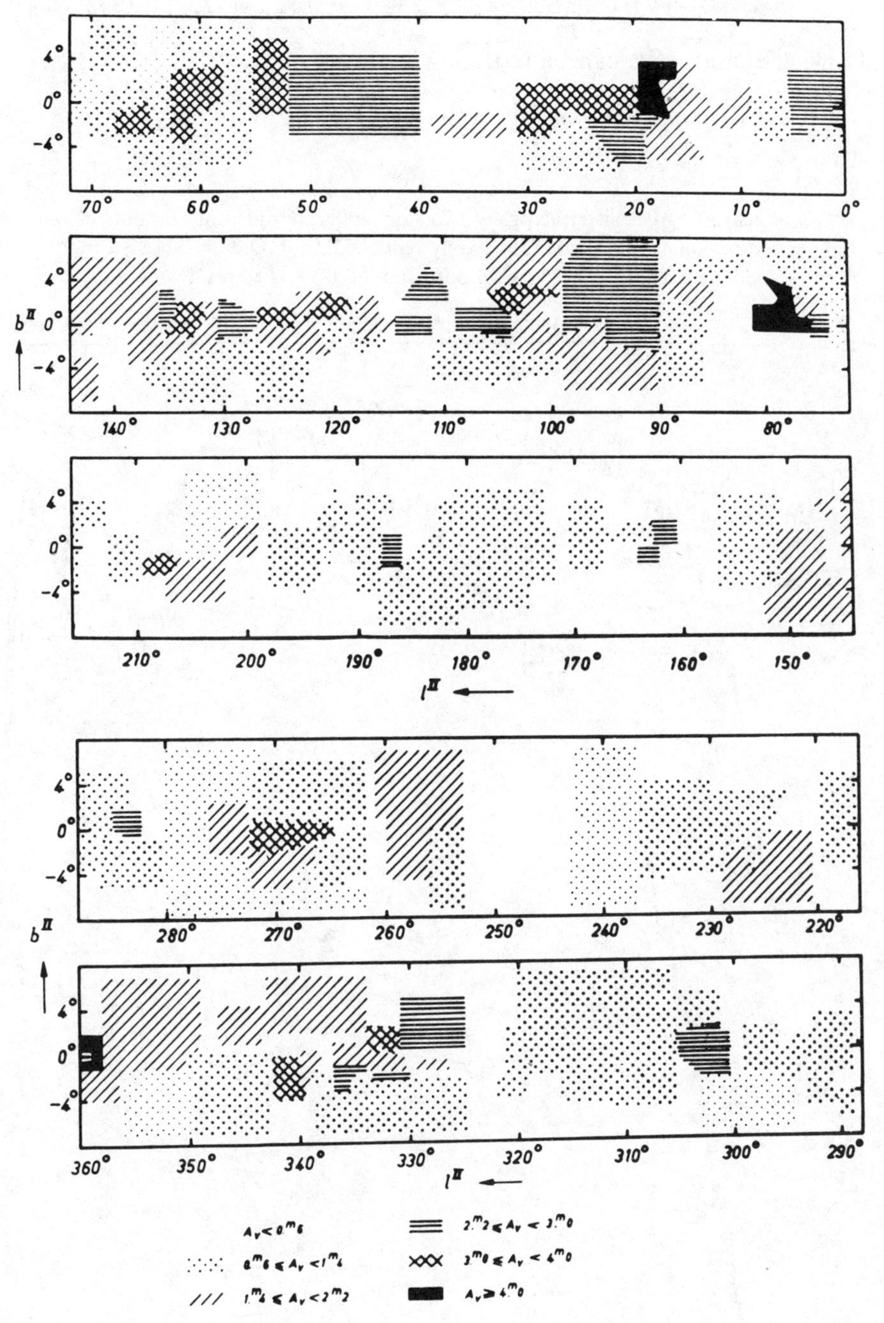

Therefore

$$A(\mathrm{B}) - A(\mathrm{V}) = A(5500) \times (1.323 - 1) = 0.323A(5500) = 0.323A(\mathrm{V})$$

Now $A(\mathrm{B})$ and $A(\mathrm{V})$ can be written as

$$A(\mathrm{B}) = m_{\mathrm{obs}}(\mathrm{B}) - m_i(\mathrm{B}) \quad \text{and} \quad A(\mathrm{V}) = m_{\mathrm{obs}}(\mathrm{V}) - m_i(\mathrm{V})$$

Figure 4.5. Density tracings of two stars suffering different amounts of extinction. (a) HD 204172 is barely reddened. (b) HD 2050196 is strongly reddened. The spectral type of both stars is B0 ib. From Divan (1954).

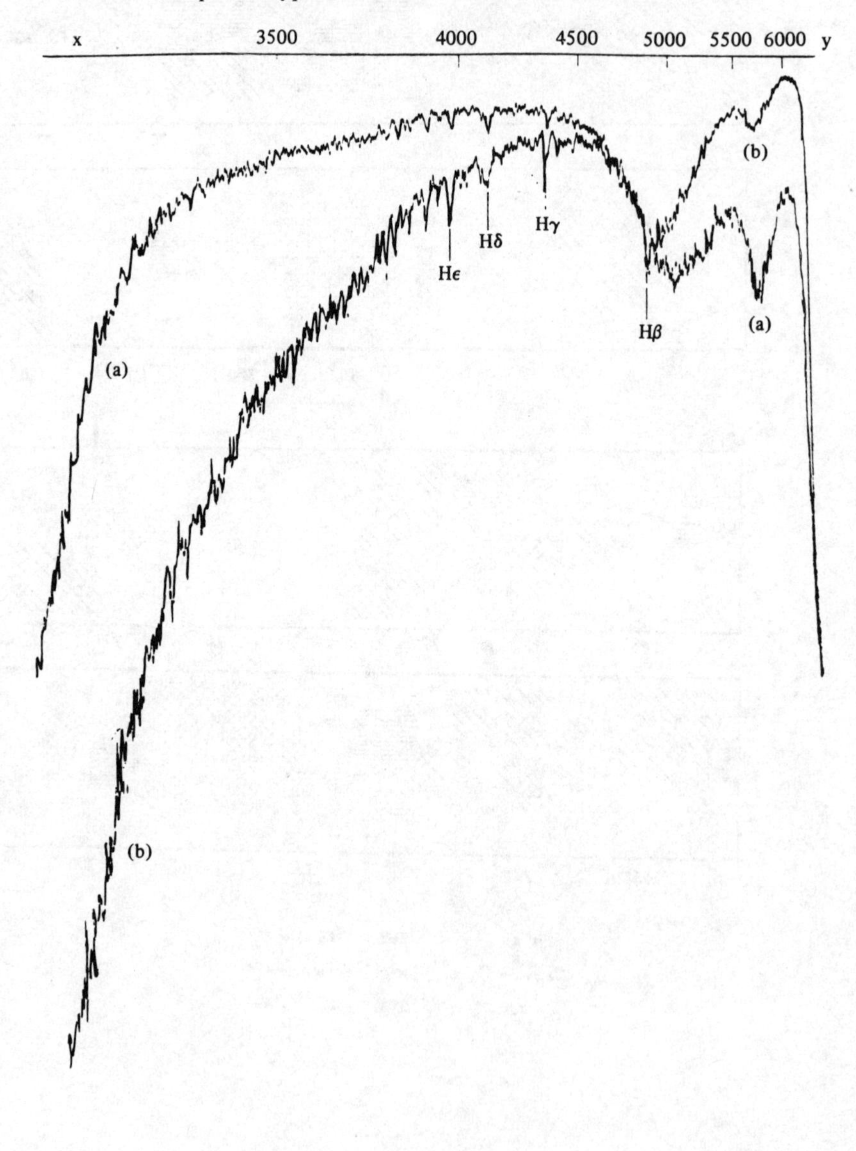

where m_{obs} are the observed magnitudes of the star including the influence of extinction, and m_i is the 'intrinsic' magnitude of the star, i.e. the magnitude it would have without extinction. Thus

$$A(\mathrm{B}) - A(\mathrm{V}) = [m_{obs}(\mathrm{B}) - m_{obs}(\mathrm{V})] - [m_i(\mathrm{B}) - m_i(\mathrm{V})]$$

Obviously $m_i(\mathrm{B}) - m_i(\mathrm{V})$ should be the same for all stars which are physically identical, because the influence of the interstellar medium is contained in the first bracket. The two parentheses are obviously color indices and we can write

$$m_{obs}(\mathrm{B}) - m_{obs}(\mathrm{V}) = (\mathrm{B{-}V})_{obs} \quad \text{and} \quad m_i(\mathrm{B}) - m_i(\mathrm{V}) = (\mathrm{B{-}V})_i$$

The difference of both is always positive (because extinction weakens the starlight), and is therefore called an *excess*, denoted by $E(\mathrm{B{-}V})$. In the present notation

$$E(\mathrm{B{-}V}) = (\mathrm{B{-}V})_{obs} - (\mathrm{B{-}V})_i = A(\mathrm{B}) - A(\mathrm{V}) \tag{4.11}$$

On the other hand we have seen that

$$E(\mathrm{B{-}V}) = A(5500) \times \text{const.} = A(\mathrm{V}) \times \text{const.}$$

and therefore we obtain the important result that

$$E(\mathrm{B{-}V}) = 0.323\, A(\mathrm{V}) \quad \text{or} \quad A(\mathrm{V}) = 3.1 \times E(\mathrm{B{-}V})$$

This reasoning can obviously be applied to any other color index. It follows also that the ratio of *two* color excesses, for example $E(\mathrm{U{-}B})$ and $E(\mathrm{B{-}V})$, is a constant. All these results are consequences of the initial hypothesis that

$$A = A_1(\lambda) A_2(r, b)$$

and are valid *only* in this case.

Table 4.5. *Interstellar absorption.*

λ(Å)	λ^{-1}	$F(\lambda)$	λ(Å)	λ^{-1}	$F(\lambda)$
∞	0	0	4000	2.50	1.419
34 000	0.29	0.052	3440	2.91	1.581
22 000	0.45	0.123	2500	4.00	2.35
12 500	0.80	0.281	2190	4.57	3.12
9000	1.11	0.484	2000	5.00	2.78
7000	1.43	0.748	1800	5.56	2.50
5500	1.82	1.000	1600	6.25	2.62
4400	2.27	1.323	1390	7.18	2.74

Complications arise if the photometric system has wide band filters. In such a case the general formula (4.2) has to be written as

$$E = \int_{\lambda_a}^{\lambda_b} E(\lambda)S(\lambda)A(\lambda)\,\mathrm{d}\lambda = A(\mathrm{V})\int_{\lambda_a}^{\lambda_b} F(\lambda)E(\lambda)S(\lambda)\,\mathrm{d}\lambda$$

With $F(\lambda)S(\lambda) = S'(\lambda)$ we regain

$$E = A(\mathrm{V})\int_{\lambda_a}^{\lambda_b} S'(\lambda)E(\lambda)\,\mathrm{d}\lambda$$

but we have in fact changed the definition of $S(\lambda)$ to $S'(\lambda)$ and therefore altered λ_0.

Thus for wide band filters, λ_0 *also* changes with $A(\mathrm{V})$. This is incidentally the reason why stars of different interstellar reddening must be included in a list of standard stars.

Let us turn back for a moment to the color index $m_i(\mathrm{B}) - m_i(\mathrm{V})$. Since this color index should be the same for all physically identical stars, it is called the 'intrinsic color index'. It is determined by choosing a group of physically identical stars and selecting the 'bluest' color of any star of the group, which is usually also the color of the nearest star. (This is because in general reddening increases with distance.) This color is then taken as the color all stars of the group should have if their colors were not reddened, i.e. it is the intrinsic color of the group. Stars with abnormal blue colors, which might be freaks of some kind, have to be excluded.

The question of what to do when even the nearest star of a group is far away – as happens for instance for the supergiants – is more difficult to answer. Usually in this case stars are considered in open clusters which contain both supergiants and main sequence stars. The intrinsic colors of the latter are known, and from them the reddening can be obtained; if all stars of the cluster suffer the same reddening, the supergiants can thus be 'de-reddened' and their intrinsic colors obtained. Of course, this assumes that the reddening is the same, no matter where the star is in the cluster.

4.7 Interpretation of measures

Once a star has been measured in a given photometric system with n colors, the result may be represented as an image point (P) in an n-dimensional coordinate system. The experimental errors of each color (ε_i) can then be imagined as defining a small volume of space centered around the image point P. For the simple case with $n = 2$, the space would be two-dimensional and the small 'volume' a circle (if $\varepsilon_1 = \varepsilon_2$) or an ellipse (if $\varepsilon_1 \neq \varepsilon_2$).

We should, however, refrain from considering this as a sphere with a hard border, since the very concept of experimental error (ε_i) produces a 'probabilistic' sphere, in the sense that any remeasurement of the star will have a certain probability p of falling within a given distance from the previous point. We should thus imagine these spheres as having a diffuse border. With this image in mind, a group of stars will be represented by a cloud of points, covering a certain domain of the (photometric) n-dimensional space. Two points shall be distinct if their distance is larger than a certain minimum value, which is of the order of 2ε. If the points are closer than the minimum distance, they cannot be distinguished from each other.

This simple idea leads us to define around each image point P an n-dimensional box with lengths $2\varepsilon_i$ ($\pm\,\varepsilon_i$ around the center). We shall call this a 'Golay box'. All stars falling into a box are identical, photometrically speaking, in the n-color system in which they were measured. Mathematically speaking, we can define boxes and distances in many different senses. For instance, in two-dimensional space we can consider a box defined by a rectangle, a circle, or an ellipse. (Whichever way is chosen depends upon the way distance Γ is defined in the n-dimensional space.) In the first case (the rectangle)

$$r = |\Delta x| + |\Delta y|$$

In the second (the circle)

$$r = [(\Delta x)^2 + (\Delta y)^2]^{1/2}$$

and in the third (the ellipse)

$$r = [a^2(\Delta x)^2 + b^2(\Delta y)^2]^{1/2}$$

Here we have denoted the two axes by x and y, Δ is the difference between two values of the coordinates, and a and b are weights arising from the different precision with which an index may be measured. Obviously other definitions of distance are possible also, and the reader is referred to Nicolet (1981) for more details.

The usefulness of photometric boxes becomes evident when we see them as a parallel to those of spectral classification. There also we consider that all objects in one box are identical from a spectroscopic point of view and in a given classification system. The difference lies in the fact that the spectroscopic system is discontinuous whereas the photometric system is continuous. We can improve the analogy by changing slightly the idea of a box. If we divide the photometric space not into boxes centered on actual stars, but into an n-dimensional grid (Egret 1982) with a mesh width $2\varepsilon_i$, we can build the perfect analog of spectral classification. These small n-dimensional parallelipipeds are called 'cells'. Note that a box can never be empty, but that a cell may well be empty. In fact in any photometric system most cells are empty – this implies that only certain regions of the photometric space are more or less densely populated.

For instance, in the UBV diagram of figure 5.1(a), (b), we immediately notice a heavily populated sequence and, in its proximity, a large region (to the right of the sequence) where the density of the points diminishes rapidly. As we shall see, this is due to the effect of interstellar extinction, which smears out the clear sequence of figure 5.1(a). Therefore, besides the physical factors intervening in the positioning of points, we also have an extraneous factor (interstellar extinction) which influences all photometric measures. The question is then what to do with the boxes and/or the cells: should we use the observed colors, or should we correct them first for interstellar extinction? This question remains open; in the last instance it depends on how much confidence we place upon the de-reddening procedure and the assumption that extinction everywhere depends upon λ in the same way; i.e. that $F(\lambda)$ is identical everywhere. Furthermore, the de-reddening procedure must be accurate within a precision of the order of ε since otherwise the grouping of stars into boxes would lose its sense because of the persistence of small differential interstellar effects in the de-reddened colors.

The Geneva school in general prefers to deal with the observed colors (i.e. without de-reddening), whereas other authors (e.g. Egret 1982) prefer de-reddened colors. In practice observed colors pose very stringent conditions for boxes – not only the physical parameters of the star, but also the interstellar effects must be identical. In the case of early type stars, where interstellar effects are almost always present, this will result in a very small number of boxes or a large number of empty cells. In such a case the use of de-reddened colors would seem to be preferable.

It would seem natural to extend the idea of 'grouping' in the photometric space to a higher order of complexity, in the sense of defining broader regions ('families of boxes') through the use of statistical classification methods. Such an extension has not been envisaged yet, although it would be very useful.

Another concept is the 'purity' of a box or a cell. If for instance we are interested in stars having a particular anomalous physical property (e.g. a certain composition), we can describe the location of these stars in the n-dimensional photometric space by indicating the length of the intervals on each of the n axes they occupy. Such a procedure does not tell us if the same (hyper) volume occupied by these stars is *only* populated by these stars. Should it turn out that there are many other kinds of stars which also fall in the same volume, the photometric system used is not very useful for singling out the objects. We may introduce the idea of 'purity':

$$p = \frac{\text{number of stars in a cell having the desired property}}{\text{total number of stars in the cell}}$$

This concept was introduced by Jaschek and Frenkel (1985).

The usefulness of introducing well-defined parameters is underlined by the results of Olsen (1979). He observed photometrically in the uvby system a large sample of southern A5–G0 stars. Based on photometry alone, he tried to predict the spectral peculiarities of stars. To do so he assumed that similar indices produce similar groups of stars. Abt, Brodzik and Schaeffer (1979) classified spectroscopically 92 of Olsen's stars and found that out of 78 stars predicted photometrically as being either more luminous than dwarfs or abnormal in composition, 56 (i.e. three-quarters) are effectively abnormal, but only 37 stars (i.e. one-third) have the specific peculiarity predicted photometrically. Using the purity parameter we could provide a figure characterizing the purity achieved by the uvby system, which is obviously not very high.

4.8 Spectrophotometry

We shall designate as spectrophotometry all studies in which the resolving power is larger than about 5 Å and smaller than 50 Å. Higher resolving powers correspond to spectroscopy and lower to narrow band photometry. In reality the difference is even less clear because in narrow band photometry filters as narrow as 10 Å are used occasionally.

Because of its intermediate position, spectrophotometry may be considered as either a multicolor narrow band photometry or a low resolution spectroscopy. This explains its advantages, inconveniences and main uses.

Spectrophotometry has until now not been used much for stellar classification; this is mainly because of the small number of spectrophotometric scans available. The most important catalogs are listed in table 4.6. All these catalogs are available from data centers.

4.9 Auxiliaries

Under this heading we group information needed to deal with various photometric systems. We start by recalling the two standard

Table 4.6. *Spectrophotometric catalogs.*

Authors	Number of stars	Wavelength region
Breger (1976)	937	$\lambda\lambda$3200–12 000
Ardeberg and Virdefors (1980)	356	$\lambda\lambda$3500–11 000
Glushneva (1982)	735	$\lambda\lambda$3225–5375
Jacoby, Hunter and Christian (1984)	161	$\lambda\lambda$3510–7427
Gunn and Stryker (1983)	175	$\lambda\lambda$3130–10 800

Table 4.7. *Photometric catalogs.*

System	Authors		Number of stars ($\times 10^3$)	Notes
UBV	Mermilliod and Nicolet	(1977)	53.0	Measures
	Mermilliod	(1983)	26.0	Update of measures
	Nicolet	(1978)	59.0	Averages
uvbyβ	Hauck and Mermilliod	(1980)	20.0	Measures and averages
Geneva	Rufener	(1981)	14.6	Averages
U_cBV	Nicolet	(1975)	7.2	Measures and averages
Vilnius	North	(1980)	1.9	Averages
DDO	McClure and Forrester	(1981)	2.2	Averages
RI	Jasniewicz	(1982)	5.7	Measures and averages
UBVRI . . N	Morel and Magnenat	(1978)	4.5	Measures and averages
UBVRI	Lanz	(1986)	8.4	Measures and averages
Infrared	Gezari, Schmitz and Mead	(1982)	30.0	Compilation for $\lambda > 1\ \mu$m
TDI	Thompson *et al.*	(1978)	31.0	4 ultraviolet bands $\lambda < 3000$
Celescope	Davis, Deutschman and Haramundanis	(1973)	5.7	4 ultraviolet bands $\lambda < 3000$
ANS	van Duinen *et al.*	(1975)	3.6	6 ultraviolet bands $\lambda < 3000$

The 'note' indicates whether only average colors or measures, or both, are provided. 'Numbers of stars' have been rounded off.

reference books on photometry, namely Golay (1974) and Straizys (1977), which discuss many (but not all) photometric systems. General discussion can also be found in the various proceedings of symposia, colloquia and conferences. We mention in particular the one on *Problems of calibration of multicolor photometric systems* (Philip Davis 1979).

We turn next to general references on individual photometric systems.

(*a*) *Response functions.* Hauck and Mermilliod (1976) tabulate response functions for 30 different photometric systems.

(*b*) *Standard stars.* Standard stars for seven photometric systems are given by Philip Davis (1979); and Philip Davis and Egret (1985) provide data for some more systems.

(*c*) *Catalogs.* Photometric catalogs are usually available in machine-readable versions, and are obtainable from data centers. Table 4.7 summarizes briefly the most important catalogs available in 1986; the reader should consult the bulletins of the data centers for updates. A general survey of photometric catalogs was made by Mermilliod (1984).

A general file called 'General catalog of photometric data' (to be published) exists at Geneva-Lausanne which provides a reference to all systems in which a given star has been measured.

References

Abt H.A., Brodzik D. and Schaeffer B. (1979) *PASP* **91**, 176

Ardeberg A. and Virdefors B. (1975) *AA* **39**, 26

Ardeberg A. and Virdefors B. (1980) *AA Suppl.* **40**, 307

Breger M. (1976) *Ap. J. Suppl.* **32**, 7

Crawford D.L. and Mander J. (1966) *A.J.* **71**, 114

Davis R.J., Deutschman W.A. and Haramundanis K.L. (1973), *S.A.O. Special Report* **350**

Deeming T.J. (1960) *MNRAS* **121**, 52

Divan (1954) *AA* **17**, 456

Ducati J. (1986) *Astroph. Sp. Sc.* **126**, 269

Egret D. (1982) Ph. D. Thesis, Strasbourg

Gezari D., Schmitz M. and Mead J.M. (1982) *NASA TM* 83819

Glushneva I.N. (1982) *Spectrophotmetric scans of stars*, Nauka, Moscow

Golay M. (1974) *Introduction to astronomical photometry*, Reidel D. Publ. Co, Dordrecht.

Goy G., Jaschek M. and Jaschek C. (1986) (in preparation)

Gunn J.E. and Stryker L.L. (1983) *Ap. J. Suppl.* **52**, 121

Hauck B. (1982) *BICDS* **22**, 67
Hauck B. and Mermilliod J.Cl. (1976) *BICDS* **10**, 28
Hauck B. and Mermilliod M. (1980), *AA Suppl.* **40**, 1
Hayes D.S., Pasinetti L.E. and Davis Philip A.G. (ed.) (1985) *IAU Symp.* **111** 'Calibration of fundamental stellar quantities', D. Reidel Publ. Co.
Jacoby G.H., Hunter D.A. and Christian C.A. (1984) *Ap. J. Suppl.* **56**, 257
Jaschek C. and Frankel S. (1985) *AA* **158**, 174
Jasniewicz G. (1982) *AA Suppl.* **49**, 99
Kraft R.P. (1960) *Ap. J.* **131**, 330
Lamla E. (1982) in Landolt–Börnstein, New Series, group 6, vol. 2b, p. 35
Lanz T. (1986) *AA Suppl.* (in press)
Lockwood (1968) *A. J.* **73**, 14
Lucke P.B. (1978) *AA* **64**, 367
McClure R.D. and Forrester W.T. (1981) *Publ. Dom. Obs.* **15**, 439
Mermilliod J.C. (1983) *BICDS* **25**, 79
Mermilliod J.C. (1984) *BICDS* **26**, 3
Mermilliod J.C. and Nicolet B. (1977) *AA Suppl.* **29**, 259
Morel M. and Magnenat P. (1978) *AA Suppl.* **34**, 477
Neckel T. (1966) *Z. f Astroph.* **63**, 221
Nicolet B. (1975) *AA Suppl.* **22**, 239
Nicolet B. (1978) *AA Suppl.* **34**, 1
Nicolet B. (1981) *AA* **97**, 85
Nissen P.E. (1974) *AA* **36**, 57
North P. (1980) *AA Suppl.* **41**, 395
Ochsenbein F. (1974) *AA Suppl.* **15**, 215
Olsen E.H. (1979) *AA Suppl.* **37**, 367
Peat D.W. (1964) *MNRAS* **128**, 435
Philip Davis A.G. (1979) '*Problems of calibration of multicolor photometric systems*', Dudley Obs. Report 14
Philip Davis A.G. and Egret D. (1985) *IAU Symp.* **111**, 'Calibration of fundamental stellar quantities', Hayes R., Philip D. and Pasinetti L. (ed.) D. Reidel Publ. Co. (microfiche added to volume)
Price M.J. (1966) *MNRAS* **134**, 135
Rufener F. (1981), *AA Suppl.* **45**, 207
Rufener F. (1982) *BICDS* **22**, 5
Savage B.D. and Mathis J.S. (1979) *Ann. Rev. AA* **17**, 73
Straizys V. (1977) *Multicolor stellar photometry*, Mokslas Publ., Vilnius
Thompson G.I., Nandy K., Jamar C., Monfils A., Houziaux L., Carnochan D.J. and Wilson R. (1978) *Catalog of stellar ultraviolet fluxes*, Science Research Council
van Duinen R.J., Aalders J.W., Wesselius P.R., Wildema K.J., Wu C.C., Luinge W. and Snel D. (1975) *AA* **39**, 159
Young A.T. (1984) *NASA Conf. Publ.* 2350

5

Photometric systems

In this chapter we shall study two photometric systems in order to show in detail some of the uses of a photometric system. We have chosen the UBV and the uvby or Strömgren system. For other photometric systems the reader is advised to consult Golay (1974) or Straizys (1977).

5.1 The UBV system

The UBV system was developed in the fifties by Johnson (see Johnson and Morgan 1953) for the photometric study of stars classified in the Yerkes system.

It uses three wide passbands, each about a thousand angstroms wide, called U (ultraviolet), B (blue), and V (visual), with λ_0 around $\lambda 3500$, $\lambda 4300$ and $\lambda 5500$ respectively. The choice of the passbands was made in part for historical reasons. V corresponds approximately to the 'visual magnitudes' handed down essentially from Ptolemy. B corresponds on the average to the 'photographic magnitudes' of the end of the nineteenth century. Finally U was chosen so as to get as much ultraviolet light as possible. The functions $S(\lambda)$ are tabulated in table 5.1, but readers should be warned that different authors use slightly different $S(\lambda)$. The system is thus not suited for very high precision. Furthermore the $S(\lambda)$ of the U color goes down to $\lambda 3000$, whereas the atmosphere usually cuts off the spectrum below $\lambda 3300$. The ultraviolet limit of the U band is therefore not fixed by the filter system, but differs from place to place (for instance with the elevation of the observatory) and sometimes from night to night. This explains the relatively large scatter in the U magnitudes of a given star when observed from different places.

With the three passbands two independent indices can be formed. We follow the majority of authors by using U–B and B–V, although U–V, B–V can also be used.

The observed color–color (or two-color) diagram is given in figure 5.1(a).

Table 5.1. *Response function for the* UBV *system.*

$\lambda \times 10^{-2}$ (Å)	$S(U)$	$S(B)$	$S(V)$
30	0.025		
31	0.250		
32	0.680		
33	1.137		
34	1.650		
35	2.006	0.000	
36	2.250	0.006	
37	2.337	0.080	
38	1.925	0.337	
39	0.650	1.425	
40	0.197	2.253	
41	0.070	2.806	
42	0.000	2.950	
43		3.000	
44		2.937	
45		2.780	
46		2.520	
47		2.230	
48		1.881	0.020
49		1.550	0.175
50		1.275	0.900
51		0.975	1.880
52		0.695	2.512
53		0.430	2.850
54		0.210	2.820
55		0.055	2.625
56		0.000	2.370
57			2.050
58			1.720
59			1.413
60			1.068
61			0.795
62			0.567
63			0.387
64			0.250
65			0.160
66			0.110
67			0.081
68			0.061
69			0.045
70			0.028
71			0.017
72			0.007

Note: This table is taken from Matthews and Sandage (1963) and is valid for one air mass.

The curve for black bodies is also drawn. Negative values of both indices are called 'blue' and positive ones 'red'. (Please observe that in the astronomical convention U–B is drawn contrary to mathematical convention.) The ranges in U–B and B–V are about 3^m and 2^m respectively; the larger range in U–B is due to the rapid disappearance of violet light in the spectra of late type stars; for B–V the situation is better because both bands are nearer to the maximum of radiation for late type stars.

Figures 5.2 and 5.3 provide the relation between U–B and B–V indices and MK types. As can be seen, the relation has a steeper slope in U–B than in B–V for U–B < 0, whereas B–V in general has a smoother relation to spectral type (and temperature) than U–B. Therefore UBV relations with spectral type (and temperature) are done with U–B for early type stars and B–V for later type stars.

Turning back to figure 5.1(a), we observe a maximum deviation between black body radiators and stars in the region around A0. Since at this type Balmer lines are also at their maximum strength (see figure 4.1) we suspect a causal relation between both. The index U–B measures the intensity ratio between λ3500 and λ4300, and it provides a measure of the Balmer discontinuity, i.e. of the hydrogen line strength.

Having thus used one color index for spectral type (or temperature), we are left with the other one for other uses. It would be tempting to see if luminosity

Figure 5.1. The (U–B, B–V) relation for unreddened main sequence stars from Johnson and Morgan (1953). The line corresponds to that for a black body.

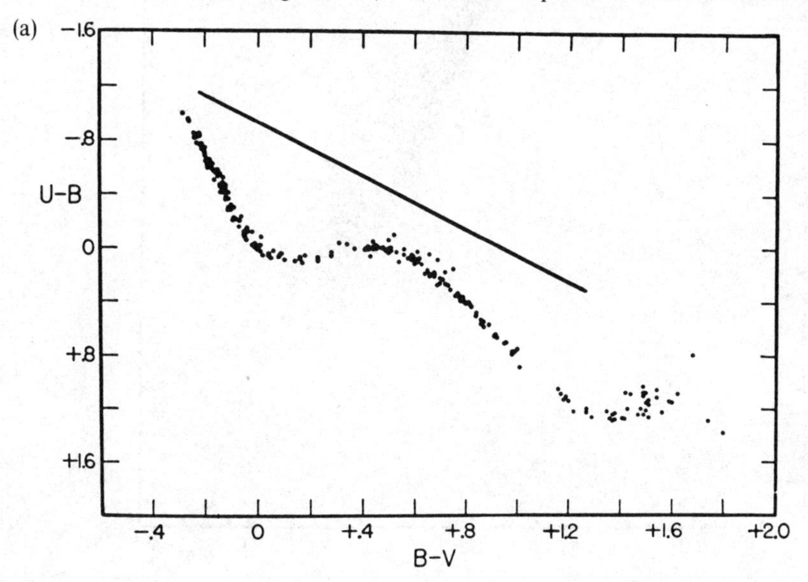

can be obtained from the UBV photometry. We have reproduced in figure 5.4 an arbitrary selection of early type stars of the same spectral class and different luminosity types, and a glance at this figure shows that B0 Ia and B0 V stars can be present at many different places in the two-color diagram, depending upon the reddening they suffer in space. Having thus two

Figure 5.1. (*Continued*)

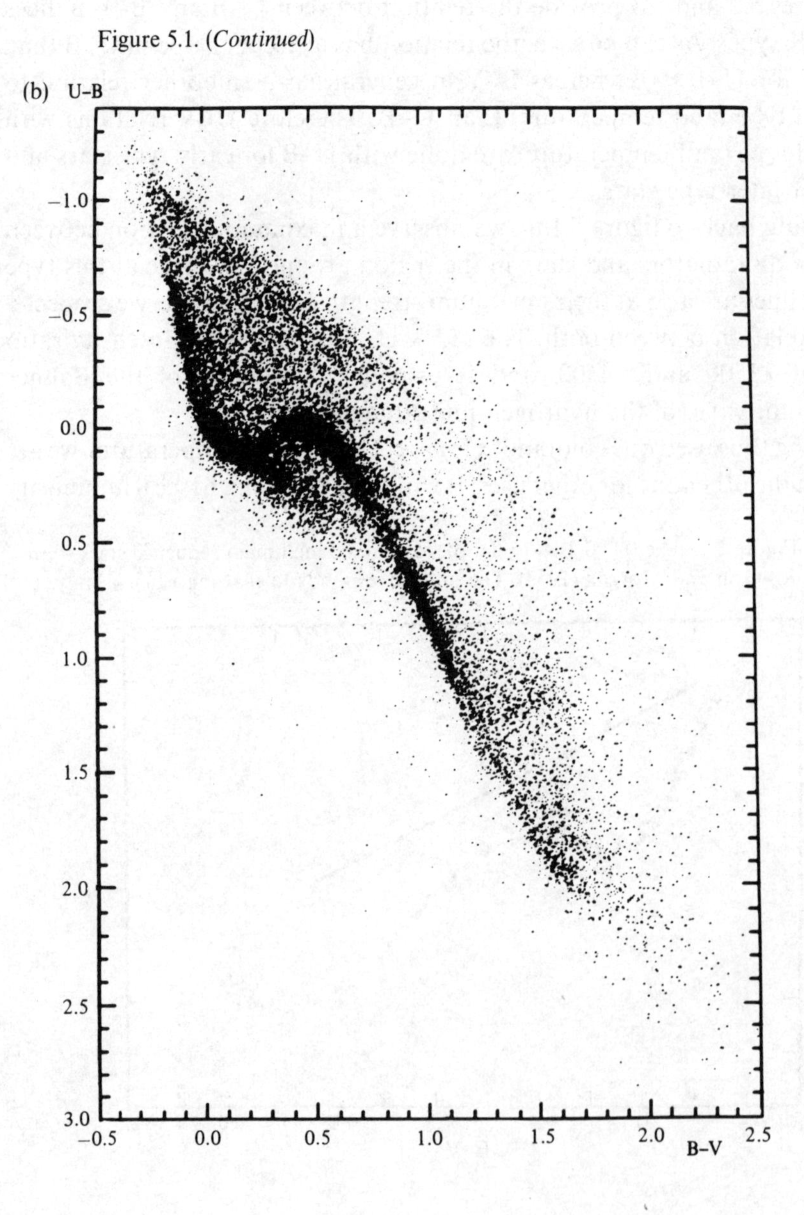

causes – reddening and luminosity – but only one more color index available, we cannot hope to determine *both.*

We consider therefore three different cases:

(*a*) *Extinction known.* If the extinction is known, or it is assumed to be negligibly small because one considers nearby stars, the observed colors are the intrinsic colors, denoted as $(U-B)_0$ and $(B-V)_0$. We have already seen in figures 5.2 and 5.3 that the intrinsic colors differ for dwarfs and giants, and similar distinctions can be made for supergiants. The observed colors can thus tell what the luminosity class of the star is. However, this procedure is in general unusable because in the solar neighborhood, where interstellar extinction is zero, there are no supergiants; so the procedure can only be used for late giants and dwarfs.

Tables of the intrinsic colors for stars of different luminosity classes are given by Golay (1974), Straizys (1977) and Schmidt–Kaler (1982).

Figure 5.2. The relation between U–B and Yerkes spectral classification. Average relations for de-reddened stars. (Sp. T. = spectral type.)

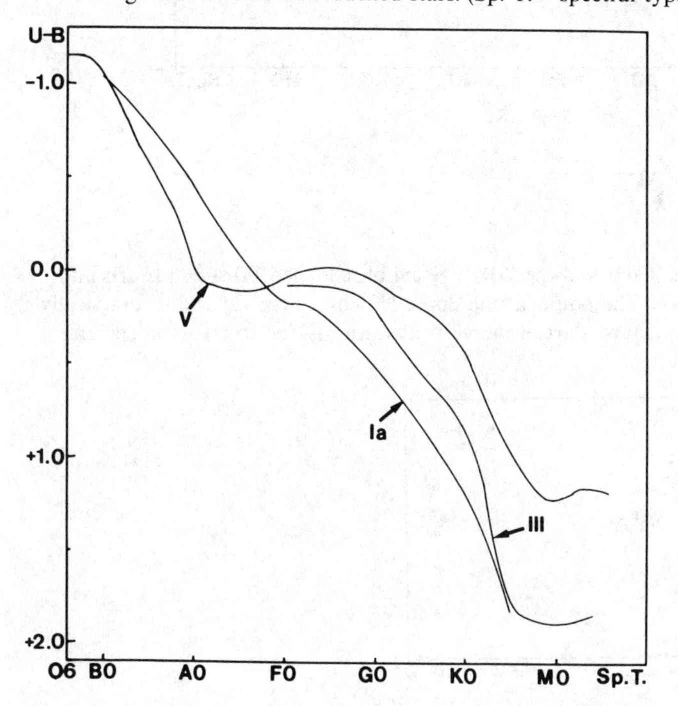

Figure 5.3. The relation between B–V and Yerkes spectral classification. Average relations for de-reddened stars. (Sp. T. = spectral type.)

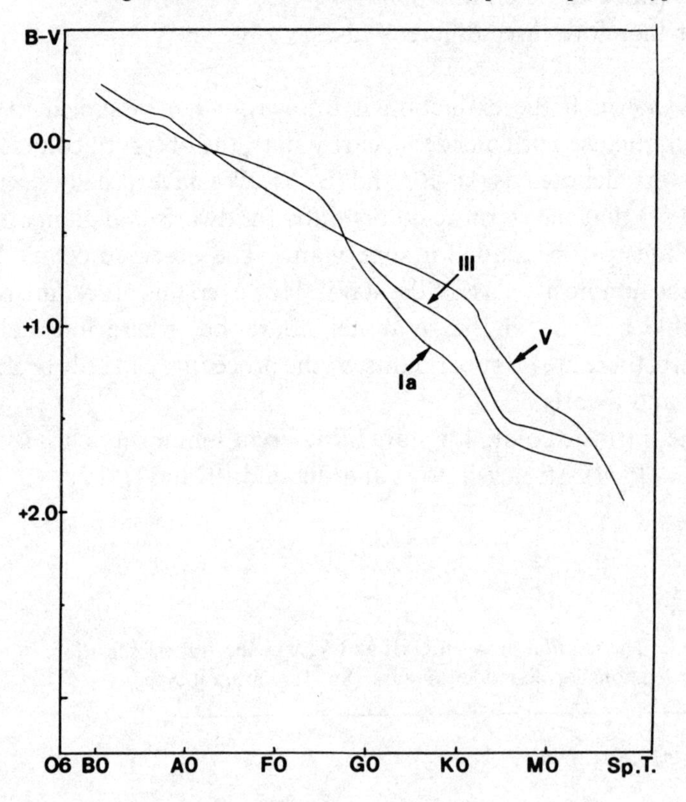

Figure 5.4. Colors in the UBV system of reddened B0 stars (dwarfs and supergiants). The points at the upper left corner correspond to practically unreddened stars. Part of the vertical scatter is due to errors in spectral classification.

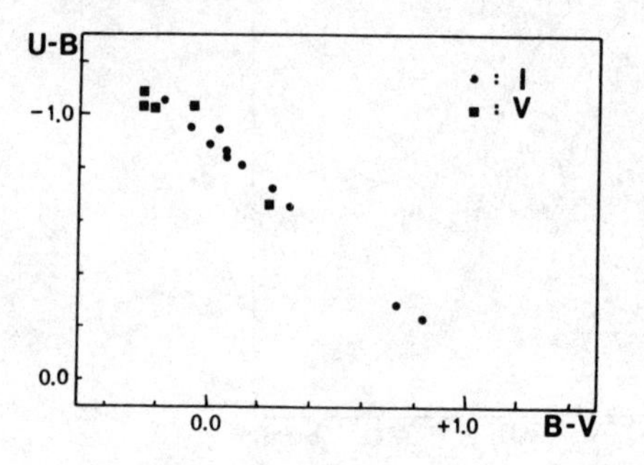

(*b*) *Luminosity class known.* In this case we use the fact found previously that the ratio of the reddenings in two colors is constant (see section 4.5). Therefore to de-redden a star, we trace a straight line with the appropriate slope through the observed color points until we encounter the main sequence. Point P_0 in figure 5.5 is the de-reddened image of P. The same procedure can be done numerically. Let

$$\alpha = \frac{E(\mathrm{U-B})}{E(\mathrm{B-V})} = \frac{(\mathrm{U-B})_o - (\mathrm{U-B})_i}{(\mathrm{B-V})_o - (\mathrm{B-V})_i}$$

Then

$$\alpha(\mathrm{B-V})_o - \alpha(\mathrm{B-V})_i = (\mathrm{U-B})_o - (\mathrm{U-B})_i$$
$$\alpha(\mathrm{B-V})_o - (\mathrm{U-V})_o = \alpha(\mathrm{B-V})_i - (\mathrm{U-B})_i = Q$$

If $(\mathrm{U-B})_i$ and $(\mathrm{B-V})_i$ are linearly related, the right-hand-side can be written as a function of $(\mathrm{U-B})_i$ alone and can thus be tabulated as a function of color and/or spectral type. This is the 'Q method' of Johnson and Morgan (1953). The numerical values are provided in table 5.2.

Obviously a similar relation can be established for supergiants, but we must always know the luminosity class of the stars beforehand.

(*c*) *Both extinction and luminosity known.* If the extinction can be assumed to be zero because the stars are nearby and the luminosity class is known, the color index is available for further use. This has been done by Sandage and

Figure 5.5 De-reddening procedure in UBV photometry. P = observations; P_0 = de-reddened image of P; dp = reddening path; I.S. = 'intrinsic' main sequence.

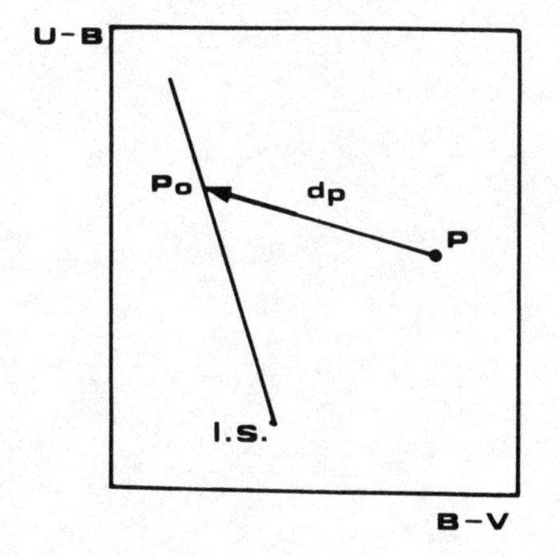

Eggen (1959) in their study of late type subdwarfs. These stars do have very weak metallic lines, and therefore their continuum should be closer to that of a black body than normal dwarfs, and according to figure 5.1 this should position them *above* the observed main sequence in the color–color diagram. Since we have seen that the blanketing increases toward shorter wavelengths, its effect should be stronger in the U band than in the B band, and stronger in B than in V. Therefore an object with very weak (or no) lines should suffer a small displacement in B–V and a larger displacement in U–B; this is illustrated in figure 5.6. Point B represents a normal dwarf. If we could gradually weaken all its metallic lines, by decreasing the metal abundance in the star's atmosphere, the star would occupy positions A_1, A_2, A_3, etc. The maximum effects observed are of the order of $0^m.3$ for BA_3.

Table 5.2. *Q values for different spectral types (Sp.t.) of dwarfs, mean relation.*

Sp.t.	Q	Sp.t.	Q	Sp.t.	Q
O9	0.90	B3	0.57	B9	0.13
B0	0.90	B5	0.44	A0	0.00
B0.5	0.85	B6	0.37		
B1	0.78	B7	0.32		
B2	0.70	B8	0.27		

Figure 5.6. Effect of blanketing in the UBV diagram. B = dwarf with normal abundance; A_1, A_2 and A_3 represent the position of B when metallic lines are progressively decreased. The curve represents part of the main sequence.

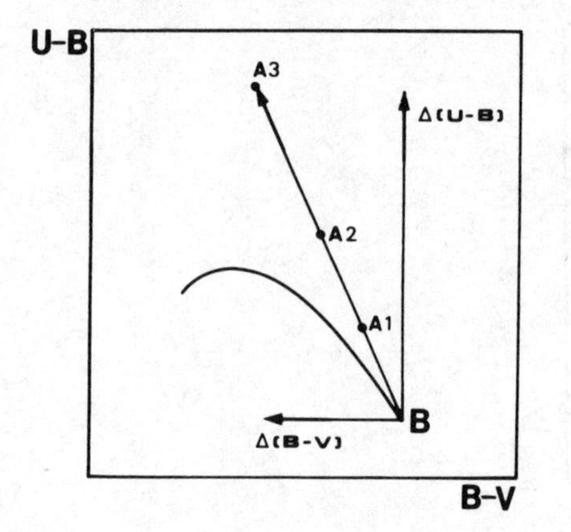

Obviously the reverse situation would happen if the star had stronger metal lines: it would appear *below* the main sequence in the two-color diagram.

We conclude with a remark on the detectability in general of spectral features by means of broad band photometry. If one works with 10^3 Å wide filters and at a precision of $\pm 1\%$, it is clear that features whose equivalent widths are less than 10 Å will not influence measures, either in emission or in absorption. Also smaller features will disappear unless they are so numerous that their total contribution becomes important as in the case discussed above. This insensitivity to spectral features is a general characteristic of broad passband systems.

We can now summarize the advantages of the UBV system. Being a broad band photometry, it is efficient even with small telescopes. It enables both a temperature assignment and a measure of the interstellar extinction to be made. It is well suited to the observation of B-type stars, the knowledge of which is important for galactic structure and cluster studies. Furthermore since it contains only two indices, the interpretation of the measures can be easily visualized. As a result of all this, a large number of stars has been measured in it; indeed, more stars have been measured in this system than in all other photometric systems taken together. It is the most widely used system for photometric classification.

To extend the usefulness of the UBV broad band system to later type stars, Johnson (1965) added more bands toward the red (see table 5.3).

There are three reasons for adding these bands. First of all, red or infrared colors are better suited as temperature indicators of cooler stars since they fall closer to the maxima of emission. Secondly, interstellar extinction effects diminish toward longer wavelengths. Thirdly, blanketing also diminishes toward longer wavelengths. We shall however not discuss these matters further. For more details, see Lamla (1982).

Table 5.3. *Arizona photometric bands.*

Band	$\lambda_0(\mu m)$	Band	$\lambda_0(\mu m)$
R	0.70	K	2.2
I	0.90	L	3.4
J	1.25	M	5.0
H	1.62	N	10.2

5.2 The uvby system

The four-color system was introduced by Strömgren in the late fifties (see Strömgren 1964). It uses four intermediate band filters centered at $\lambda_0 = 3500$, 4100, 4700 and 5500 Å. These filters are called in order u (ultraviolet), v (violet), b (blue) and y (yellow). The response functions of the filters are given in table 5.4, taken from Lamla (1982).

Table 5.4. *S(λ) for the uvby system.*

u		v		b		y	
λ[Å]	S(λ)	λ[Å]	S(λ)	λ[Å]	S(λ)	λ[Å]	S(λ)
3150	0.00	3750	0.00	4350	0.00	5150	0.00
3175	0.26	3775	0.16	4375	0.38	5175	0.42
3200	2.96	3800	0.35	4400	0.91	5200	1.02
3225	6.61	3825	0.83	4425	1.50	5225	1.58
3250	11.04	3850	1.46	4450	2.16	5250	2.28
3275	15.95	3875	2.26	4475	3.33	5275	3.90
3300	20.97	3900	3.02	4500	4.60	5300	5.51
3325	24.99	3925	4.78	4525	7.28	5325	7.90
3350	28.14	3950	7.81	4550	11.16	5350	11.70
3375	31.26	3975	12.97	4575	17.80	5375	15.84
3400	33.00	4000	20.00	4600	26.58	5400	19.27
3425	33.92	4025	29.88	4625	35.35	5425	19.98
3450	34.39	4050	39.92	4650	39.92	5450	20.13
3475	34.30	4075	47.04	4675	40.08	5475	20.05
3500	33.37	4100	48.98	4700	38.06	5500	19.84
3525	32.14	4125	47.52	4725	31.71	5525	18.99
3550	30.43	4150	43.00	4750	22.83	5550	15.90
3575	27.82	4175	36.64	4775	14.09	5575	11.48
3600	24.33	4200	27.78	4800	8.80	5600	7.44
3625	20.38	4225	17.76	4825	5.44	5625	4.33
3650	16.07	4250	10.88	4850	3.06	5650	2.70
3675	11.29	4275	6.54	4875	1.94	5675	1.72
3700	7.45	4300	3.88	4900	1.15	5700	1.00
3725	4.55	4325	2.59	4925	0.92	5725	0.57
3750	2.11	4350	1.92	4950	0.67	5750	0.29
3775	0.86	4375	1.33	4975	0.51	5775	0.19
3800	0.39	4400	0.68	5000	0.28	5800	0.14
3825	0.00	4425	0.34	5025	0.14	5825	0.07
3850	0.00	4450	0.00	5050	0.00	5850	0.00

Choice of these particular passbands was dictated by the following reasons.

- **y** was positioned as close as possible to the V band of UBV photometry, to preserve a link with the 'V' magnitudes. It lies in a region free from strong lines.
- **b** was positioned in a region including Hβ, which apart from this line does not contain too many strong lines.
- **v** is centered on Hδ and contains many additional lines in stars later than F; it lies *after* the Balmer jump.
- **u** is positioned between the atmospheric cut-off (λ3200) and the Balmer jump.

Furthermore the interstellar extinction for the different λ_0s is (table 4.5; section 4.6)

$$A(3500) = 1.579\,A_v \qquad A(4100) = 1.394\,A_v \qquad A(4700) = 1.223\,A_v$$
$$A(5500) = 1.0\,A_v$$

The differences between consecutive indices are thus 0.185 A_v, 0.171 A_v and 0.223 A_v, implying that the indices are evenly spaced with regard to interstellar extinction.

Besides these four, we normally add a narrow band (interference) filter to measure the intensity of Hβ. There is some confusion in the literature as to what the 'Strömgren' system is, i.e. with or without β. We shall use it in the sense that excludes the β filter, but nevertheless we shall discuss its use.

We start the discussion by noticing that with four passbands we have six color indices, three of them independent. The three color indices give raise to at least three diagrams with two indices. A spatial representation of the four-color system thus requires a three-dimensional set-up. We start by considering the (u–b, b–y) diagram (figure 5.7), which can be compared with figure 5.1 which gives (U–B, B–V). Although the shapes are rather similar, the inflection is different. By a similar procedure to that followed in UBV photometry, we can also establish the reddening lines. This is illustrated in figure 5.8 for the case of O-type stars.

Two-color diagrams with observed colors are, however, seldom used in uvby photometry. The only color index used as a temperature indicator is (b–y). Strömgren has introduced two parameters (instead of the indices u–v and b–v) defined as

$$c_1 = (u - v) - (v - b)$$
$$m_1 = (v - b) - (b - y)$$

Figure 5.7. uvby photometry. The (u–b, b–y) diagram. The curve provides the average relation for non-reddened dwarfs.

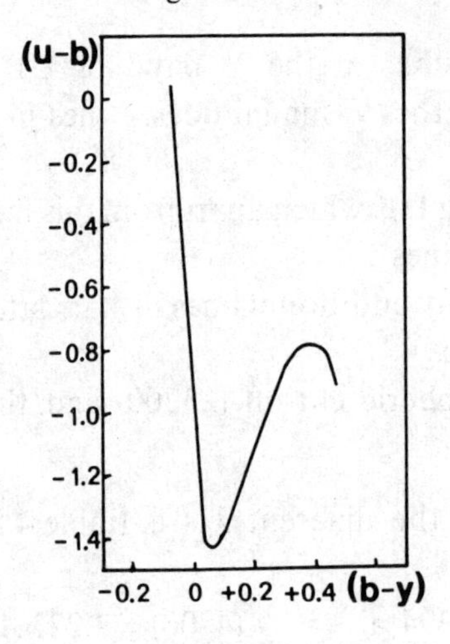

Figure 5.8. The reddening for O-type stars in the ubvy system. The almost vertical line is part of the sequence of non-reddened dwarfs. The positions of the reddened stars cluster around an average line (middle). The two other lines represent possible reddening paths for small regions of the sky (for instance, Cygnus). From Crawford and Mandwewala (1976).

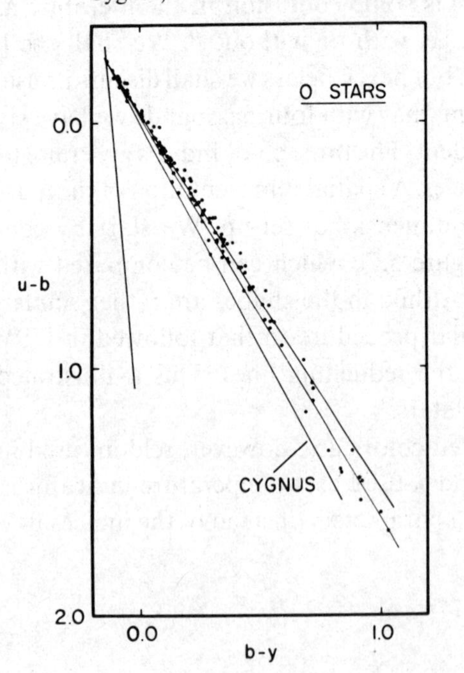

The first measures the Balmer discontinuity and the second the line blanketing. This can be seen in figure 5.9. Since the blocking is about twice as large in u as in v, the difference $c_1 = u - 2v + b$ is practically independent of the blocking and thus measures mainly the Balmer discontinuity.

The two indices are still affected by interstellar extinction, although the effect should not be very great because, as we have observed before, the extinction *differences* for the different passbands are small. If we assume that a universal $F(\lambda)$ 'law' applies we can introduce

$$(c_1)_o - (c_1)_i = E(c_1) \qquad (i = \text{intrinsic}; \quad o = \text{observed})$$

$$(b-y)_o - (b-y)_i = E(b-y)$$

Furthermore, because of the 'universal' form for interstellar extinction,

$$E(c_1) = \alpha E(b-y)$$

where α is best determined observationally. Crawford and Mandwewala (1976) obtain $\alpha = 0^m\!.19$. Then

$$(c_1)_o - (c_1)_i = 0.19(b-y)_o - 0.19(b-y)_i$$

and the quantity

$$(c_1)_o - 0.19(b-y)_o = (c_1)_i - 0.19(b-y)_i = [c_1]$$

is a de-reddened quantity. Similarly we can introduce

$$[m_1] = (m_1)_o + 0.33(b-y)_o$$

and

$$[u-b] = (u-b)_o - 1.53(b-y)_o$$

In general, representations in Strömgren photometry are made with three de-reddened parameters. We show in figures 5.10, 5.11 and 5.12 some of the important ones taken from Philip and Egret (1980). Figure 5.10 is a diagram

Figure 5.9. Scheme to explain the c_1 and m_1 indices. For explanation, see text.

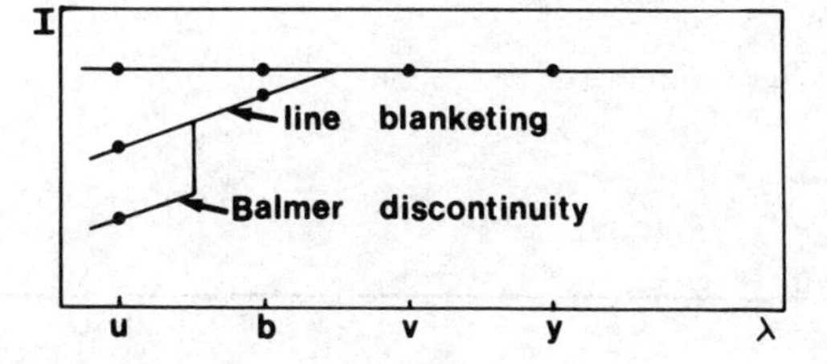

Figure 5.10. uvby photometry: β versus (b–y) diagram. The general direction of reddening (**R**) and the luminosity effect (**L**) are indicated by arrows. From Philip and Egret (1980).

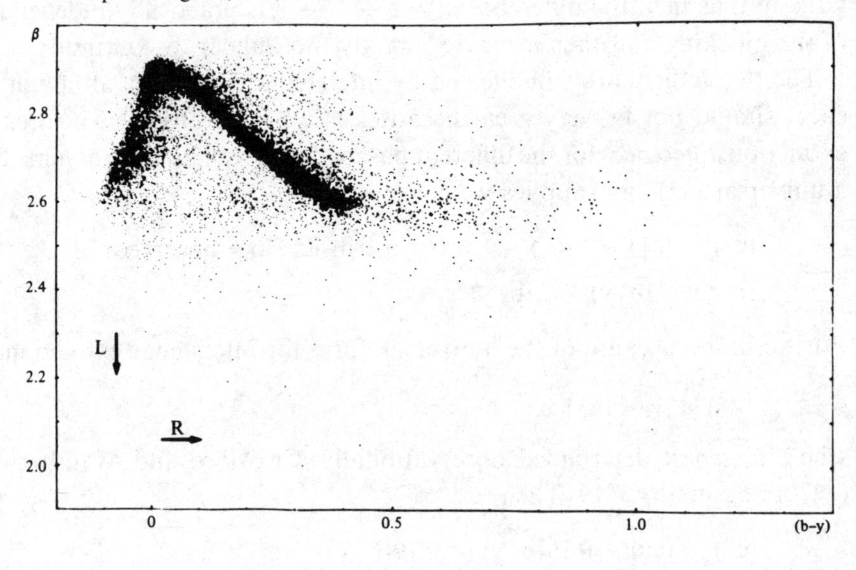

Figure 5.11. uvby photometry: c_1 versus (b–y) diagram. The general direction of reddening (**R**) and the luminosity effect (**L**) are indicated by arrows. From Philip and Egret (1980).

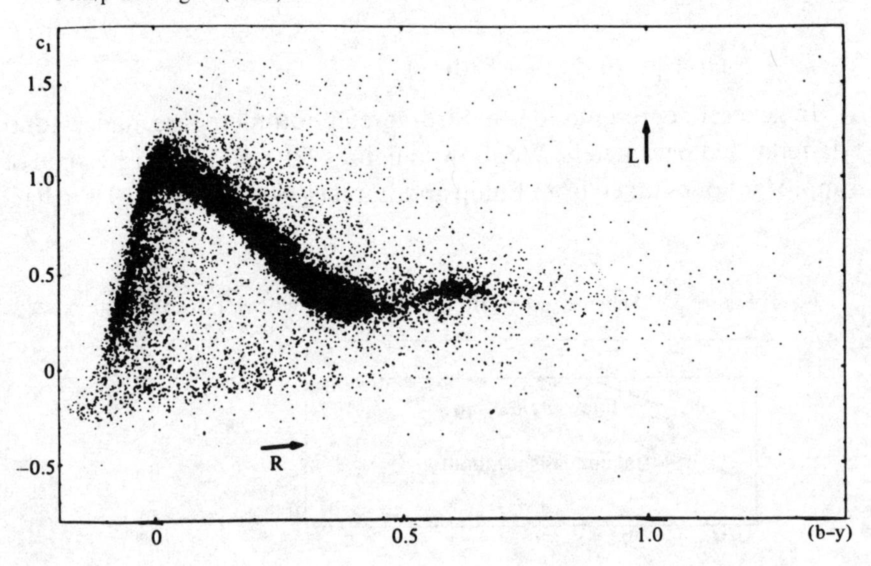

of β vs. (b–y), where (b–y) is a temperature parameter and β indicates the Hβ line strength. Since Balmer lines are strongest around A0(b–y $\sim$ 0.0), there is a hill-shaped main sequence with decreasing β-strength toward both early and late type stars. The scattered points which 'fill' the 'hill' are supergiants, which at a given temperature do have weaker hydrogen lines. Stars scattered toward the right are reddened stars.

One possible use of the diagram is obvious: if stars of different luminosity are well separated in β, the diagram can be used to determine absolute magnitudes. This requires of course that interstellar reddening be corrected beforehand.

Similar effects are to be expected in a diagram of (c_1, b–y), which is given in figure 5.11. Here c_1 represents the Balmer jump and (b–y) the temperature. We expect to find a maximum at b–y $\sim$ 0 and, if observed values of (b–y) are used, some scatter toward larger values of b–y due to reddened stars. The effect of de-reddening is shown very well in figure 5.12. The narrow left side represents of course the suppression of reddening; the scatter to the left on the right side represents the effect of stellar luminosity. More luminous stars have smaller Balmer jumps, and so they fall higher on the band.

Figure 5.12. uvby photometry: c_0 versus $(b-y)_o$ diagram. c_0 and $(b-y)_o$ represent de-reddened c_1 and (b–y) values.

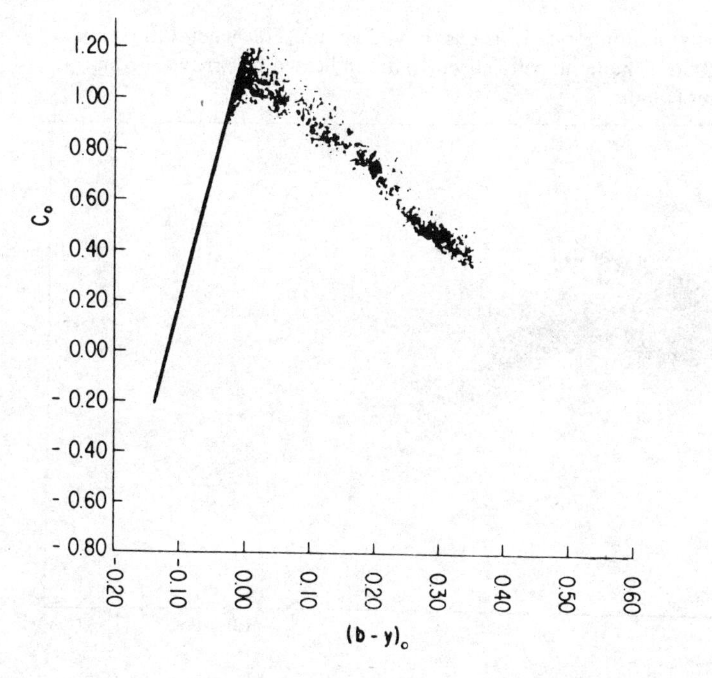

Figure 5.13 gives the relation between (m_1, b–y). Roughly, points occupy a V-shaped area, with its vertex at about b–y ~ -0.1. Early type stars occupy the left part (b–y < 0) but can be displaced (parallel to vector **R**) because of interstellar reddening. Later type stars follow more or less the lower part of the diagram, but show a significant scatter. This scatter is due partly to reddening and luminosity effects, but it primarily reflects the intensity of the metallic lines. Since early type (reddened) stars which have *few* lines are found upwards, it is clear that the diagram implies that there are more stars with *weaker* metallic lines than with stronger metallic lines. The latter should lie in the lower part of the diagram. Figure 5.14 shows the situation when reddening is taken out.

Oblak, Considere and Chareton (1976) have produced average uvby β colors for different spectral types and luminosity classes. Some of their values are illustrated in figures 5.15, 5.16 and 5.17 which permit an easy passage from the Yerkes system to the Strömgren system. An extension toward later spectral types has been given by Ardeberg and Lindgren (1985).

Although it is impossible to review here all the finer uses of the Strömgren system, figure 5.18 shows just what kind of peculiarity we can hope to distinguish (Kilkenny and Hill 1975). We should however remember what was said before on the lack of 'purity'; so an uncommon combination of color

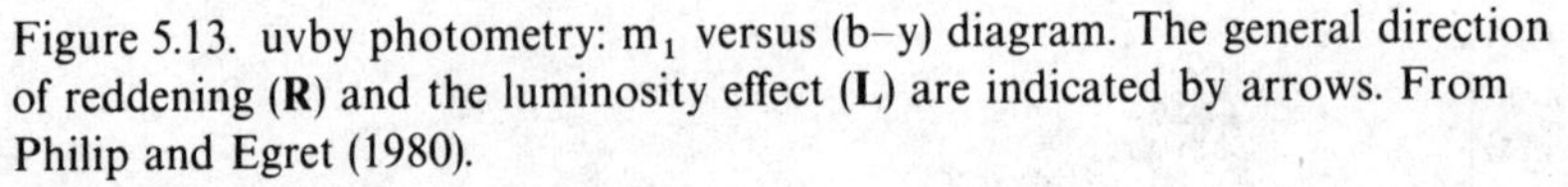
Figure 5.13. uvby photometry: m_1 versus (b–y) diagram. The general direction of reddening (**R**) and the luminosity effect (**L**) are indicated by arrows. From Philip and Egret (1980).

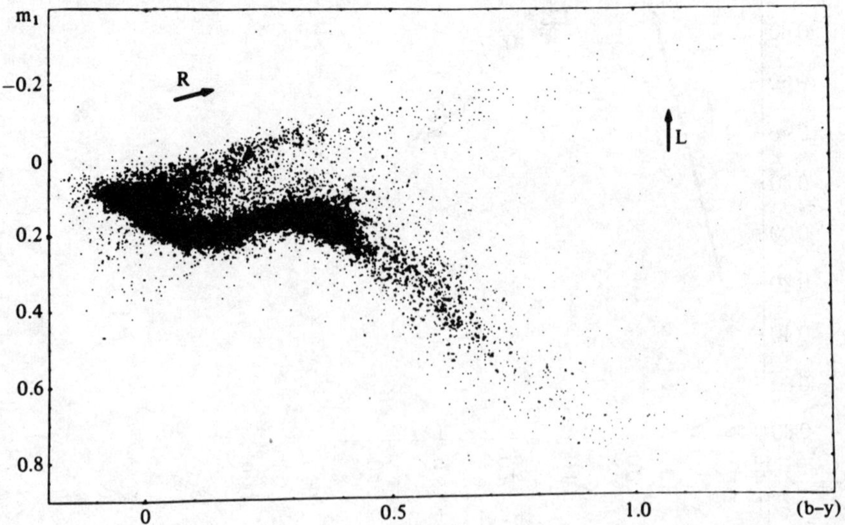

Figure 5.14. uvby photometry: m_0 versus $(b-y)_o$ diagram. m_0 and $(b-y)_o$ represent de-reddened m_1 and $(b-y)$ values.

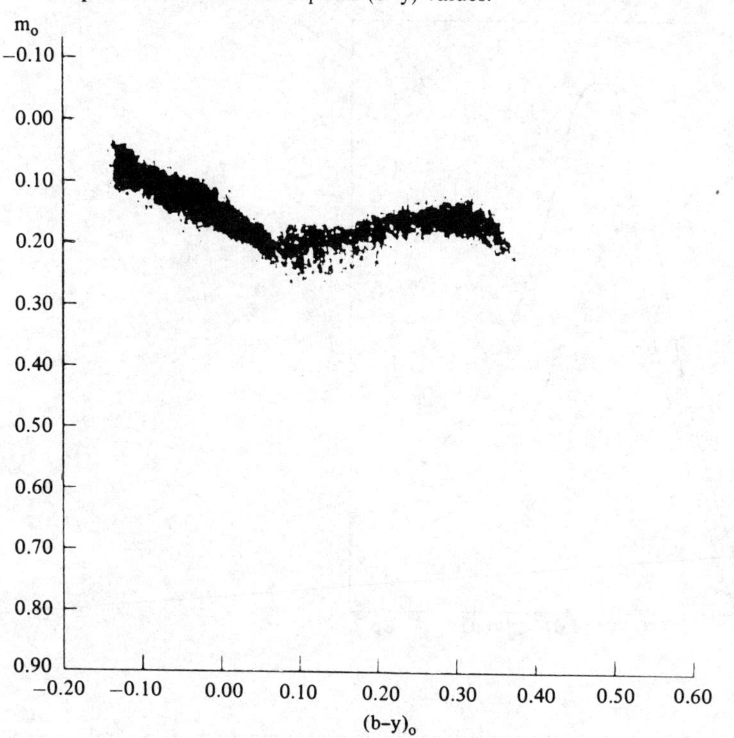

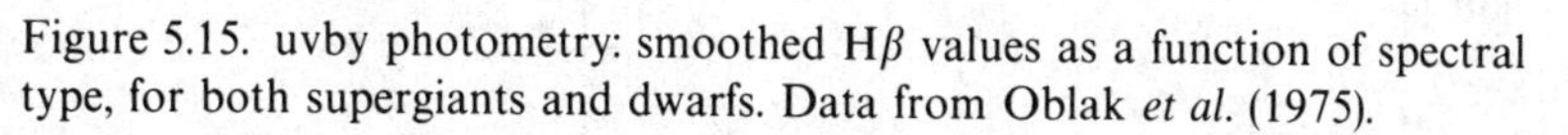

Figure 5.15. uvby photometry: smoothed Hβ values as a function of spectral type, for both supergiants and dwarfs. Data from Oblak *et al.* (1975).

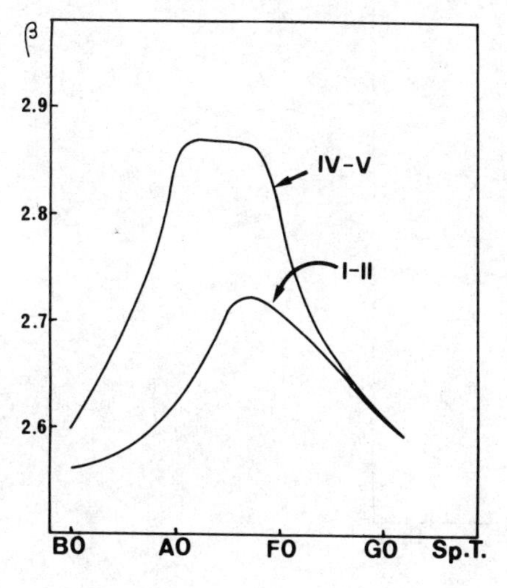

Figure 5.16. uvby photometry: smoothed values of $[c_1]$ as a function of spectral type for two luminosity groups. Data from Oblak *et al.* (1975).

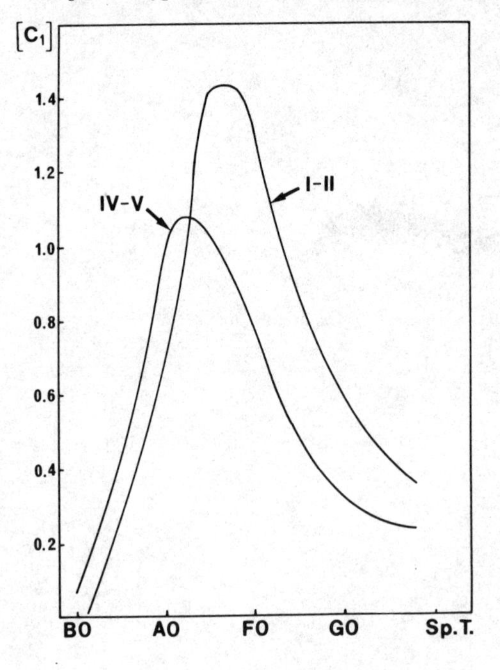

Figure 5.17. uvby photometry: $[m_1]$ values as a function of spectral type for two luminosity groups. Data from Oblak *et al.* (1975).

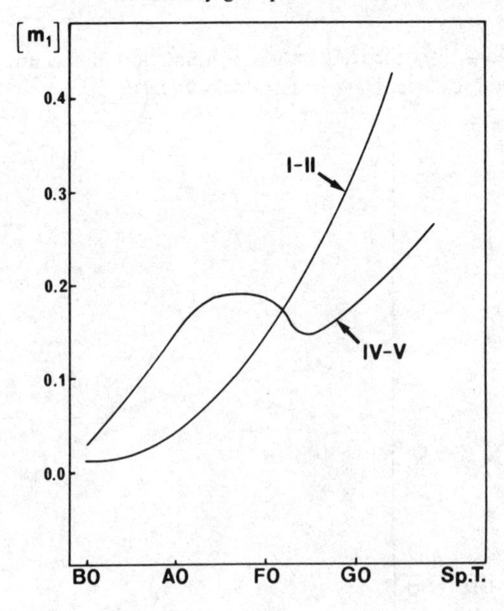

indices tells only that the star is abnormal, but not what kind of spectroscopic abnormality is in fact causing the photometric abnormality.

For more details we refer the readers to Strömgren (1966), Golay (1974) Philip, Miller and Relyea (1976), Philip and Egret (1980), as well as to the meeting on *Multicolor photometry and the theoretical HR diagram*, edited by Philip and Hayes (1975), and the various papers in the meeting in honor of Strömgren (Reiz and Andersen 1978).

When compared with the UBV system previously discussed, we see that the uvby system uses well-defined narrower filters. It is thus well reproducible and accordingly its precision is high. Errors are better than $0^m.01$ and are generally quoted as being of the order of $\pm 0^m.008$; in this respect the system is clearly superior to the UBV. Due to the narrowness of its filters, the measures are more easily interpretable from the point of view of theory, since we average over a smaller number of spectral features; moreover the passbands were deliberately chosen so as to extract a maximum amount of information from the stellar spectra. It should be kept in mind that the

Figure 5.18. uvby photometry: schematic m_1 versus (b–y) diagram. The location of some peculiarity groups is indicated: sd = subdwarfs; D = degenerate; hor. br. = horizontal branch; Am = metallic line stars. The arrow marks the reddening path.

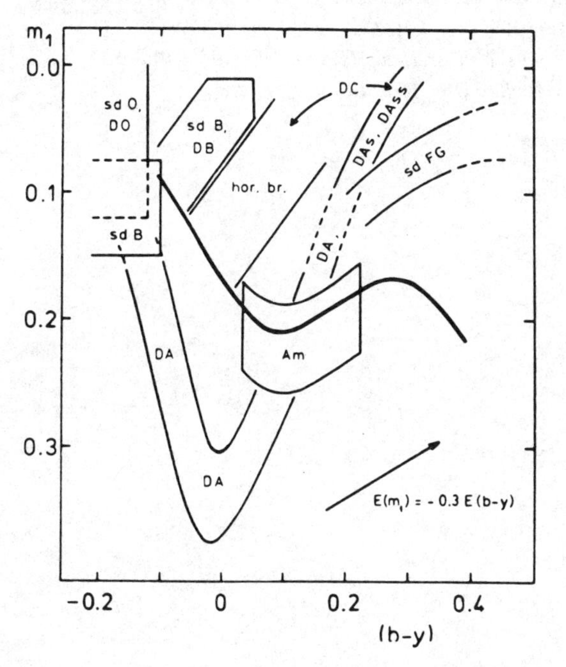

system was specifically conceived to deal with nearby stars earlier than the sun and these limitations are slightly more restrictive than those for the UBV system.

References

Ardeberg A. and Lindgren H. (1985) *IAU Symp.* **111**, 509, Reidel D. Publ. Co., Dordrecht

Crawford D.L. and Mandwewala N. (1976) *PASP* **88**, 926

Golay M. (1974) *Introduction to astronomical photometry*, Reidel D. Publ. Co, Dordrecht

Johnson H.L. (1965) *Com. L. Planet Lab.* **3**, 73

Johnson H.L. and Morgan W.W. (1953) *Ap. J.* **117**

Kilkenny D. and Hill P.W. (1975) *MNRAS* **173**, 625

Lamla E. (1982) in Landolt–Börnstein, *Numerical data and functional relationships in science and technology.* New series, group 6, vol. 2b, p. 35.

Oblak E., Considere S. and Chareton M. (1976) *AA Suppl.* **24**, 69

Philip A.G.D., Miller T.M. and Relyea L.J. (1976) in *Dudley Obs. Rep.* **12**

Philip A.G.D. and Egret D. (1980) *AA Suppl.* **40** 199

Reiz A. and Andersen T. (1978) *Astronomical papers dedicated to Bengt Strömgren*, Copenhagen University Observatory

Sandage A. and Eggen O. (1959) *MNRAS* **119**, 278

Schmidt-Kaler T. (1982) in Landolt–Börnstein, *Numerical data and functional relationships in science and technology.* New series, group 6, vol. 2b, p. 35.

Straizys V. (1977) *Multicolor stellar photometry*, Mokslas Publ. Vilnius

Strömgren B. (1964) *Ann. Rev. AA* **4**, 433

Multicolor photometry and the theoretical HR diagram (1975) Philip D. and Hayes D.S. (ed.), *Dudley Obs. Rep.* **14**

6

Comparison of classification methods

Having seen in some detail spectroscopic and photometric classification methods, we shall compare them in this chapter. We shall examine first their 'problem solving capability' and 'information content'. Finally we shall discuss the relation between classification and physical parameters. (For more details, see Jaschek 1982.)

6.1 Problem solving capability

In the chapter on spectral classification we have seen that in the Yerkes system there are two parameters, according to which stars can be arranged. If a star cannot be assigned a unique place in the scheme, it is called 'peculiar'. We have also seen that in some cases abbreviations are needed for stars with varying degrees of rotation and that in some cases magnetic fields can be detected by the inspection of spectrograms. Therefore the list of parameters which can be ascertained from spectrograms is:

spectral type
luminosity class
spectral peculiarity
rotation
magnetic field.

Without going into details we may say that the spectral type corresponds to stellar surface temperature, luminosity class to stellar luminosity and spectral peculiarity to either abnormal atmospheric structures or anomalies in the abundance of chemical elements.

Our next question is whether these parameters can only be determined spectroscopically, or if photometry is able to do the same or better. This is a crucial question because it will determine the choice of instrumentation to attack a given problem.

We have already seen that in the UBV system both U–B and B–V indices are correlated with spectral type. Between B0 and M0 there are roughly 50 spectral subdivisions (see chapter 5) whereas we have in B–V an interval of $1^{m}70$, from $-0^{m}30$ to $+1^{m}40$. If we assume that the error of a B–V measure is of the order of $\pm 0^{m}02$ we arrive at the 'figure of merit' $f(=\text{interval} \div \varepsilon)$ equal to 85, better by almost a factor of two. So, overall, photometry is as good as or better than spectroscopy. A similar figure applies to U–B measurements, and similar arguments can also be made with other photometric systems.

Turning now to luminosity class, we see that due to interstellar extinction UBV cannot predict it except in the hypothetical case when we deal with de-reddened observations (see figures 5.2 and 5.3) and chose the region where the sequences of dwarfs and giants are most widely separated.

Next, in the ubvy system, we have already seen that a luminosity assignment is possible between O- and G-type stars if an Hβ index is measured in addition to uvby photometry, and that a precision of $\pm 0^{m}3$ is possible in M_v. Such a result is certainly better than that offered by spectroscopy. Jaschek and Mermilliod (1984) have shown that in the B9–F5 range the M_v of stars classified spectroscopically as dwarfs is about 2^{m}, which is much larger than the uncertainty for Hβ photometry.

From these considerations we see that both temperature and luminosity can be determined with better precision by photometrists, and if we add that interstellar reddening can be determined *only* by photometry, we could be led to think that spectral classification may be superseded by photometry. Let us note, however, that this result applies only if the stars are known to be earlier than G and to be normal, and their interstellar extinction can be handled in the normal way.

The situation changes if spectral peculiarities are considered. Here the spectroscopists have the advantage; they can quickly recognize a large number of peculiarities, whereas photometrists are more or less helpless. Consider just one case: the Hβ measurements which are central for luminosity determinations of early type stars. Since photometrists measure an equivalent width, but do not see the profiles, they cannot know if the star is a supergiant with intrinsically small $W(\beta)$, or a dwarf with a central emission in Hβ simulating a small $W(\beta)$. This is clearly a case for the spectroscopists.

The same is true for many of the forty groups of peculiarity which we shall consider later on in this book. Photometrists can of course overcome these limitations by constructing special systems for special types of stars, but although this approach is feasible, it is certainly not economic.

We should also remember the problem of 'purity'. If 'purity' cannot be ascertained, the photometric attack is almost certainly bound to run into difficulties.

These considerations can be complemented with short remarks on other parameters. Rotation is almost impossible to handle for a photometrist, but can be established easily on suitable spectrograms. For magnetic fields the answer is more difficult. If the average field is of the order of a few kilogauss, high resolution plates are needed to detect it; if the field is larger than 10^5 gauss it can be easily seen from the split Balmer lines. Photometrists on the other hand have developed a technique to measure fields up to 5×10^3 gauss (Cramer and Maeder 1981) usable for stars in the range B5–A1. We see once again that we can do things photometrically if we sufficiently narrow down the problem.

In conclusion, part of this discussion can be summarized by observing that spectral classification is a good, rapid technique to distinguish normal from abnormal stars. Once the star is classified as normal, photometrists can obtain more accurate information than spectroscopists. In this sense both techniques are complementary and should be used this way.

We consider next the information content of both systems.

6.2 Information content

A very simple argument could start with the consideration that a spectrogram obtained for MK classification covers a spectral range of 1200 Å ($\lambda\lambda$3600–4800) with a resolution of about 2 Å. This makes 600 elements of information, of which many are not independent in normal stars, but might be vital in the case of a peculiar star. For instance, a central emission in Hβ concerns just one information element, but is sufficient to classify the star as an emission line object.

Photometries with n bands can be likened to a receiver with n information elements, and we see clearly that as far as detection of peculiarity goes, the case is unfavourable for photometrists since n is always smaller than 10. Nevertheless we may also argue that in a broad sense these 600 bits are not truly independent, so that we have in reality a much smaller number of information elements. This is certainly true for physical parameters like temperature and luminosity, and in this sense spectroscopy is not vastly superior to photometry.

A different argument has been given by Nicolet (1982). He starts with the number of luminosity classes (15) and spectral types (71) which furnish 6.1 + 3.9 = 10 bits. For UBV photometry he admits possible errors of $\pm 0^m02$ in B–V and $\pm 0^m03$ in U–V. Next he considers two stars S_1 and S_2 measured in UBV and evaluates the probability p that they will be distinguishable in UBV. The number of bits will be $-\log_2(p)$. He then uses a sample of 160 stars (N_e) to calculate the number of pairs N_p which differ by more than the pre-specified errors. He obtains $N_p = 48$ and therefore $p = N_p/N_e(N_e - 1) =$ 0.001 89 and thus 9.05 bits. Therefore the UBV system and the MK system do have the same information content. This is true since from MK we can determine spectral type and luminosity class (T and M_v) and from UBV temperature and interstellar extinction. But we have assumed that the stars are normal, and the decision as to whether stars are normal or not was based spectroscopic criteria.

The reasoning of Nicolet can also be applied to other types of photometric systems. For the Geneva system (four colors) he obtains an information content of 12.67 bits; this is still less than that of the MK system if we add to the previous 10 bits (see above) the 5.3 bits coming from the forty groups of peculiarity distinguishable with MK classification. We can conclude that even with four colors we are still below the information content of the MK system.

From this general reasoning we can only say again that both ways of classifying stars are complementary rather than conflicting.

We should be aware that if photometry in discontinuous bands is substituted by spectrophotometry of medium and high resolution the advantages of both systems can be combined. Up to now the number of stellar spectrum scans published is certainly two orders of magnitudes smaller than both the number of two-dimensional spectral classifications and the number of photoelectric photometric measures, and we can only hope that the use of better detectors will improve the present situation.

6.3 Physical parameters and classification

The most frequent criticism of stellar classification methods comes from people interested in the physics of stellar atmospheres or from theoreticians. Crudely, the question they pose can be stated: 'What is the classification good for?' And usually critics add that to call a star A2 v is useless because it does not tell them anything about the effective luminosity, the surface gravity, the chemical composition, atmospheric structure or other physical variables. The conclusion is then drawn that stellar classification methods belong to the past and are to be banned from serious consideration.

To answer such arguments, we start by recalling that classification methods are ways of ordering objects, and ways of reducing the infinite number of astronomical objects to a smaller number of typical objects. Labelling a star A2 v means that we can avoid the analysis of 10^5 similar objects in our galaxy in favor of the analysis of just *one* A2 v star. Once we have derived the physical parameters for one A2 v star, we can apply them to *all* similar stars. A classification takes a few minutes whereas a detailed atmospheric analysis of any star takes some $10^4 - 10^5$ minutes, which is a rather sobering fact. Of course the classification is not an answer in itself and it can never substitute physical analysis. But it is an essential first step, because it locates roughly the object with regard to others. Because this is a general framework, it is independent of the actual (physical) parameters we assign to the object. An A2 v star is cooler than an A0 v star, independently of the actual value of the temperature of the A0 v and the A2 v stars. Whereas the temperature scale of the stars has changed many times in the past, the spectral type has not changed. This, incidentally, is why books are written on 'B-type stars', rather than 'normal stars with surface temperatures 25 000 to 10 000 K'.

The same is true for luminosity classes. Whereas Arcturus (α Boo) is still undisputedly spectroscopically a giant, the opinions about its surface gravity (see for instance Bell, Edvardsson and Gustafsson 1985) are still in a process of flux – an astonishing fact for a well-studied object. Up to now absolute magnitudes of stars are mostly derived from their luminosity classes. Here again when we speak of luminosity class II we allude to something which can be seen in the spectrum, independently of the calibration of the luminosity class in terms of absolute magnitude, which has changed several times over the years.

In much the same way this is true for stars with peculiar spectra. Here also the classification designation is used because it is a short summary of a long description. Obviously the names we give to a group should avoid, as much as possible, any theoretical interpretation. For instance, the term 'shell stars' should have been banned from the start, although it is probably too late to change it now (see the corresponding section), because it suggests an interpretation. Another group which should be rebaptized is that of 'super metal rich stars'. The name affirms that the abundances are anomalous, and this is an interpretaion. What we *see* are strong metal lines, and that is how the group should be called. Although this seems an academic discussion, we feel it most important to keep the facts we observe well separated from their interpretation – all sciences are full of discussions arising from wrong designations.

The grouping of stars into boxes allows us to select convenient 'representative members'. It seems obvious that if we want to study a group in detail without analyzing all its members, it is convenient to select one or several 'representative members', and this is clearly a task for the classifier. Now a perusal of the literature – old and new – shows that very often this rule was not followed, with ensuing confusion. The confusion arises because if the star is not representative of the group, we cannot apply its properties to all group members.

A few examples will clarify this point. The star τ Sco was very often selected as a typical early type dwarf. However, this star is atypical in the sense that it has very sharp lines – although this facilitates analysis, the physical parameters obtained for this star should not be generalized because early type stars with sharp lines are in a small minority.

τ UMa was analyzed in detail by Greenstein (1948), and the results of the analysis applied to all Am stars. This caused trouble, because although τ UMa is an Am star, it is one of the latest Am stars known, and is atypical as such.

Be stars occur most frequently among B2 stars (Jaschek and Jaschek 1983) and it would seem obvious that if we want to learn something about the Be star group, we should study stars of this type rather than of types B5 or B8. Many authors still study exceptional Be stars – which is acceptable – but then they generalize their results to *all* Be stars. Of course, Einstein was a man, but that does not mean that we should call him a 'typical man'.

In short classification is a step which must always be accomplished before we study an object in detail. If we omit this, we study *one isolated object*, interesting in itself, but one which does not tell us much about the properties of other stars.

A last example may be quoted to show the usefulness of classification methods. This is the case of spectroscopic studies based upon satellite observations, in the wavelength range $\lambda\lambda$1000–3000. As a typical case let us consider the observations carried out with the 'International Ultraviolet Explorer' (IUE) from 1978. Because IUE is a typical mission satellite, each observer could ask for it to observe a certain number of interesting objects. No precaution was taken to observe systematically a large set of reference stars. But when a star with an interesting spectrum is observed, we usually also want to know how it compares with other, normal stars. Thus observers were forced to include in their programs some stars assumed to be normal. But how can it normally be ascertained, if not through classification? The result was that in many cases comparisons turned out to be wrong because the star the observer had assumed to be normal was in fact abnormal in the ultraviolet. Recently efforts have been made to classify UV spectra so that a set of reference stars is available (Heck *et al.* 1984).

We shall conclude this chapter by recommending the lecture of Mihalas (1984) which discusses the relevance of (spectroscopic) classification methods to physical theory.

References

Bell R.A., Edvardsson B. and Gustafsson B. (1985) *MNRAS* **212**, 497

Cramer N. and Maeder A. (1981) *23ème Coll. Inst. d'Astrophys. Liège*, p. 61

Greenstein J. (1948) *Ap. J.* **107**, 151

Heck A., Egret D., Jaschek M. and Jaschek C. (1984) *IUE low dispersion spectra reference atlas.* Part I. Normal stars, ESA SP-1052

Jaschek C. (1982) *Mitt. Astron. Gesell.* **57**, 167

Jaschek C. and Jaschek M. (1983) *AA* **117**, 357

Jaschek C. and Mermilliod J.C. (1984) *AA* **137**, 538

Mihalas D. (1984) in *The MK process and stellar classification*, Garrison R.F. (ed.), Toronto, p. 4

Nicolet B. (1982) Thesis, Geneva Observatory

Part 2

7

Introduction

The second part of this book is devoted to the description of the different groups of stars which have been defined over the years. We shall start with the hottest stars and work toward the cooler ones, each chapter dealing with one type of object. The material is arranged broadly into 'families' (chapters), so that stars which are similar appear together.

Stars with peculiar spectra are dealt with in sections added to the relevant chapter. Since many groups are not limited to one spectral class, the place of a given group in the book is usually the one corresponding to the first appearance of the group. So, for example, Am stars are discussed in section 10.1 of the A-type stars, despite the fact that some Am stars are found among F-type objects.

As far as possible, each chapter has the following structure. First, we define each group, taking in as much history as we need, but without trying to write the whole history of the group. Then we describe the spectrum of the type of star, usually in the classical region, but, if information exists, also in the ultraviolet, infrared and radio regions. Rotation is also included here. Next we examine what photometry can tell us about the group; then we consider the absolute magnitude, whether any of the group are binaries and finally statistical properties like presence in clusters, distribution on the sky and the frequency of the stars of the group. The chapters conclude with a mention of atlases and material for further reading.

We repeat here what we said in the preface, namely that this is a book on *classification* and not on the physics of the stars. Further, we provide general descriptions and are not attempting to replace the many excellent monographs which have been written about the particular groups.

We have already introduced in the first part of this book the spectroscopic and photometric aspects about which we shall comment in each section. Here we shall add only a few short paragraphs concerning the statistical properties which we discuss in each group.

Spectra

Unless otherwise stated, the spectra reproduced in the book come from an *Atlas of spectrum tracings* by G. Goy, M. Jaschek and C. Jaschek (to be published). The original spectra were taken at the Observatoire de Haute-Provence, with the 120 cm telescope and a spectrograph providing a plate factor of 65 Å/mm. The spectra were then traced at Geneva. The reduction procedure is described by Goy (1981, 1984, 1985), and consists essentially of filtering the fast Fourier transform of the register, eliminating both the highest frequencies (noise) and the lowest ones. The latter causes the spectrum to become flat, instead of showing the usual curvature seen in figure 4.5.

We have chosen to provide in this book only tracings of spectra, because of the well-known difficulties (and high costs) of reproducing spectrograms accurately. The reader can compare tracings and spectra of the same types of stars by comparing the two parts of figure 4.1. The choice of tracings over real spectra is however *only* dictated by printing costs and we would like to emphasize again that whenever possible real spectra, not tracings, should be looked at, as explained in detail in chapter 2.

Membership of stars in groups

The membership of stars in a given group comes from the *catalog of stellar groups* (CSG) Jaschek and Egret (1982, 1986), which lists about 25 000 stars in more than 50 peculiarity groups. Being a compilation catalog taken from what the authors considered to be the best observing lists, the catalog has its limitations, as far as individual members are concerned, because the authors did not see the original spectra on the basis of which the classification was made. There is therefore the possibility that a star might have been placed in the wrong group or might not be peculiar at all. We hope that the number of such misclassifications is small. The CSG, as far as we know, represents the only effort made to group data on spectrum peculiarities originally scattered over hundreds of publications.

For normal stars (which are not contained in the CSG), the data were taken from the SIMBAD database of the Centre de Données Stellaires. For a description of the content of SIMBAD, see Ochsenbein (1984); it is the largest existing astronomical database and contains about 6×10^5 stars.

Absolute magnitude

Because of their importance for evolutionary considerations and for studies of the spatial and kinematic distributions of stars, we have discussed the absolute magnitude of each group. A vast literature exists on average absolute magnitude determinations. Summaries of this work can be found in four papers in the Landolt–Börnstein series: Schmidt–Kaler (1982) deals with normal stars of all luminosity types, Seitter and Dürbeck (1982) look at stars of peculiar groups, protostars are discussed by Appenzeller (1982), and white dwarfs by Weidemann (1982).

Calibration problems are not yet solved even for well-known groups; see, for instance, Grenier *et al.* (1985), who deal with a calibration of B5 to F5 main sequence and giant stars, showing that existing calibrations can still be improved. Even in this case r.m.s. errors of M are still of the order of $\pm 0^{m}.5$. Since the sample used for the derivation of the absolute magnitude in this paper was of the order of 900 stars, it is clear that the average absolute magnitude for groups composed of smaller numbers of stars will have correspondingly larger errors.

The difficulty is worse in general for many of the peculiarity groups containing few stars since we find that:

(a) no reliable trigonometric parallaxes exist;
(b) few objects of this type are known in clusters or external galaxies with well-determined distances;
(c) the number of objects in the group is too small to derive a reliable statistical parallax.

Because of this we have indicated in each case what seems to us to be the best value for the average absolute magnitude, so that the group can be placed in the HR diagram. For reasons of space we have not included detailed discussions, which must be sought in the papers quoted.

A general description of the methods used for the distance determination is given by Heck (1978).

Binaries

It is a well-known fact that most of the stars form part of binary or multiple systems. We do not know if this is true for all kinds of stars, and we have therefore devoted a paragraph, whenever possible, to assemble present knowledge on whether the objects of a group are binaries. The interested reader can find more details in the proceedings of the meeting *Binary stars in the HR diagram* (1983), and in Halbwachs (1983) and Abt (1983).

Presence in clusters

We have devoted one paragraph to the presence of stars (of a given group) in clusters or in external galaxies, because of the obvious relevance to studies of absolute magnitude, frequency, evolutionary considerations, and relations to other groups.

The interested reader may find more information in Lynga and Lundström (1980) and in Alter, Balazs and Ruprecht (1982).

Distribution on the sky

Unless otherwise stated, the distribution is derived from the SIMBAD data base. Since existing data are severely biased because of the use of different discovery techniques in different parts of the sky (see for examples Egret and Jaschek 1984), it is useless to go beyond the simplest characteristics of the sky distribution.

Frequency of group members

For future work it is important to have an idea of how frequent the objects of a given group are. Therefore we have tried to quote the number of group members with $V < 6^{m}.5$, or the percentage or group members with respect to the total number of stars brighter than $6^{m}.5$, which is 9110. We selected the limit $6^{m}.5$ because *The bright star catalogue* (Hoffleit and Jaschek 1982) provides a convenient source of reference.

We also quote the number of objects with $V < 9^{m}$. This limit was chosen to allow comparison with stars included in the *Henry Draper catalogue*, which number 225 000. Since the magnitude limit of the HD varies however over the sky, the proportions with regard to HD stars are to be taken as approximate only.

List of group members

Interested readers may obtain lists of group members from the authors of the above-mentioned *Catalog of stellar groups* (Jaschek and Egret 1982, 1986). Bibliographic references to papers in which a given star is studied can also be obtained from the Strasbourg data center.

References

Abt H.A. (1983) *Ann. Rev. AA* **21**, 343

Alter G., Balazs B.A. and Ruprecht J. (1970) *Catalogue of star clusters and associations*, Budapest; first supplement, CDS (1982)

Appenzeller I. (1982) in Landolt–Börnstein, Group VI, vol. 2b, p. 357

Egret D. and Jaschek M. (1984) in *Cool stars with excesses of heavy elements*, Jaschek M. and Keenan P.C. (ed.) Reidel Publ. Co., p. 135

Goy G. (1981) *Arch. Sc. Genève* **34**, fasc. 2, 251

Goy G. (1984) *Arch. Sc. Genève* **37**, fasc. 2, 229

Goy G. (1986) (in press)

Goy G., Jaschek M. and Jaschek C. *Atlas of spectrum tracings* (to be published)

Grenier S., Gomez A.E., Jaschek C., Jaschek M. and Heck A. (1985) *AA* **145**, 331

Halbwachs J.L. (1983) *AA* **128**, 399

Heck A. (1978) *Vistas in Astronomy* **22**, 221

Hoffleit D. and Jaschek C. (1982) *The bright star catalogue*, 4th edition, Yale University Observatory, New Haven, Conn., USA

Jaschek M. and Egret D. (1982) *Catalog of stellar groups*, part I, Publ. Spéciale CDS no. 4

Jaschek M. and Egret D. (1986) *Catalog of stellar groups*, part II, (to be published)

Lynga G. and Lundström I. (1980) *IAU Symp.* **85**, 123

Ochsenbein F. (1984) *BICDS* **26**, 75

Schmidt-Kaler T. (1982) in Landolt–Börnstein, Group VI, vol. 2b, p. 1

Seitter W. and Dürbeck H.W. (1982) in Landolt–Börnstein, Group VI, vol. 2b, p. 269

Weidemann V. (1982) in Landolt–Börnstein, Group VI, vol. 2b, p. 373

Binary stars in the HR diagram (1983) Jaschek C., Jaschek M. and Florsch A. (eds.), Strasbourg

8

O-type stars

8.0 Normal stars

An O-type spectrum is characterized by the simultaneous presence of lines of neutral and ionized helium. If ionized helium is absent, the spectrum corresponds to a B-type. The lines of ionized helium strengthen toward earlier stars, whereas neutral helium lines and hydrogen lines decline toward earlier types. Table 8.1, taken from Conti (1973), illustrates this by way of the equivalent widths of some of the stronger lines.

Because of this antagonistic behavior of He II and He I, we can use them for classification. The MK system uses the ratio He II λ4541/He I λ4471 as the criterion for spectral type for the interval 03–09 (for an illustration, see figure 8.1). The He II lines are grouped in the so-called Pickering series (analogous to the Bracket series of hydrogen) with lines at λλ5414, 4850, 4542, 4339, 4200, 4100, 4025 and 3968,..., and the Fowler series (analogous to the Paschen series of hydrogen) with lines at λλ4656, 3204,...

Besides helium, some other elements are also present in O-type spectra. Among these we have Si IV(λ4089) and C III(λλ4068, 4647 and 4651) in late O-type stars. N III(λλ4634, 4640) if present is only seen in emission.

In these stars emission lines are very often present and can appear with varying degrees of intensity, from barely perceptible to rather strong, although they are never as broad and intense as in Wolf–Rayet stars. They can appear in different objects as a weak central emission on a wider absorption, or as a complete fill-in, or as a moderately strong emission. If a star shows emission lines in the Balmer series, it is called Oe. If the emissions appear in N III(λ4634, λ4640) *and* in He II(λ4686) the object is called Of (see sections 8.1 and 8.2). Besides these groups there are some peculiar objects which combine diverse emission features (see section 8.2). Since, however, all three groups (O, Of and Oe) are closely related, we will deal in this section with their common properties.

Luminosity effects can also be distinguished. In the MK system, the authors use in early O-type stars a criterion found by Botto and Hack (1962). At O6, supergiants and dwarfs differ in various features. He II λ4686 is in emission in supergiants and in absorption in dwarfs. N III $\lambda\lambda$4634–42 is in emission in both classes, but is stronger in supergiants. N IV $\lambda\lambda$3479–85 on the other hand is in absorption in both and is stronger in supergiants.

Table 8.1. *Equivalent widths (in Å) of some strong lines.*

	Hγ λ4340	He I λ4471	He II λ4541
O4	1.5	0.15	0.8
O6	1.8	0.4	0.8
O8	2.0	0.8	0.6
B0	2.5	1.0	0.2

Figure 8.1. Spectra of O-type stars – spectral types and luminosity classes. For explanation see text.

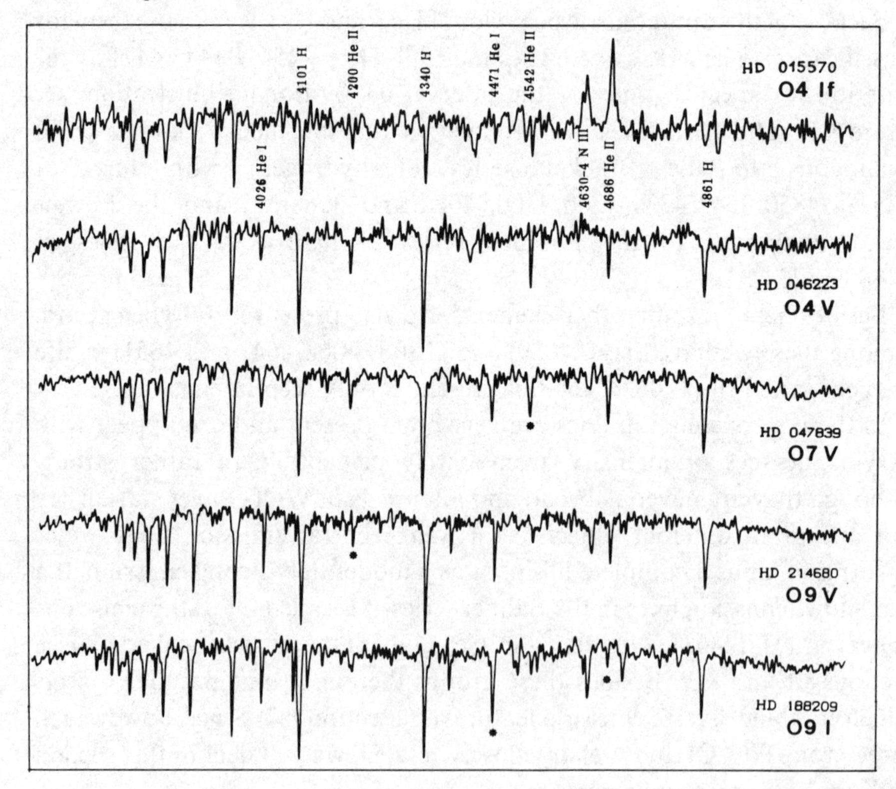

Nevertheless special care must be taken with regard to spectral types, since some lines (i.e. $\lambda\lambda$3479–85) vary with both spectral type and luminosity class over a short interval.

In late O-type stars the ratio C III λ4649/He II λ4686 is used. In supergiants λ4649 is stronger than λ4686, whereas in dwarfs the reverse is true. Also the intensities of the Balmer series show a well-marked negative luminosity effect (see figure 8.1).

Conti and Alschuler (1971) used the ratio Si IV λ4089/He I λ4143, the Si IV being stronger in supergiants. But this criterion is valid only at high dispersion ($A \sim 20$ Å/mm) in the interval O6–O9.

However, it is not yet certain that the luminosity assignments of O-type stars are final, because a plot of the luminosity class against the luminosity of early O-types, independently determined, does not show a clear pattern. This is illustrated in figure 8.2, taken from the data of Underhill (1982). Similar criticisms have been voiced by Conti (1974) and Goy (1976).

Underhill also finds that effective temperatures do not agree very well with spectral types; this is shown in figure 8.3 taken from her paper. The disagreements are seen in the earliest stars, and disappear toward later types. If we consider the fact that temperatures are derived mostly from the

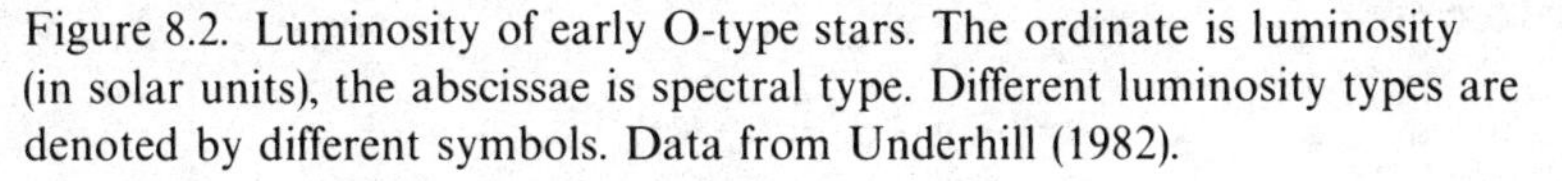

Figure 8.2. Luminosity of early O-type stars. The ordinate is luminosity (in solar units), the abscissae is spectral type. Different luminosity types are denoted by different symbols. Data from Underhill (1982).

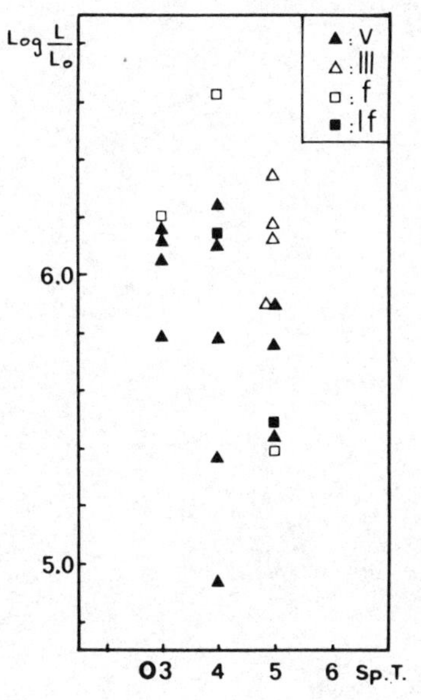

ultraviolet, whereas spectral types are derived from the classic region, such disagreements are not too serious. The discrepancies disappear when we consider spectral types based upon ultraviolet criteria.

Beyond the classic region, the visual ($\lambda\lambda$5400–6000) has also been explored by Conti (1974). Walborn (1980) has provided an illustration of the temperature and luminosity effects observable in the form of an atlas. Among the atomic species easily observable are O III and C IV. Both of these are poorly represented in the classic region and reach their maximum in the visual at O6–7. With regard to luminosity effects he finds (the same as in the classic region) a positive luminosity effect for certain emission lines, such as C III, λ5696, which parallels the behavior of N III and He II.

The near infrared region has been studied by Andrillat and Vreux (1979). They found that the He I λ10 830 line is often seen in emission and they state that as a rule the emission is stronger in stars of high luminosity, whereas it is

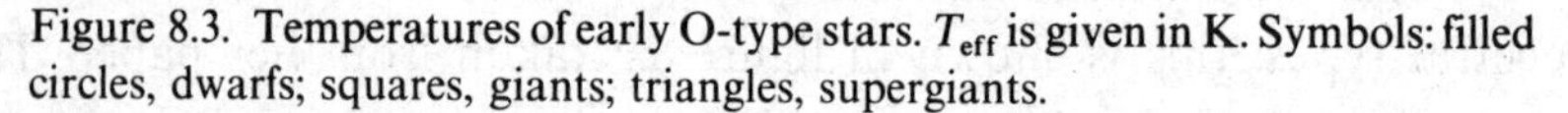

Figure 8.3. Temperatures of early O-type stars. T_{eff} is given in K. Symbols: filled circles, dwarfs; squares, giants; triangles, supergiants.

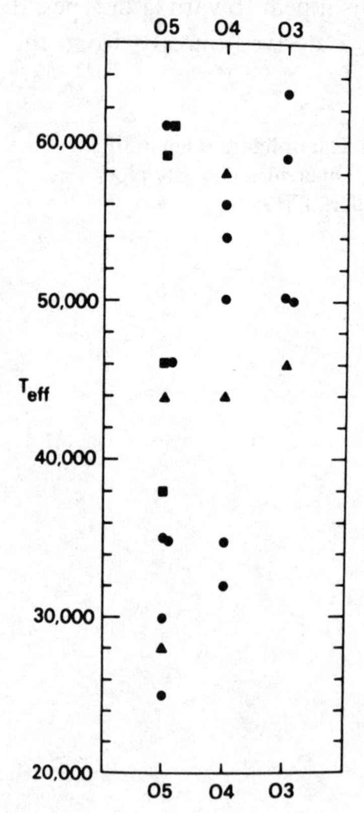

weaker or absent in dwarfs. In several stars they found emission line strength variations.

The ultraviolet spectral region has been analyzed, from the point of view of classification, by Jaschek and Jaschek (1984). It has been shown that stars can be ordered by spectral type and luminosity, as in the classic region.

A list of the usable lines is given in table 8.2 (Heck *et al.*, 1984).

Based upon the classification of many UV spectra, the result was obtained that spectral types derived from UV criteria agree rather well with MK types. The agreement is less satisfactory for luminosity and some discrepancies exist.

Equivalent widths. A small number of O-type stars has been studied at high dispersion, providing among others line identification and equivalent width measurements. Table 8.3 lists the stars studied by Peterson and Scholz (1970). Other studies were made by Underhill (1958), Underhill and de Groot (1965), Hutchings (1968), Buscombe (1969) and Simon *et al* (1983). Equivalent widths of some strong lines only, are given by Conti and Alschuler (1971) and Conti (1973, 1974).

Table 8.2. *Some UV lines characteristic of O-type stars.*

λ	Identification	Comments
1175	C III	Increases from O4 toward B1, where it has its maximum
1255		Decreases from O3 to O7
1300	Si III	Decreases from O3 to O9
1333		Decreases from O3 to O9, where it disappears
1371	O V	Decreases from O3 to O7, where it disappears
1394, 1403	Si IV	Becomes well visible from O7 onwards, and has a strong positive luminosity effect
1428	C III	Increases from O4 toward B1, where it has its maximum
1453		Has its maximum at O4 and disappears at B0
1548	C IV	Decreases from O_3 toward B-type, it disappears at B2; it has a positive luminosity effect
1718	N IV	Has a maximum at O4; at O7 it is blended with Al III λ1721 present in B-type stars

Radio observations have shown that in a few cases radio emission does exist. However at present the observations are limited, because of instrumental reasons, to a couple of bright stars (Barlow 1978). Most of the emitters are Of stars (Abbot *et al.* 1980).

Rotation. Slettebak (1956) measured the line widths of a number of O stars. His conclusion is that the broadening is only partially due to axial rotation, since macroturbulence and other atmospheric phenomena also contribute. The only thing to be said is that *if* line broadening is due to rotation alone, $V \sin i$ lies between 130 and 220 km/s. The latter value is thus an upper limit for rotation, and shows that in general O-type stars do not have sharp lines.

Spectral peculiarities. The O-type stars are rich in subgroups characterized by peculiar spectral features. We shall describe them in separate sections and only list them here. We have already mentioned the Of and Oe stars, characterized by emission lines. Then we have the WR stars, closely related to the O-type stars, with very strong emission lines.

At fainter luminosity we find the subluminous WR stars, the UV bright stars and the subdwarf O-type stars.

Some of these groups are not restricted to O-type stars, but continue into early B-type; among these we have the subdwarfs, the emission line stars, the CNO stars and the 'run aways', which shall be discussed in the chapter on B-type stars. O-type white dwarf stars shall be described in the chapter on 'degenerates'.

Photometry. We start with the general remark that the maximum emission of these stars lies in the far ultraviolet because of their high temperature; at

Table 8.3. *Some O-type stars analyzed at high dispersion.*

HD	Sp. type
34 078	O9.5 v
57 682	O9 v
36 861	O8
47 839	O7
54 662	O6
164 794	O5

$\lambda > 3000$ Å we are dealing with the tail of the energy distribution. For this reason in any photometric system in the classic region, O-type stars are confined to a narrow range of colors. In UBV for instance

$$-1.0 \geqslant U - B \geqslant 1.15$$

$$-0.30 \geqslant B - V \geqslant -0.33$$

The colors in broad band photometry are shown in figure 8.4; for the uvby system, see section 5.2.

Since colors are insensitive to spectral type, any classification error has little influence on the color. This is a unique situation, which happens with no other object. Since on the other hand O-type stars are objects of high luminosity, they can be seen from great distance. Both reasons justify the extended use of O-type stars for interstellar extinction studies. The only difficulty is that the intrinsic colors are also difficult to establish (Burnichon 1975).

Figure 8.4. Broad band photometry of O-type stars: U, B, V,..., L colors for dwarfs. V differs by 0^m06 between O5 to O9, 5, B by 0^m02 and the other colors by less than 0^m01.

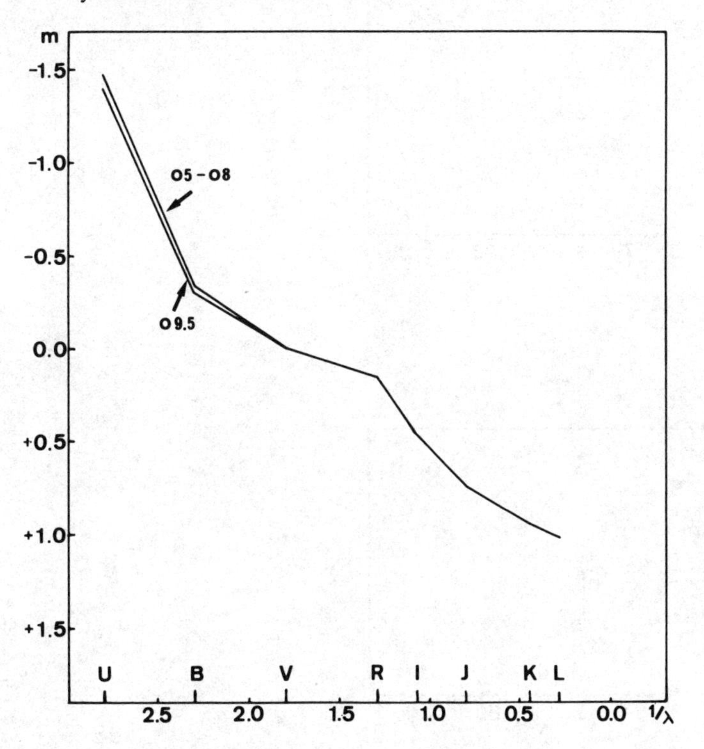

There is some evidence that many O-type stars are variables, by amounts of a few tenths of a magnitude (Hutchings 1979).

Narrow band systems are difficult to use on O-type stars because of the frequent appearance of emission lines and because most of the lines are rather weak (see table 8.1). Landolt (1970) used a narrow band system centered on the lines λ4637 (N III) and λ4686 (He II). This permitted him to separate O and Of stars, as would be expected (see figure 8.5). As in spectroscopy, he does not find a close correlation of this index with luminosity.

Infrared colors of O-type stars were studied by Castor and Simon (1983). They observed J . . . M colors for a large sample of stars and concluded that essentially all Oe and Of type stars show infrared excesses, which are of the order of a few tenths of a magnitude at 4.7 μm.

Figure 8.5. Narrow band indices in O-type stars. The *n* index measures N III λ4637 and the *h* index He II λ4686. Notice the separation between O- and Of-type stars. Data from Landolt (1970).

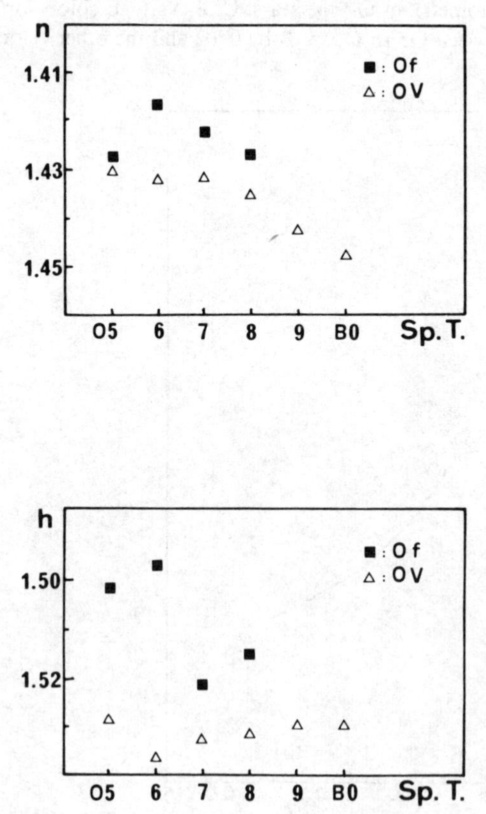

Radial velocities are known for many O-type stars and the majority of them belongs to population I. The remainder ($\sim 10\%$) have velocities greater than 40 km/s and are called 'run-aways' (Blaauw, 1960). These velocities can be either due to abnormal causes (ejection of an O star by its exploded companion) or indicate that some O-type objects belong to the old disc population.

Garmany, Conti and Massey (1980) found that 36% of the stars are binaries. A surprising fact pointed out by Bohannan and Garmany (1978) is that there is a large number of double lined spectroscopic binaries (SBs) whereas both single line systems and low amplitude systems are scarce. Since observational selection is one possible explanation, Garmany *et al.* (1980) report on a re-examination of the brighter northern stars. They find that among 76 stars about half are single, one-quarter double lined SBs, and the rest single line binaries and triple systems, so that the above-mentioned lack of single systems seems to be real. Goy (1976) has stressed that many O stars do form part of multiple systems, but that only the brightest (and nearest) systems are recognized as such. The fact that for fainter O-type objects this is not the case, could explain many contradictory observations. O-type stars appear very frequently in associations and open clusters. In associations they are closely linked to the surrounding gas and dust, so that it seems preferable to speak of O complexes rather than of O-type stars (see for instance Pecker 1976). A good example is Orion (Goudis 1982). About 60% of all O-type stars belong to clusters and associations, according to Garmany, Conti and Chiosi (1982).

Absolute magnitude. The luminosities of O-type stars are determined from objects belonging to clusters and associations. We have already mentioned the fact that many O-type stars belong to multiple systems, a fact which is often difficult to recognize, but which falsifies the position of the star with respect to the cluster sequence. Secondly, we have alluded to the fact that some minor disagreements exist on luminosity classification. With this in mind, we quote in table 8.4 the values taken from Schmidt-Kaler (1982). These values are still somewhat uncertain, especially for the early types.

The distribution of O-type stars closely follows the galactic plane. We have $|b| = 2°$ and $\beta \simeq 60$ pc. For a discussion of the distribution in the galactic plane, see Massey (1985).

Frequency. The number of O stars is given in table 8.5. Despite their low frequency, O-type stars are extremely important because they constitute the 'hot' tip of the main sequence. Evolution proceeds very quickly and the stars are thus found near their birthplaces.

There are two general catalogues of O-type stars. The first is by Goy, of which the latest edition is 1980. The second is by Garmany, Conti and Chiosi (1982).

8.1 Of stars

An Of star is an O-type star exhibiting emissions at both $\lambda\lambda$4630–34 of N III and λ4686 of He II. The emissions are distinct but not very strong, in contrast to WR objects where the emissions are always very strong. (For illustration, see figure 8.6.)

Of stars were defined by Pearce (1930), and can be recognized at classification dispersion, although in cases when emission is faint, a higher dispersion is more convenient. The spectral range of Of stars is from the earliest types up to O9; in general the phenomenon is more frequent in early types.

Besides the two emissions mentioned above, there are very often weak emissions at Hα, Hβ and at C III λ5696 (Conti 1974). In some stars other emissions may appear, but as a general rule these emissions are faint and can only be seen at high resolution.

Table 8.4. *Absolute magnitudes of O-type stars.*

	V	III	Ia
O3	−6.0		−6.8
O5	−5.7	−6.3	−6.8
O7	−5.2	−5.9	−6.8
O9	−4.5	−5.6	−6.8

Table 8.5. *Proportion of O-type stars.*

m	Number of stars	% of total
6.5	46	~5‰
9	310	~1.5‰

In recent years, two independent studies of the group were carried out by Conti and associates, and by Walborn. Conti used calibrated 16 Å/mm plates, whereas Walborn used 65 Å/mm plates, widened to 1.2 mm. Both authors have introduced new designations for these objects, which differ between authors and papers. The reader should therefore look for the latest paper of each author (Conti and Leep 1974; Walborn 1973) where the authors converge toward a common notation, without however reaching complete agreement. The basic idea behind the scheme is that there is no gap between the O and the Of stars; an idea previously expressed by Underhill (1966). They use a refined subdivision:

O((f)): N III in emission, He II strongly in absorption
O(f): N III in emission, He II weakly in absorption or emission
Of: N III in emission, He II strongly in emission

An ordinary O star comes in this scheme before O((f)).

As a further step both Conti and Leep (1974) and Walborn (1973) suggest that Of stars are O-type supergiants. This was first proposed by Underhill (1949) and Slettebak (1956) produced the first plot linking the intensity of the λλ4630–34 line to absolute magnitude. Slettebak also mentions the existence of possible emission line intensity variations. There are some well-documented cases like λ Cep (Conti and Frost 1974) and HD 108 (Andrillat *et al.* 1973). For more references, see Grady, Snow and Timothy (1983). Nevertheless such evidence was challenged by Conti (1976).

Figure 8.6. Emission lines in O-type stars: Oe and Of.

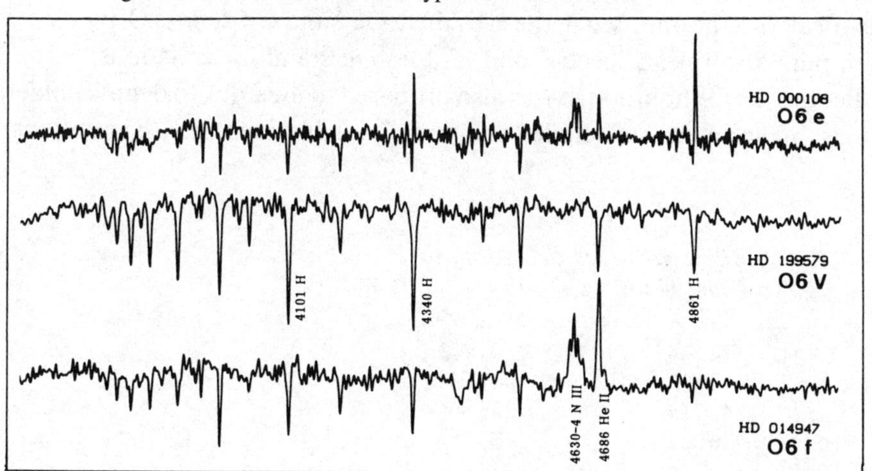

Many of the stars with strong emission also show P Cygni type profiles with expansion velocities of the order of several hundred (400–500) km/s, infrared excesses and radio emission.

Walborn (1977) called attention to the fact that in some stars the N IV λ4058 emission is stronger than the N III λλ4634–40–42 emission. He denotes such a star as 'f*'; when λ4058 is simply in emission, the notation is 'f$^+$'. At present only three 'f*' are known, namely HD 269810, HD 93129 A and VI Cyg 7, all located in the Cygnus OB2 association. Vreux and Andrillat (1974) found that HD 93129 A has variable features.

From what has been said in this section it seems clear that even within the group there is a great variety of spectral features, ranging for instance (for a given line) from a pure absorption to a pure emission profile. A classification scheme trying to accommodate all these complexities rapidly becomes complex in itself, and we can ask where the borderline between a classification (i.e. a short description) and a description of the spectrum should be established. Probably notation like 'f*' is already beyond the borderline. In any case much more attention needs to be paid to the variability of the spectral features than has been done until now.

8.2 Oe stars

Oe stars are O stars exhibiting emissions in the Balmer series without being accompanied by emissions in N III λλ4630–34 or He II λ4686 (see figure 8.6). Sometimes there are emissions in He I λ5876. The emissions in the Balmer lines are roughly centered on the absorption profiles and central reversal is often visible. The Oe notation was introduced by Conti and Alschuler (1971), but it should be observed that this classification is *not* identical with the one used in the HD; there Oe 5 and Od denote O-type stars with pure absorption spectra, and 'e' does *not* stand for emissions.

Besides the Oe notation, it was also proposed to use O(e), to denote milder effects. In the absence of detailed analysis of possible line strength variations,

Table 8.6. *Oe stars according to Frost and Conti (1978).*

HD	24 534	HD	60 848
	39 680		149 757
	45 314		155 806
	46 056		203 064

it seems best to use such notation very cautiously. In table 8.6 we give the known Oe stars according to Frost and Conti (1978).

Conti and Leep (1974) introduced another subgroup, that of Oef stars which is called 'Onfp' by Walborn (1973). We prefer the former notation. The few stars in this group are characterized by a double emission structure in the He II λ4686 analogous to the one described in the Oe group.

Both types (Oe and Oef) are characterized by broad absorption profiles suggesting fast rotation, analogous to that in Be stars.

Frost and Conti (1978) have analyzed three Oe stars in more detail and they conclude that Oe stars are the prolongation of the hotter Be stars in the O-type star domain. The following analogies between both groups are noted.

(1) Balmer lines and He I are seen in double emission.
(2) The stars have a large rotational velocity.
(3) They lie close to the main sequence.
(4) Emissions are variable in time.

Finally they find that the proportion of Oe to O stars is about 16%, which is close to the percentage of hot Be to B stars.

Andrillat, Vreux and Dennefeld (1982) concluded similarly that a close relation with Be stars exists in the near infrared region. In particular Paschen lines are found in emission and sometimes He I λ6678, Fe II and the infrared Ca II triplet.

8.3 Wolf–Rayet stars

A Wolf–Rayet spectrum is characterized by wide emission lines, standing out distinctly from the continuous spectrum, which corresponds to a hot star.

The first objects of the group were discovered by the French astronomers Wolf and Rayet (1867), during a visual spectroscopic survey, before the systematic use of photographic plates. The first systematic study of the group was made by Plaskett (1924), who provided also the first atlas of these objects. Further progress became possible when Edlen (1933) showed that the emissions were due to highly ionized elements and almost immediately Payne (1933) was able to show that WR stars can be arranged in two sequences, one the so-called carbon sequence and the other the nitrogen sequence (see figure 8.7). Table 8.7 provides the ions found in each sequence.

From a glance at table 8.7 it is clear that the emissions represent a wide range of ionization – whereas C II has an ionization potential of 24.4 eV, those of O VI and N V are 138.7 eV and 97.8 eV. The emission lines are very broad, having widths from 4 to 100 Å. Further, two rules are observed; first, for each species the quantity $\Delta\lambda/\lambda$ is constant ($\Delta\lambda$ being the emission line width) and, second, within each star the emission lines of a more ionized element are narrower than those of a less ionized element.

Hydrogen lines are not always seen. When they are present, they are usually in emission; the Paschen series is often seen in emission even if no emission is visible in the Balmer lines (Sahade 1981).

Later studies showed that the stars of each branch can be arranged in an order of excitation. Such a scheme was proposed by Beals and adopted by

Figure 8.7. Spectra of two WR stars.

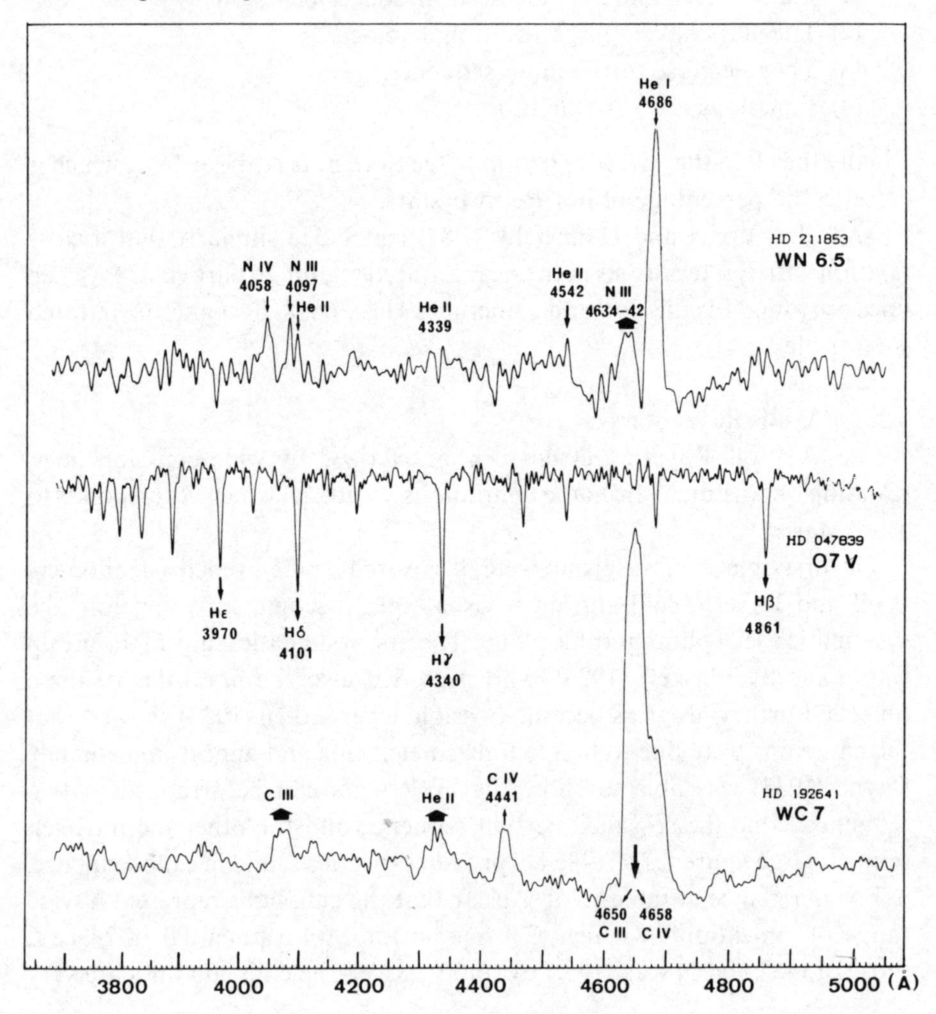

the IAU in 1938 (IAU 1938). It is still used with further refinements, and we shall follow closely the definitions used by Smith (1968).

Smith keeps the division into a carbon sequence (WC) and a nitrogen sequence (WN). She introduces further subdivisions using for the WN sequence the ionization stage of nitrogen which predominates, and for the WC sequence the fact that the emission widths ($\Delta\lambda$) grow steadily toward earlier types. These criteria are given in tables 8.8 and 8.9.

Table 8.7. *Emission lines in WR stars.*

Carbon sequence		Nitrogen sequence	
He I		He I	
He II	$\underline{4686}$	He II	$\underline{4686}$
C II	4267	N III	$\underline{4097}, \underline{4640}, \underline{5314}$
C III	3609, 4187, 4325, $\underline{4650}$, $\underline{5696}$	N IV	3483, $\underline{4057}$
C IV	4441, $\underline{4658}$, 4786, $\underline{5805}$	N V	$\underline{4605}$, $\underline{4622}$
O II	4134, 4317, 4349, 4366, 4414, 4417		
O III	3714, 3760, 3961		
O IV	3385, 3405, 3412, 3562, 3725, 3736		
O V	$\underline{5592}$		
O VI	3815, 3835		

Table 8.8. *Classification of WN spectra.*

Class		Criteria
WN8	N III $\gg$ N IV	He I strong with violet absorption edges, N III λ4640 $\approx$ He II λ4686, N III λ5314 present
WN7	N III $\gg$ N IV	He I weak, N III λ4640 $<$ He II λ4686.
WN6	N III $\approx$ N IV	N V present but weak, N III $\lambda\lambda$4634–41 band present
WN5	N III $\approx$ N IV $\approx$ N V	N III λ4634–41 band present
WN4.5	N IV $>$ N V	N III very weak or absent
WN4	N IV $\approx$ N V	N III very weak or absent
WN3	N IV $\ll$ N V	N III absent

Source: From Smith (1968).

It should be stressed that the division into two sequences is not exclusive, since in most stars of either sequence – carbon or nitrogen – the other element is also seen, for example C IV in WN stars or N V in WC stars. What is important is that in each sequence the element that characterizes it is considerably stronger. No real dichotomy exists, but rather predominance of one of the elements. It seems further that there are a few stars whose main characteristic is the great enhancement of oxygen (Barlow and Hummer 1982); these may be denoted as WO. These authors note the great strength of O IV, O V and O VI and introduce further subdivisions of the group, as with the WC and WN stars.

We may naturally try to quantify the criteria used, by measuring the equivalent widths of the lines and plotting them (or their ratios) against spectral type. Figure 8.8 reproduces one of the results of Conti, Leep and Perry (1983a).

Obviously, for a given type there is great variation in line ratios. In fact the variation is so great that we may ask if the subdivisions are really meaningful. Although in *all* spectral groups there is some scatter around the mean, in the WR stars the scatter seems to be too large, and the conclusion of Conti *et al.* (1983a) that 'the WN sequence, while primarily an ionization one, does have other physical elements influencing the appearance of the spectra' is justified.

Infrared region. The near infrared (λλ5700–10 300) was studied in detail by Vreux, Dennefeld and Andrillat (1983) and they found the species listed in table 8.10. Figure 8.9 illustrates two typical spectra. The important fact we learn from this region is that the separation into WC and WN sequences may be kept.

Table 8.9. *Classification of WC spectra.*

New notation	C III λ5696 / O V λ5592	C III λ5696 / C IV λ5805	Width of C III, IV λ4650	Beals' notation
WC5	< 1.0	0.3	85 Å	WC6
WC6	> 1.0	0.3	45 Å	WC6
WC7	8.0	0.7	35°	WC7
WC8		1.0		WC7–8
WC9		3.0	10 Å	WC8

Source: From Smith (1968).

Photometry. Any broad band photometry is exposed to contamination by strong emission line features. If a line has an equivalent width of 40 to 100 Å, it may considerably disturb a broad band measurement like U, B or V. Pyper (1966) has evaluated the correction by microphotometring different spectra

Table 8.10. *Elements present in the $\lambda\lambda$5700–10 300 region.*

WN	WC
He II	C II
He I	C III
sometimes H I (Hα)	C IV
N III	He I
N IV (very strong)	
N V	
C IV	

Figure 8.8. Line ratios for nitrogen in different states of ionization, as a function of spectral type and luminosity class. Data from Conti *et al.* (1983a).

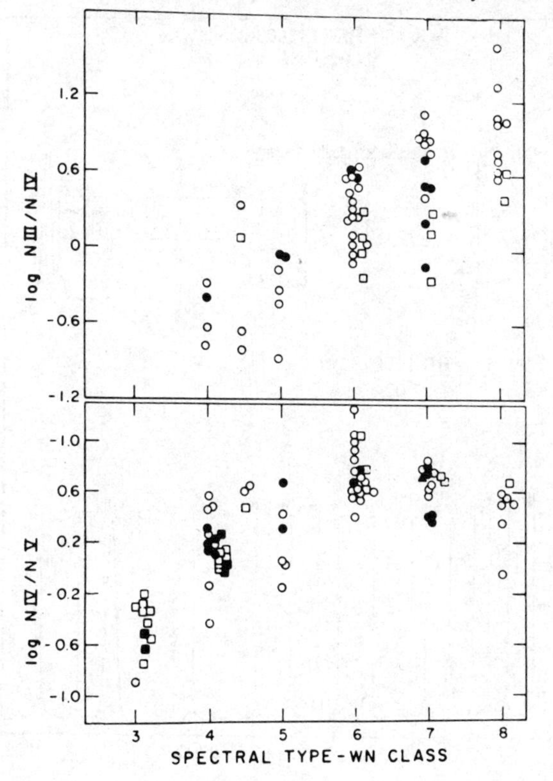

of WR stars, and measuring the equivalent width of the emission lines. If we call the flux in the emission lines F_L and the flux in the continuum F_C, the correction due to the presence of emission lines is

$$\Delta S = 2.4 \log \frac{F_L + F_C}{F_C}$$

and this can be evaluated for each color. In table 8.11 the corrections for two stars are given. Figures 8.10 and 8.11 illustrate how these changes affect the location of the stars in the UBV diagram. When reddening is taken into account, the $(U-B)_o$ colors for WR stars are less than -1^m00. In the uvby system, the $(b-v)_o$ lie in the interval -0.21 to -0.55.

Another way of overcoming the difficulties introduced by the emission lines is to observe colors in regions free from emission. This was done by Smith (1968) who chose narrow band filters about 100 Å wide positioned at

Figure 8.9. Spectra of two WR stars in the near infrared. Data from Vreux, Dennefeld and Andrillat (1983).

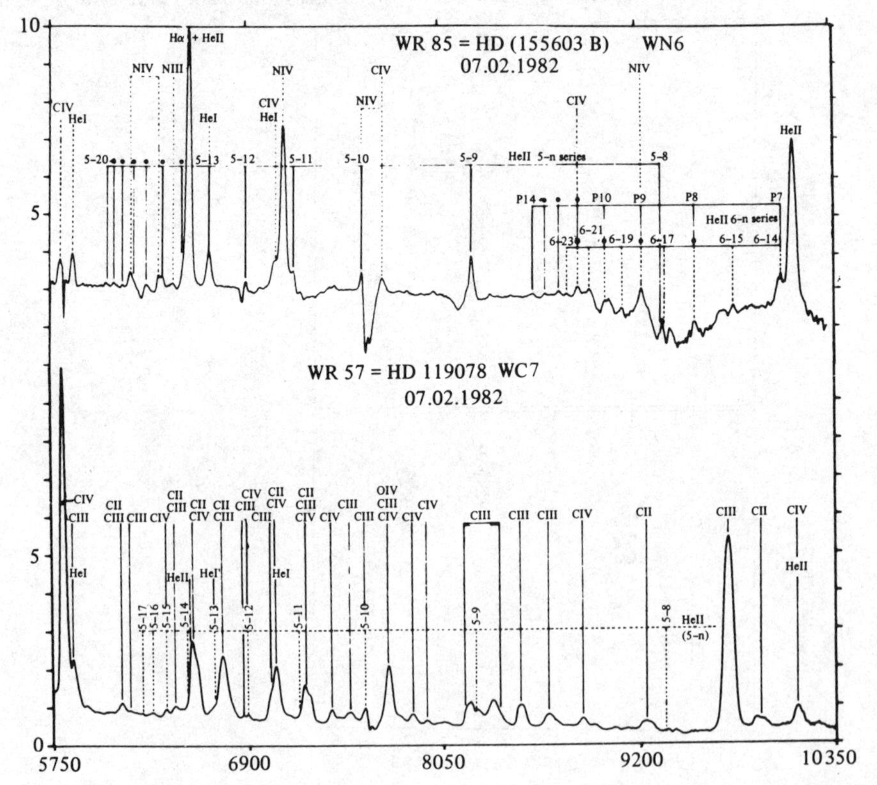

$\lambda_0 = 3500, 3650, 4270, 5160$ and 5500 for the WN stars. Other types of stars can also be measured in this system to calibrate it, and to produce de-reddened indices.

Smith hoped that by using narrower filters she could suppress the scatter between the (de-reddened) colors of WN stars of the same spectral subtype. Although there is a general tendency in this sense, the scatter is still rather large and persists even using still narrower filters, as Massey (1982) did. He used a band width of less than 10 Å, and still found differences of $\pm 0^m1$ in the color indices for stars of the same spectral subgroup.

The drawback of the systems is that they cannot be applied to WC stars since their emission lines differ from the WN stars.

Table 8.11. *Photometric corrections.*

HD	Type	Δ(U–B)	Δ(B–V)
165 688	WN5	− 0.24	+ 0.30
165 763	WN6	− 0.42	+ 0.12

Figure 8.10. UBV photometry of WR stars without correction for emission lines. Square, WN; triangle, WC; circles, binaries. Open figures represent possible binaries. Colons indicate uncertain values. The solid line represents the main sequence. Data from Pyper (1966).

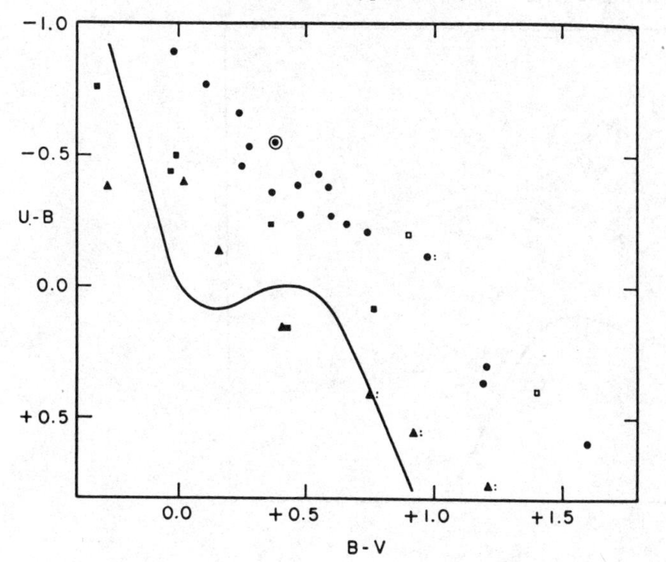

Another possibility which has been explored is the use of infrared colors. As we have seen, the near infrared is full of emission and the same happens at longer wavelengths. Figure 8.12, taken from Pitault *et al.* (1983), shows that a large scatter can also exist within the spectral groups, and therefore the use of infrared colors does not solve the problem. Light variability (not due to stellar eclipses) has also been analyzed, but somewhat contradictory data is found in the literature. Cherepashuk (1974) observed two stars, one a WC and the other a WN, over five years and found real variations in the average colors, although small variations over time scales of days occurred, with amplitudes of less than $0^{m}02$.

Binaries. The question of whether WR stars are binaries has been much debated over the years. Those WR stars that show absorption lines besides the emissions were all thought to be binaries, of the type WR + OB, where OB stands for O- or B-type companions, and the percentage of such assumed binaries tended to increase over the years. However, Niemela (1976) showed that absorption lines could originate on the WR stars themselves. Therefore the most cautious procedure is to accept only those SBs for which orbits exist, and these are only a small percentage of all WR stars (Pedoussaut, Ginestet and Carquillat 1983). Among the 78 WR stars

Figure 8.11. UBV photometry of WR stars corrected for emission lines. Colons indicate uncertain values. Data from Pyper (1966). This figure should be compared with figure 8.10.

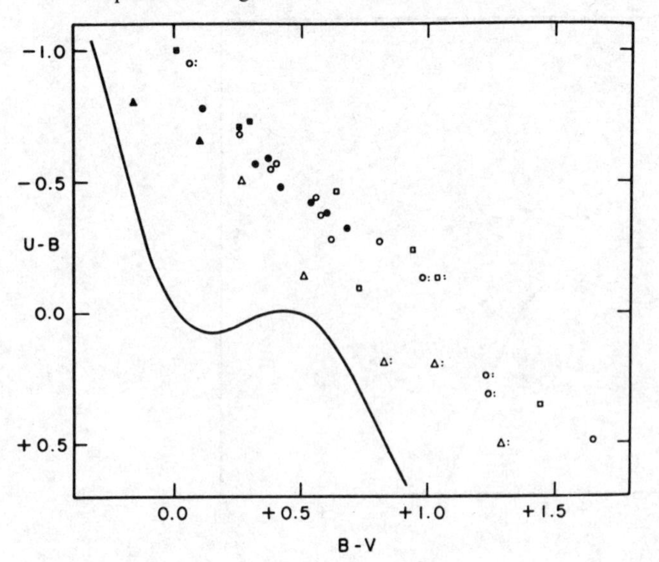

with HD numbers, at present 20 have orbits, but of course not all WRs have been examined in detail for radial velocity variations.

It is rather surprising to find that the majority of these spectroscopic binaries are double lined. In a way this is a selection effect because line shifts are only easily detectable in stars in which both absorption and emission lines are present. From a more complete survey, Lamontagne and Moffat (1982) find, out of 23 stars, eight systems, five candidates and ten probably single stars. Among the known binaries there are a few eclipsing binaries (van der Hucht *et al.* 1981).

Absolute magnitude. WR stars being luminous objects, their absolute magnitude is best established from membership in open clusters or in other galaxies. This has been done several times in the literature and three general difficulties emerge. The first is that in different stellar systems not all subtypes are equally well represented (see below, under 'distribution'). The second is that a large dispersion is present in the absolute magnitudes. This dispersion

Figure 8.12. Infrared colors of WR stars. The colors are corrected for emission lines. ∧, WC5; ∨, WC6; ∗, WC7; +, WC8 and 8.5; ×, WC9; ∘, WN3 to 5; □, WN6; ■, WN7; ●, WN8; underlined symbol, binary system or star with absorption lines; number between parentheses from van der Hucht *et al.* (1981). Data from Pitault *et al.* (1983).

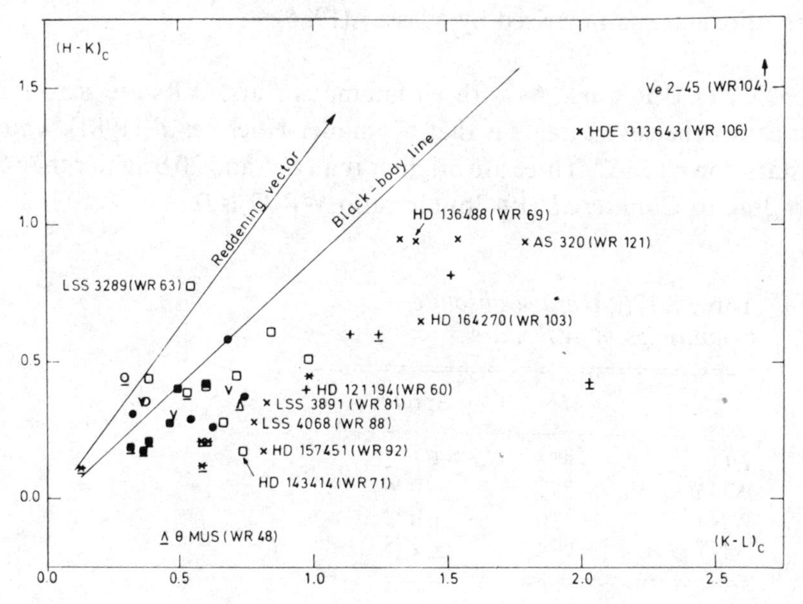

does not diminish even when objects are grouped by spectral type, which is not surprising since we have seen that spectral type is poorly correlated with physical characteristics. Finally, the third difficulty is that care must be taken with regard to the magnitude system, since magnitudes are extremely sensitive to emission lines. With these reservations in mind we quote the results of van der Hucht *et al.* (1981), Conti (1982), and Conti *et al.* (1983b) in table 8.12.

Distribution in the galaxy. In general WR stars are strongly concentrated toward the galactic plane; in this plane they follow closely the spiral arms, so that they can be used as spiral tracers. They appear also in associations and young clusters (Reddish 1967) and at least 40% of the WR stars are associated with H II regions (van der Hucht *et al.* 1981).

One very striking phenomenon is the absence of WR stars in the anti-center quadrant ($140 < l < 220$). This effect is probably real since WR stars are easy to detect because of their emission lines. If we accept that surveys are complete within 2 kpc around the sun, we find that their distribution is similar to that of the earliest O-type stars (Conti *et al.* 1983b). This is used as an argument to show that WR stars come from O-type stars. Within the solar neighborhood, the WR stars are divided about evenly between WN and WC types (Hidayat, Supelli and van der Hucht 1982).

WR stars were also detected in external galaxies. Both their number and the proportion of WC/WN stars varies from one galaxy to another. The observations are summarized by Massey (1985).

Frequency of WR stars. As with all luminous stars, WRs are scarce. The most comprehensive catalog is that of van der Hucht *et al.* (1981), who list 159 stars down to 15^{m}. Three are brighter than $6^{m}_{\cdot}5$ and 20 brighter than $9^{m}_{\cdot}0$. According to Conti *et al.* (1983b) the ratio WR/O is 0.14.

Table 8.12. *Average absolute magnitudes of WR stars.*

	$\langle M \rangle$	Spread
WC	$-4^{m}_{\cdot}3$	$\pm 1^{m}_{\cdot}2$
WN3	$-3^{m}_{\cdot}9$	$\pm 0^{m}_{\cdot}6$
WN4	$-4^{m}_{\cdot}0$	$\pm 0^{m}_{\cdot}8$
WN7–9	$-6^{m}_{\cdot}4$	$\pm 2^{m}_{\cdot}5$

Atlases. Spectra of WRs were published by Smith (1968) and Hiltner and Schild (1966). Lundström and Stenholm (1984) have provided a set of tracings of spectra in the region $\lambda\lambda$4400–5900.

8.4 Subluminous WR stars

About 10% of the WR stars are central stars of planetary nebulae (PN). These stars are similar to other WR stars, but van der Hucht *et al.* (1981) call them [WR]. We shall retain this concise notation.

The difference between WR and [WR] is that their absolute magnitudes differ considerably. Although the scale of absolute magnitudes of PN is still rather uncertain, values between -2 and $+3$ are found for the [WR] stars, which make them definitely fainter than the WR stars.

Although PN belong to the old disc population, [WR] are often called 'Pop II WR' stars. The spectral classification of the objects is difficult because the [WR] spectra are contaminated by the lines from the planetary nebulae. If these lines are eliminated the resulting spectra are similar to ordinary WR spectra. No clear difference can be found between them; the emission line width which was thought to be smaller in [WC] spectra was found to be comparable, at least in some cases, to those in WCs (Smith 1973).

Thus the classification scheme for [WR] stars is essentially the same as that for WR stars. Heap (1982) has proposed some extensions, which are summarized in table 8.13.

Groups WC 10 and WC 11 were introduced by Carlson and Henize (1979), and follow the WC 9 type. In WC 10, C II is seen in emission, but C III λ5696 and C IV λ5801 are absent. In WC 11, C II is still in emission, but all emissions are weaker than in WC 10. At present only four objects of these two types are known.

Table 8.13. *Classification criteria for [WR] stars.*

	O VII λ5670	O VI λ3811 / C IV λ4658	O VI λ5292 / C IV λ5806
WC2	Present		
WC3	Absent	>0.5	>0.3
WC4		<0.5	<0.3
WC5–WC9		As in WC stars	
WC10–WC11		See text	

One important fact is that no spectra of the WN sequence exist in [WR] spectra. A list of [WR] objects is given by van der Hucht *et al.* (1981).

8.5 UV bright stars = AHB stars = HL stars

Strom *et al.* (1970) called attention to a group of globular cluster stars lying $1^m - 2^m$ above the horizontal branch, in the spectral range O9–F0. They coined the name AHB (above horizontal branch) stars. In most cases hydrogen lines are unusually sharp and in many cases lines of elements other than hydrogen and helium are weak.

The next analysis was made by Zinn, Newell and Gibson (1972) who observed photographically globular clusters in the U and V colors. They confirmed the existence of a number of stars lying above the horizontal branch (see figure 8.13). Since these objects were bright in UV, they called the group 'UV bright stars'.

Newell (1973) studied blue halo stars and introduced the term 'HL', denoting high luminosity population II stars. He used a photometric approach through four narrow band filters positioned at $\lambda\lambda$3535, 4272, 4340(Hγ) and 5305 which provide a measure of the Balmer discontinuity, the slope of the continuum and the equivalent width of Hγ.

Figure 8.13. A sample of UV bright stars. Circles and triangles denote respectively stars on the blue and on the red side of the RR Lyrae stars. The sequences correspond to M3. Data from Zinn, Newell and Gibson (1972).

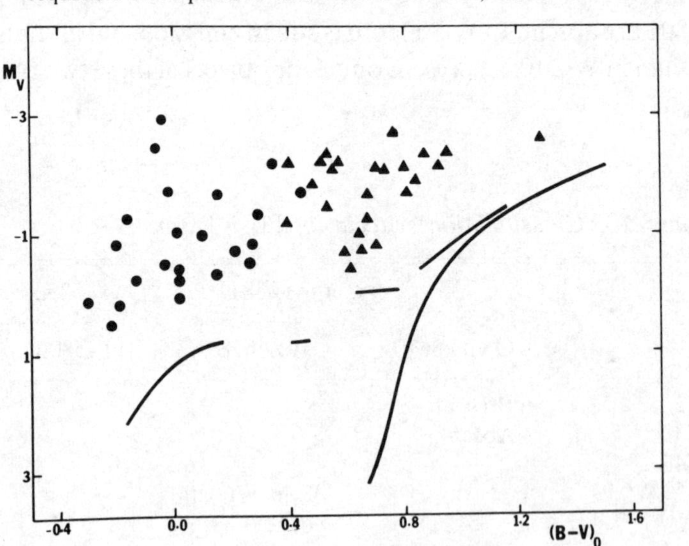

Probably the most important contribution was that he analyzed a large number of candidates and put together a fairly large sample of HL stars. Since most of them are faint, only a few, listed in table 8.14, were analyzed in detail.

No clear pattern has emerged from the stars analyzed in detail, either for the behavior of the helium or for the metallic lines. Both may be weak or normal for their color or spectral type. The fact that stars are defined partly by their luminosity (above the horizontal branch) complicates things further, because distances of field candidate stars are rather uncertain.

From what has been said it is clear that this is an ill-defined group. To add to this confusion we must talk of another group of stars for which the same name – UV bright stars – was used, but in a very different sense.

Carnochan and Wilson (1983) presented a catalogue of objects observed from the satellite TD1 having a color index bluer than that corresponding to their HD spectral type. It is however known that the HD catalogue contains a number of misclassified early B-type stars, taken with objective prism plates in poor seeing conditions – if the star is an early B-type and the helium lines are 'washed out', the star is classified as late B-type. Thus the fact that satellite ultraviolet colors do not agree with the HD spectral type is not alarming, were it not for the very large number of stars for which this happens. Carnochan and Wilson call such objects 'UV bright stars', where UV now stands for (satellite) ultraviolet.

When these groups are analyzed in detail (Barbier *et al.* 1978; Dworetsky, Whitelock and Carnochan 1982a, b; Wade and Smith 1985), most of the candidates turn out to be misclassified normal stars and only 10% or so are horizontal branch stars or subdwarfs. The designation UV bright stars can thus be abandoned.

Table 8.14. *UV bright stars analyzed.*

Barnard 29	M 13
v. Zeipel 1128	M 3
HD 93 521	Field
BD + 33°2642	Field
HZ 22	Field
HD 137569	Field

Source: Data from Tobin and Kaufmann (1984).

8.6 O-type subdwarfs

An O-type subdwarf is a star exhibiting very broad and shallow Balmer lines and a very strong HE II λ4686 line when compared to a star of similar spectral type. Because of the broadness of the Balmer lines and consequently their confluence at lower numbers, the number of Balmer lines seen distinctly is smaller than in dwarfs ($n \sim 10$–12 in subdwarfs and $n \sim 18$ in dwarfs). Usually N IV(λ3479 for instance) is also enhanced.

Members of this group can be detected at dispersions as low as 180 Å/m if the Balmer discontinuity is visible (Sargent and Searle 1968). The group is an extension toward earlier types of the B-type subdwarfs first described by Greenstein and Münch in the fifties, and is usually denoted sdO.

Figure 8.14. Spectra of some sdO stars. The upper spectrum corresponds to sdO, the following to sdOp, the third is an interesting composite object, sd + F, and the last an O6 V star. From Berger and Fringant (1978).

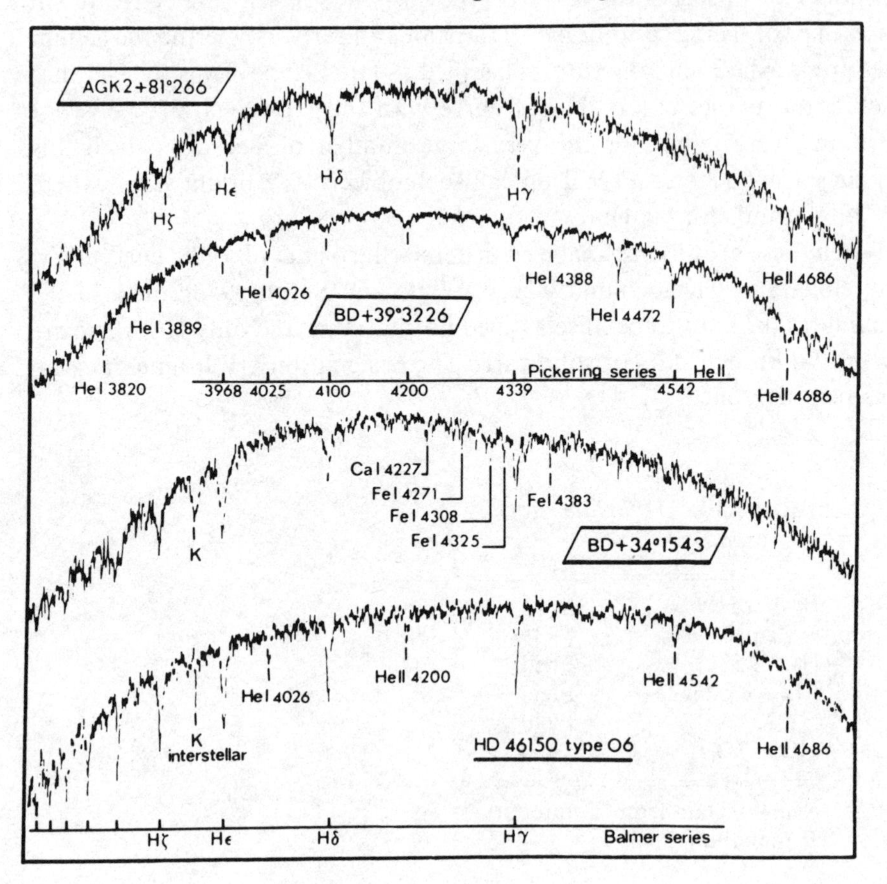

Baschek (1981) uses the notation sdOB to denote stars which have weak He I lines and strong He II; he proposes HD 149382 as a prototype of the group. Since there is some danger of confusion, we prefer to maintain the classic notation of sdO and sdB; in this case HD 149382 is sdO.

Figure 8.15. Spectra of some sdO stars. The spectra are arranged according to the hydrogen–helium ratio. The top star is a hydrogen deficient star, and the last one a helium deficient star. The abscissa shows wavelength in Å. From Hunger *et al.* (1981).

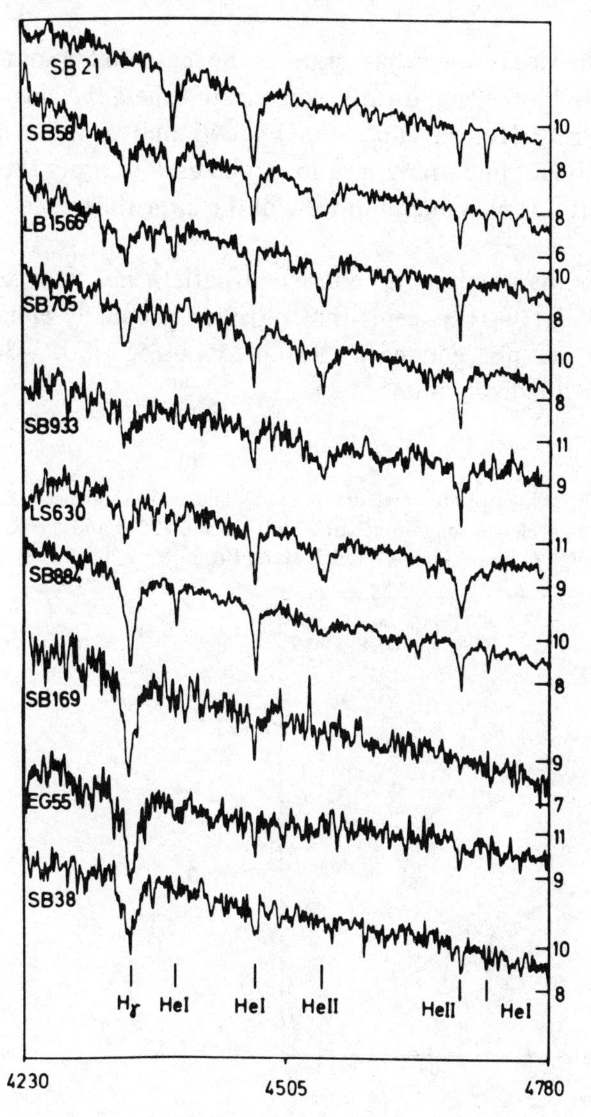

When the spectra of sdOs are examined in detail, spectroscopic anomalies are found in the helium series. In the He II series the first anomaly is the behavior of the Pickering series as compared to the Fowler series. $\lambda4686$ is very strong, as we have mentioned already, and varies little with spectral type. In normal O-type stars the intensity of $\lambda4686$ is comparable to the strength of $\lambda4542$ and $\lambda4200$; in subdwarfs $\lambda4686$ is comparatively stronger, and $\lambda4542$ and $\lambda4200$ are weaker or even absent. According to Sargent and Searle (1968) the ratio $\lambda4686/\lambda4542$ might be luminosity dependent.

Some of these features can be seen in figure 8.14 taken from Berger and Fringant (1978).

The behavior of the He I is somewhat erratic in the sense that in some stars the lines are like those in normal dwarfs, whereas in others they are weak. Figure 8.15, taken from Hunger *et al.* (1981) shows the variation in line strength. The He II to He I line strength ratio can be used to order the stars, but it is doubtful if this is meaningful in view of the large differences in the He II line series.

Of elements other than helium, no consistent pattern has emerged yet. According to Baschek (1981) it seems that nitrogen is usually enhanced, whereas C and O are not enhanced. S III($\lambda4255, \lambda4355$ and $\lambda4362$) is sometimes present with strong lines.

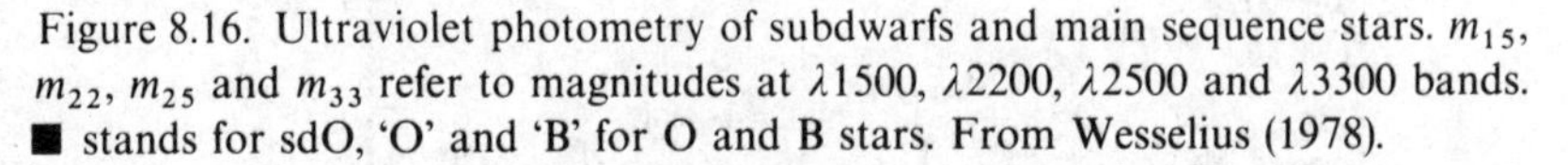
Figure 8.16. Ultraviolet photometry of subdwarfs and main sequence stars. m_{15}, m_{22}, m_{25} and m_{33} refer to magnitudes at $\lambda1500$, $\lambda2200$, $\lambda2500$ and $\lambda3300$ bands. ■ stands for sdO, 'O' and 'B' for O and B stars. From Wesselius (1978).

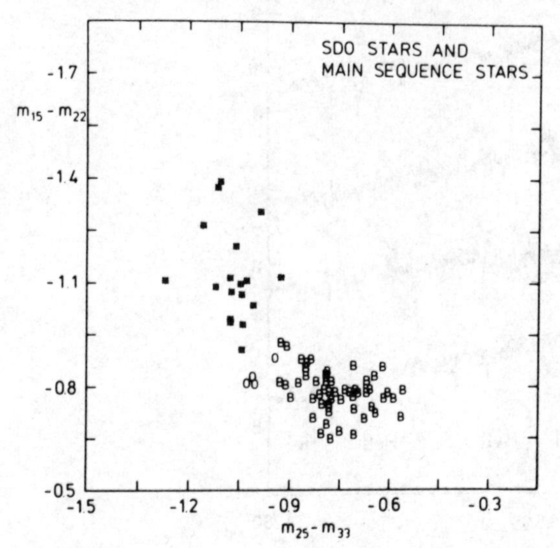

Photometric studies of these objects show that they are extremely blue, but in most ground-based work it is difficult to distinguish them from hot white dwarfs, O-type dwarfs and pure 'black bodies'. They stand out only in satellite photometry because of their enormous ultraviolet radiation. Figure 8.16 shows a graph taken from Wesselius (1978) where two sdO stars stand out as the 'bluest' stars in the diagram. They are therefore easily detectable in the wavelength range ($\lambda < 3000$), and provide a welcome addition to the search criteria; up to now they were mostly found through spectroscopic observation of blue large proper motion stars, or from high latitude objective prism surveys.

In uvby, the sdO stars occupy the region $b - y < -0.12$, $m_1 < +0.10$ and $c_1 < 0.00$.

The radial velocity (r.v.) of these stars is usually large; Greenstein and Sargent (1974) quote $-120 < \text{r.v.} < +89$. These values imply that the stars do not belong to population I.

The absolute magnitude is quite uncertain. Greenstein and Sargent (1974) conclude $\bar{M}_V \sim +2^{\text{m}}3$, i.e. six or seven magnitudes below the main sequence. In view of the large differences in the line spectrum it is possible that the sdO stars cover a quite large range in absolute magnitude, in which case the average value is not very meaningful. Glaspey *et al.* (1985) summarize the absolute magnitudes derived from sdO stars in globular clusters and find $\bar{M}_V \sim 0$ with a range between -1.7 and $+1.0$. It is however uncertain whether both types of stars are identical.

Some sdO stars were found to be central stars of planetary nebulae (Mendez and Niemela 1977; Heap 1977) and others were found in globular clusters (Remillard, Canizares and McClintock 1980), as we have mentioned already.

The number of known sdOs is small. Only 14 sdO and sdB stars are brighter than $V = 10$.

References

Abbot D.C., Bieging J.H., Churchwell E. and Cassinelli J.P. (1980) *Ap. J.* **238**, 196

Andrillat Y. and Vreux J.M. (1979) *AA* **76**, 221

Andrillat Y., Vreux J.M. and Dennefeld M. (1982) *IAU Symp.* **98**, 229

Andrillat Y., Fehrenbach C., Swings P. and Vreux J.M. (1973) *AA* **29**, 171

Barbier R., Dossin F., Jaschek C., Jaschek M., Klutz, M., Swings J.P. and Vreux J.M. (1978) *AA* **66**, L9

Barlow M.J. (1978) *IAU Symp.* **89**, 119, D. Reidel Publ. Co., Dordrecht

Barlow M.J. and Hummer D.G. (1982) *IAU Symp.* **99**, 387

Baschek B. (1981) *Comptes rendus 3ème Journée de Strasbourg*, p. 3
Berger J. and Fringant A.M. (1978) *AA* **64**, L9
Blaauw A. (1960) *Bull. Astron. Inst. Netherlands* **15**, 265
Bohannan B. and Garmany C.D. (1978) *Ap. J.* **223**, 908
Botto P. and Hack M. (1962) *Mem. Soc. Astron. Italiana* **33**, 159
Burnichon M.L. (1975) *AA* **45**, 385
Buscombe W. (1969), *MN* **144**, 1
Carlson E.D. and Henize K.G. (1979) *Vistas in Astronomy* **23**, 213
Carnochan D.J. and Wilson R. (1983) *MN* **202**, 317
Castor J.L. and Simon T. (1983) *Ap. J.* **265**, 304
Cherepashuk A.M. (1974) *Astrofisika* **10**, 347
Conti P.S. (1973) *Ap. J.* **179**, 161
Conti P.S. (1974) *Ap. J* **187**, 589
Conti P.S. (1976) *20ème Coll. Intern. Liège*, 193
Conti P.S. (1982) *IAU Symp.* **99**, 3
Conti P.S. and Alschuler W.R. (1971) *Ap. J.* **171**, 325
Conti P.S. and Burnichon M.L. (1975) *AA* **38**, 467
Conti P.S. and Frost S.A. (1974) *Ap. J.* **190**, L137
Conti P.S. and Leep E.M. (1974) *Ap. J.* **193**, 113
Conti P.S., Leep E.M. and Perry D.N. (1983a) *Ap. J.* **268**, 228
Conti P.S., Garmany C.D., de Loore C. and Vanbeveren D. (1983b) *Ap. J.* **274**, 302
Dworetsky M.M., Whitelock P.A. and Carnochan D.J. (1982a) *MN* **200**, 445
Dworetsky M.M., Whitelock P.A. and Carnochan D.J. (1982b) *MN* **201**, 901
Edlen B. (1983) *Z. f. Astroph.* **7**, 378
Frost S.A. and Conti P. (1978) *IAU Symp.* **70**, 139
Garmany C.D., Conti P.S. and Chiosi C. (1982) *Ap. J.* **263**, 777
Garmany C.D., Conti P.S. and Massey P. (1980) *Ap. J.* **242**, 1063
Glaspey J.W., Demers S., Moffat A.F.J. and Shara M. (1985) *Ap. J.* **289**, 326
Goudis C. (1982) *The Orion complex*, D. Reidel Publ. Co., Dordrecht
Goy G. (1976) *AA* **48**, 87
Goy G. (1980) *AA Suppl.* **42**, 91
Grady C.A., Snow T.P. and Timothy J.C. (1983) *Ap. J.* **271**, 691
Greenstein J. and Sargent A.I. (1974) *Ap. J. Suppl.* **28**, 157
Heap S.R. (1977) *Ap. J.* **215**, 864
Heap S.R. (1982) *IAU Symp.* **99**, 423
Heck A., Egret D., Jaschek M. and Jaschek C. (1984) *IUE low-dispersion spectra reference atlas. Part I. Normal stars.* ESA SP-1054
Hidayat B., Supelli K. and van der Hucht K.A. (1982) *Contr. Lembang* **68**
Hiltner W.A. and Schild R.E. (1966) *Ap. J.* **143**, 770
Hunger K., Gruschinske J., Kudritzki R.P. and Simon K.P. (1981) *AA* **95**, 244
Hutchings J.B. (1968) *MN* **141**, 219
Hutchings J.B. (1979) *IAU Symp.* **83**, 3
Jaschek M. and Jaschek C. (1984) in *The MK process and stellar classification*, Garrison R.F. (ed.), Toronto, p. 290
Lamontagne R. and Moffat A. (1982) *IAU Symp.* **99**, 283

Landolt A.U. (1970) *A.J.* **75**, 337
Lundström I. and Stenholm B. (1984) *AA Suppl.* **56**, 43
Massey P. (1982) *IAU Symp.* **99**, 121
Massey P. (1985) *PASP* **97**, 5
Mendez R.H. and Niemela V.S. (1977) *MN* **178**, 409
Newell E.B. (1973) *Ap. J. Suppl.* **26**, 37
Niemala V.S. (1976) *Physique des mouvements dans les atmosphères stellaires*, CNRS Coll. 250, Cayrel R. and Steinberg M. (ed.) p. 467
Payne C.H. (1933) *Z. f. Astroph.* **7**, 1
Pearce J.A. (1930) *Publ. DAO Victoria* **5**, 110
Pecker J.C. (1976) *20ème Coll. Intern. Liège*, p. 319
Pedoussaut A., Ginestet, N. and Carquillat, J.M. (1983) *Etoiles binaires dans le diagramme HR*, p. 63, 5ème Journée de Strasbourg
Peterson D. and Scholz, M. (1970) *Ap. J.* **163**, 51
Pitault A., Epchtein N., Gomez A.E. and Lortet M.C. (1983) *AA* **120**, 53
Plaskett J.S. (1924) *Publ. DAO Victoria* **2**, 287
Pyper D. (1966) *Ap. J.* **144**, 13
Reddish V.C. (1967) *MN* **135**, 251
Remillard R.A., Canizares C.R. and McClintock J.E. (1930) *Ap. J.* **240**, 109
Sahade J. (1981) *The Wolf-Rayet stars*, Collège de France, Paris
Sargent W.L.W. and Searle L. (1968) *Ap. J.* **152**, 443
Schmidt-Kaler T. (1982), in Landolt–Börnstein, group VI, vol. 2b, p. 1
Simon K.P., Jones G., Kudritzki R.P. and Rahe J. (1983) *AA* **125**, 44
Slettebak A. (1956) *Ap. J.* **124**, 173
Smith L.F. (1968) *MN* **138**, 109
Smith L.F. (1973) *IAU Symp.* **49**, 126
Strom S.E., Strom K.M., Rood R.T. and Iben I. (1970) *AA* **8**, 243
Tobin W. and Kaufmann J.P. (1984) *MN* **207**, 369
Underhill A. (1949) *MN* **109**, 562
Underhill A. (1958) *Publ. DAO Victoria* **11**, 143
Underhill A. (1966) *The early type stars*, Reidel Publ. Co, Dordrecht
Underhill A. (1982) *Ap. J* **263**, 741
Underhill A. and De Groot M. (1965) *Rech. Astr. Obs. Utrecht* XVII, no. 3
van der Hucht K., Conti P.S., Lundström I. and Stenholm B (1981) *Sp. Sc. Rev.* **28**, 227
Vreux J.M. and Andrillat Y. (1974) *AA* **34**, 313
Vreux J.M., Dennefeld M. and Andrillat Y. (1983) *AA Suppl.* **54**, 437
Wade B.R. and Smith L.F. (1985) *MN* **212**, 77
Walborn N.R. (1973) *Ap. J.* **180**, L35
Walborn N.R. (1977) *Ap. J.* **215**, 53
Walborn N.R. (1980) *Ap. J. Suppl.* **44**, 535
Wesselius P.R. (1978) *IAU Symp.* **80**, 125
Wolf C.J. and Rayet, G. (1867) *Comptes Rendus Acad. Sci.* **65**, 292
Zinn R.J., Newell E.B. and Gibson J.B. (1972) *Astron. Astrophys.* **18**, 390
Transactions of the Intern. Astron. Union (1938) **6**, 248 (ed. J.H. Oort) Cambridge Univ. Press

9

B-type stars

9.0 Normal stars

A B-type star is an object exhibiting neutral helium lines in its spectrum, but no ionized helium lines. The latter are characteristic of O-type stars. Neutral helium lines are invisible in A-type stars. (For illustration, see figure 9.1.)

The maximum strength of the He I lines is reached in early B subclasses, around B2. Hydrogen lines on the other hand have their maximum strength at A2 and therefore along the B-type star sequence hydrogen and helium exhibit an opposite trend. Table 9.1 provides the equivalent widths of the stronger lines, taken from Didelon (1982). All lines of elements other than hydrogen in the region $\lambda\lambda$3600–4800 are less intense than 1.3 Å.

Table 9.1 shows that for quantitative classification we can use in principle H and He I line strengths alone. However, the Balmer lines are too intense to use for visual classification and we must look for other, weaker lines in the $\lambda\lambda$3600–4800 region. Elements having weaker lines are listed in table 9.2. As can be seen from the table, the number of elements visible diminishes toward later B-types. For stars between B5 and A0 only a few lines are left and so all have to be used for classification. Equivalent widths for most of these lines are given by Didelon (1982).

To determine the spectral type, the Yerkes system uses ratios of intensities of lines belonging to He I, Si II and Mg II. A list of the commonly used ratios is given in table 9.3.

The use of lines of different elements in table 9.3 becomes objectionable if elements which behave abnormally are included among them. This does happen and the reader is referred to the sections on Ap stars (anomalies in Si), on the helium stars (anomalies in He) and on CNO objects (anomalies in C, N and O). If therefore an element is known to behave abnormally in a certain number of objects, the element should not be used for classification.

Table 9.1. *Equivalent widths (in Å) of some strong lines in dwarfs.*

	H I λ6562	H I λ4340	H I λ4860	He I λ4026	He I λ4471	Mg II λ4481	C II λ4267
B0	3.5	3.5	3.8	1.0	1.0	0.1	0.1
B2	5	5.1	6.0	1.5	1.4	0.2	0.2
B5	6	6.7	7.5	0.8	0.8	0.3	0.2
B8	8	10.0	10.0	0.3	0.3	0.4	0.1
A0	10	13.6	14.0	0.1	0.1	0.4	

λ6562 = Hα, λ4860 = Hβ, λ4340 = Hγ

Table 9.2. *Lines visible in B-type spectra in the region λλ3500–4800.*

B0–B2	H,	He I,	C II,	C III,	N II,	N III,	O II,	Si III,	Si IV,
B2–B5	H,	He I,	C II,	Mg II,	Si II,				
B5–A0	H,	He I,	Mg II,	(Si II),					
A0	H,	Mg II,							

Figure 9.1. Stars of spectral type B. Observe the increasing strength of the Balmer lines toward late B-type and the decreasing strength of the He I after a maximum at B2.

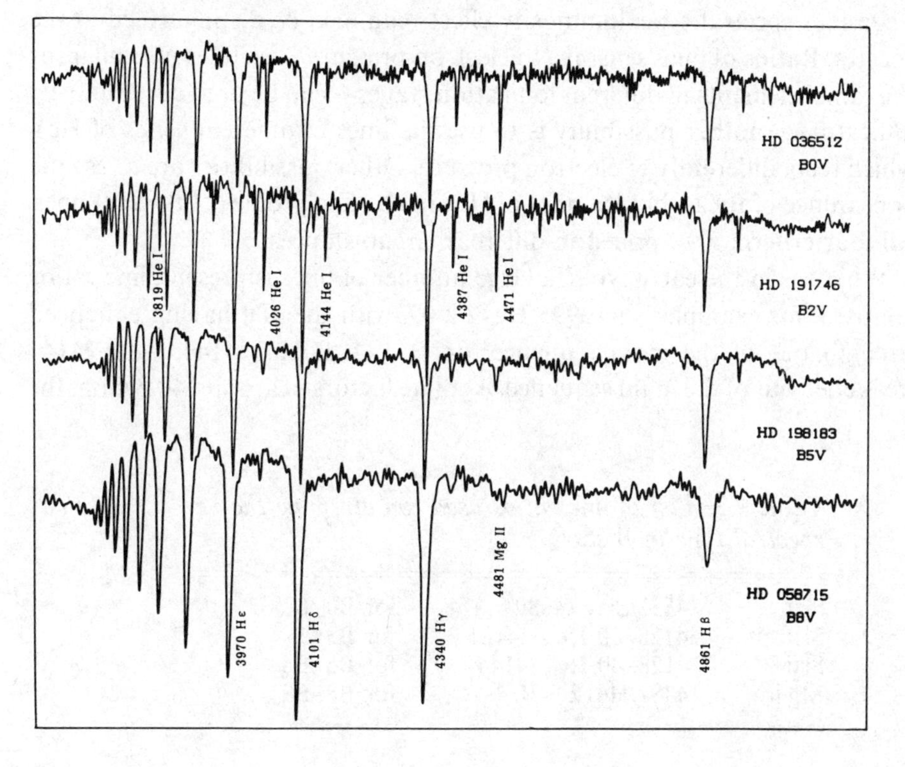

Walborn (1971b) thinks that for early type stars the best procedure is to use only He and Si, leaving aside C, N, O and Mg. Although no early type star with Si line anomalies is known, the same is not true for He, which behaves abnormally in some stars. The best procedure seems therefore to use all available lines as a check upon consistency.

In the MK system about twelve subtypes can be distinguished. Sometimes the use of higher dispersions has been proposed to refine the classification, but it does not necessarily lead to an improvement because B-type stars usually have large rotational velocities which blur the lines and make all fainter lines indistinguishable from the background. On the other hand, if rotation is low, a large number of faint lines appear at high dispersion which suggests spectral peculiarities which do not exist in reality.

At very large plate factors (200–300 Å/mm) only the strong H and He lines remain visible. All stars between O7 and B2 'collapse' into a single group conveniently called 'OB'. Stars in which He I is weaker are called 'OB^-' and are usually B3–B5 stars. Between B5 and A2 only the hydrogen lines are seen; they increase in strength toward a maximum at A2. Schemes of this type were described by Schalen (1926) and by Slettebak and Stock (1957). See the section on natural groups, 3.6.

Besides spectral type, luminosity effects can also be distinguished in the spectra. Ratios of lines sensitive to electron pressure – i.e. lines of similar (or the same) elements in different ionization stages – can be used as luminosity indicators. Another possibility is to use the lines of different series of He I, which react differently to electron pressure. Other possibilities are to use the appearance of the forbidden lines of He I or the line profiles of the He I lines. All four criteria are applied in different circumstances.

When, as in the early types, a large number of lines is present, line ratios are used; for example N II λ3995/He I λ4009, with the N II having enhanced strength out of the main sequence, or He I λ4121/He I λ4144, with λ4144 weakened out of the main sequence. At plate factors of around 40 Å/mm, the

Table 9.3. *List of line ratios used to determine the spectral type in B stars.*

Si III	λ4552/Si IV λ4089	for B0–B2
Si II	λ4128–30/He I λ4121	for B3
Si II	λ4128–30/He I λ4144	for B5–B8
Mg II	λ4481/He I λ4471	for B8–B9

forbidden He I λ4469 lines are visible only in dwarfs and can be used as luminosity discriminants around B3.

Finally the line profiles are used as a criterion in the late B-types (narrower profiles out of the main sequence) when nothing else is left. The main luminosity criteria are summarized in table 9.4. For illustration, see

Figure 9.2. Luminosity effects in B-type stars. The luminosity effects are illustrated at spectral types B2, B5 and B8. See text for criteria.

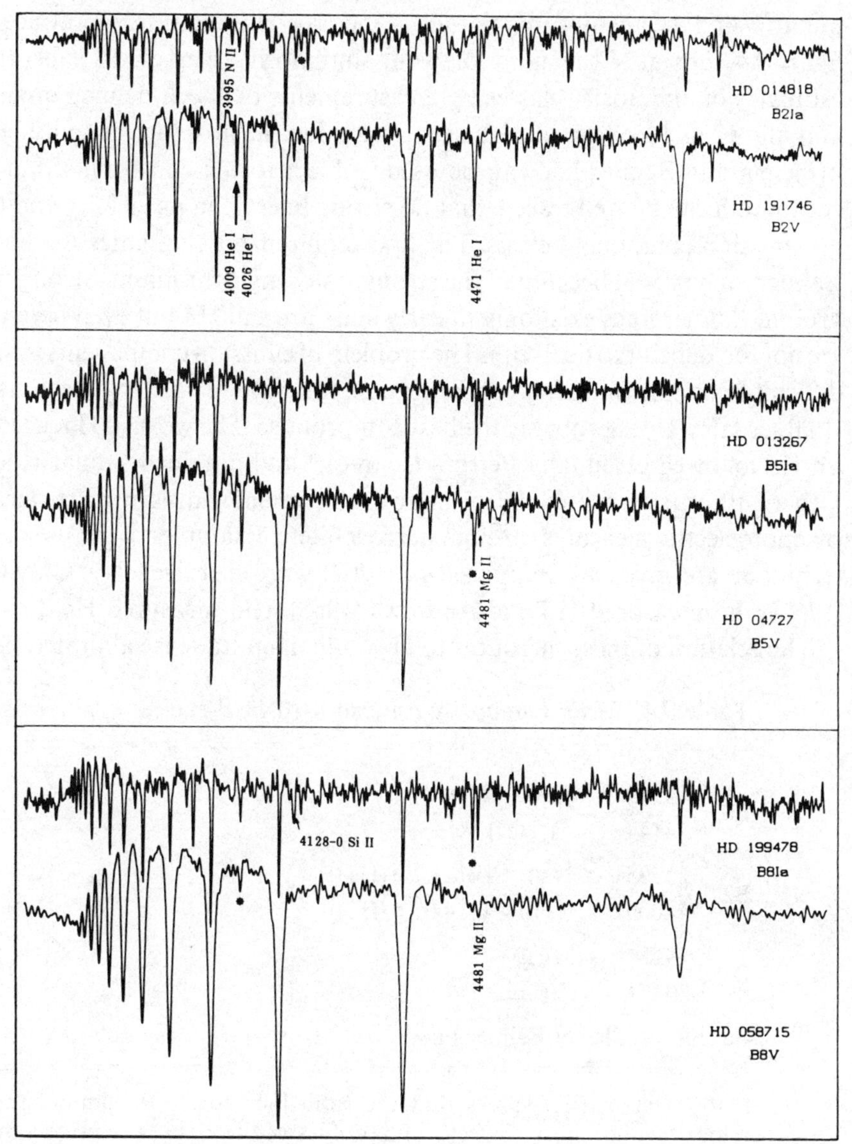

figure 9.2. It should be noticed that all lines used for determination of luminosity are weak lines. None of the lines in table 9.4 has $W > 0.5$ Å.

The dependence of the profiles of the hydrogen lines upon luminosity was discovered by the Swedish school in the twenties. Figure 9.3 shows the profiles of Hγ for three stars of the same spectral type but different luminosity: it is clear that the equivalent width (W) of the line changes with luminosity. Petrie (1952) made a long series of careful measurements of W(Hγ) and was able to show that a smooth relation exists between this quantity and the absolute magnitude of the star. This relation, reproduced in figure 9.4, obviates the need of using luminosity classes. It substitutes estimates of luminosity classes by measurements of a well-defined quantity and allows us to obtain a higher accuracy for the absolute magnitude. In principle any Balmer line can be used, subject to the constraints that the continuum can be well traced, that no strong lines lie in its vicinity and that no emission component exists. The first requirement eliminates the higher Balmer lines ($n > 5$) because of the strongly sloping continuum. Strong lines around Balmer lines exist only in early stars around Hδ but even here they are not too difficult to deal with. The problem of emission components is more difficult to handle, since faint emission components are hard to detect. As we shall see later, Hγ is probably the best compromise. The work of Hack (1953) on Hδ followed essentially Petrie's technique and obtained similar results. Other authors substituted the measure of equivalent widths in spectrograms by photoelectric measures through narrow filters. Examples of the use of this technique are given by Bappu *et al.* (1962) who measured Hγ, Crawford (1958) who measured Hβ and Andrews (1968) who measured Hα.

The relation of these measures to absolute magnitude is calibrated using

Table 9.4. *Main luminosity criteria used in B-type stars.*

B0: $\dfrac{\lambda 4119}{\lambda 4144} = \dfrac{\text{I(Si IV + He I)}}{\text{I(He I)}} : +$

B2: $\dfrac{\lambda 3995}{\lambda 4009} = \dfrac{\text{I(N II)}}{\text{I(He I)}} + \dfrac{\lambda 4121}{\lambda 4144} = \dfrac{\text{I(He I)}}{\text{I(He I)}} +$

B5: $\dfrac{\lambda 4481}{\lambda 4471} = \dfrac{\text{I(Mg II)}}{\text{I(He I)}} +$

B6–B9: profiles of Balmer lines

Note: The + sign shows that the ratio increases with luminosity – a 'positive luminosity effect'. The profiles of the Balmer lines become narrower toward higher luminosities.

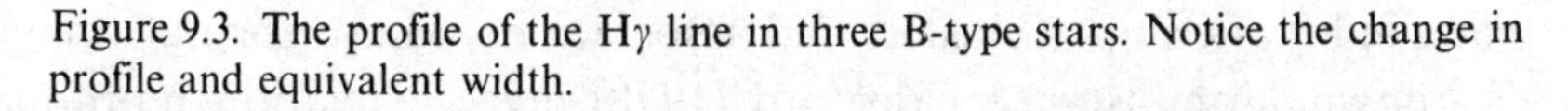

Figure 9.3. The profile of the Hγ line in three B-type stars. Notice the change in profile and equivalent width.

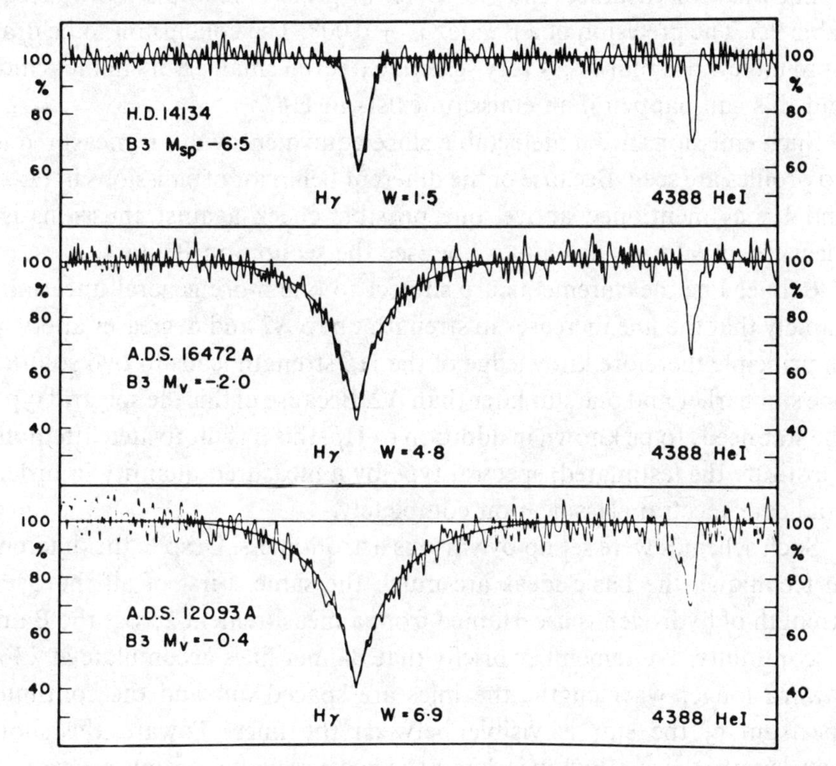

Figure 9.4. Absolute magnitude as a function of Hγ equivalent width. From Petrie (1952).

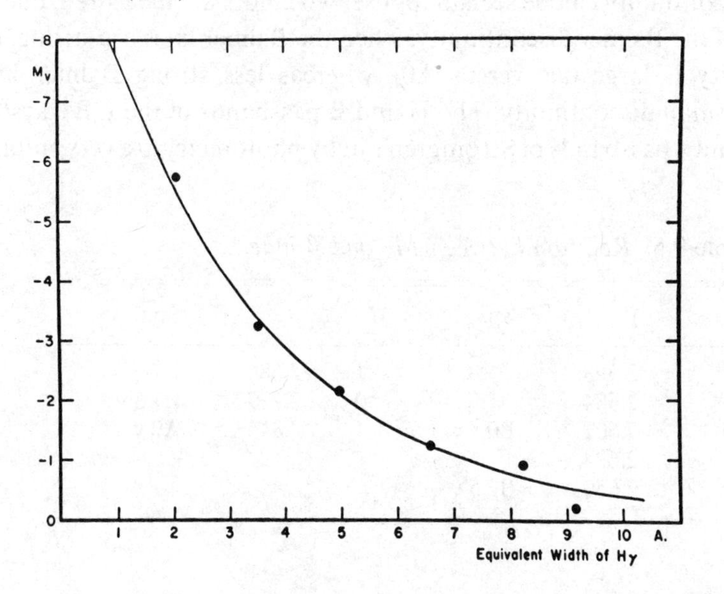

stars in binaries, open clusters and associations, whose absolute magnitude is known. For instance Crawford (1978) gives a relation reproduced in table 9.5. The precision of a β index is ± 0.008. The conclusion to be drawn from table 9.5 is that M_v is very sensitive to even small errors in the β index, and this can happen if an emission exists in Hβ.

Such emissions are undetectable since equivalent width is measured and no profiles are seen. Because of the different behavior of emissions in Hα, Hβ and Hγ as mentioned above, one possible check against emissions is to measure an additional Balmer line (see the section on Be stars).

Balmer line measurements are subject to one more general uncertainty, namely that the line increases in strength up to A2 and decreases afterward. In principle therefore knowledge of the Hβ strength leads to two solutions, one star earlier and one star later than A2. Because of this the spectral type of the star needs to be known in addition to Hβ; this has motivated attempts to substitute the (estimated) spectral type by a measured quantity in order to eradicate spectral classification completely.

Such schemes were set up by various astronomers. Despite the differences in technique, the basic ideas are much the same. First of all the overall strength of hydrogen is ascertained from a measurement across the Balmer discontinuity. We remember briefly that Balmer lines accumulate at λ3647; toward longer wavelengths the lines are spaced out and the continuous spectrum of the star is visible between the lines. Toward the shorter wavelengths the continuum is depressed because of the continuous hydrogen absorption. Figure 9.5 reproduces the spectra of stars (B0 to B8) where the Balmer discontinuity can be seen. Suppose two colors are measured, one on each side of the Balmer discontinuity: when the Balmer lines are strong, the discontinuity is large (i.e. versus A0), whereas less strong Balmer lines produce a small discontinuity. The U and B passbands of the UBV system and the u and v passbands of Strömgren's uvby photometry are very suitable

Table 9.5. *Relation between M_v and β index.*

M_v	β	Sp.t.	M_v	β	Sp.t.
−6	2.568		−1	2.685	
−5	2.584		0	2.737	B7.5 v
−4	2.602	B0.5 v	+1	2.845	A0 v
−3	2.624				
−2	2.650	B2.5 v			

for this and in fact correlate very well with spectral types, as we have seen in the chapters on photometry. Values of either u–v or U–B (when corrected for reddening) can thus substitute for the spectral type, in the same way as measures of the Hβ strength can replace the luminosity class.

The advantages of such a scheme are obvious – something directly related to the physics of the object is measured accurately and only one chemical element is used. In fact for some time it was thought that spectral classification could be replaced entirely by photometry. However a photometric system is completely blind to emission components in hydrogen lines and to anomalies in elements other than hydrogen. Because of this a spectrum must be examined first to decide if anomalies are absent, before the object is observed photoelectrically.

Besides photometric systems using filters, there is another method based upon the same idea, but using photographic plates. Since this entails much additional work because of the non-linear plate response, the technique is not widely used. We mention the Chalonge–Barbier system. The interested reader can find details in Fehrenbach (1958) and Underhill and Doazan (1982), for instance.

We mentioned above the confluence of the Balmer lines toward the discontinuity. Obviously the confluence will also depend upon the character

Figure 9.5. Intensity tracings of B-type dwarfs. From Jacoby, Hunter and Christian (1984). Observe the increase of the Balmer discontinuity toward late subclasses.

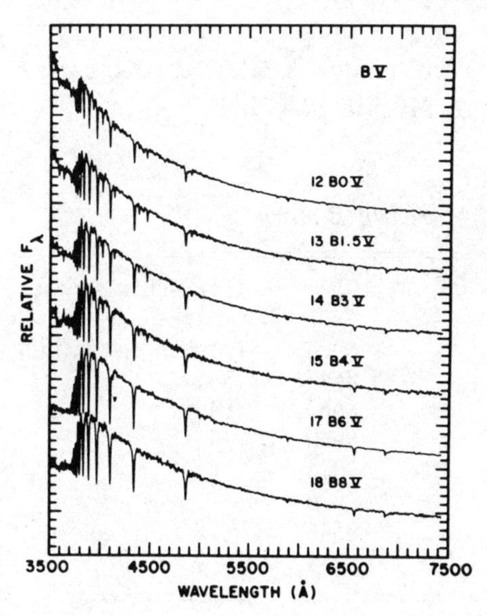

of the lines. If the lines are narrow (as in supergiants, for example), the wings will overlap less than if the wings are broad. Thus by simply counting the number of the last Balmer line visible ($n = 3$ for Hα, $n = 4$ for Hβ, etc.) in the spectrum, we can distinguish a dwarf from a supergiant. At 40 Å/mm for instance, supergiants show hydrogen lines up to $n = 24$, whereas giants show them up to $n = 18$ and dwarfs up to $n \sim 14$. These numbers depend upon the dispersion and the resolution of the spectrograph used and are to be taken only as indicative.

Infrared. The near infrared spectrum is dominated by the Paschen series and the few strong lines present are listed in table 9.6. In general the lines of the Paschen series increase in strength toward late B, as the lines of the Balmer series do. The lines of He I peak at B2. The infrared Ca triplet on the other hand increases steadily toward type A. Usually some of the lines appear in emission in Be stars; if this happens, there is usually an infrared color excess also.

Ultraviolet. In the range $\lambda\lambda$1150–3000, B-type stars can be arranged in an order which closely parallels the order of the Yerkes system based upon the $\lambda\lambda$3600–4800 region. The overall correspondence between both classifications is to be expected but it is also evident that this correspondence need not be perfect.

We give in tables 9.7 and 9.8 the principal lines which can be used for spectral classification in this wavelength range, taken from Jaschek and Jaschek (1984). For more details the reader is referred to the *IUE low dispersion spectra reference atlas* by Heck *et al.* (1984).

Table 9.6. *Lines in near infrared of B stars.*

λ	Species	λ	H (Paschen series)
7065	He I	8204	Paschen discontinuity
7772	O I	8545	
8446	O I	8665	P13
8498	Ca II	8860	11
8542	Ca II	9015	10
8662	Ca II	9229	9
8680	N I	9546	8
10830	He I	10049	7

Emission lines. Up to now we have mainly discussed absorption lines, but emission lines also exist in the spectra of a number of B-type stars. Usually the lines in emission are the hydrogen lines, with the general rule that $E(\mathrm{H}\alpha) > E(\mathrm{H}\beta) > E(\mathrm{H}\gamma)$ and so on.

A star with Balmer lines in emission is called Be (see section 9.1). About one-tenth of all B-type stars exhibit emissions in the Balmer lines at one time or another. Often when Balmer lines are seen in emission Paschen lines are also found in emission (Andrillat and Houziaux 1967).

Besides hydrogen, the element most often found in emission is Fe II. He I on the other hand is seldom seen in emission. Usually these lines appear only in stars having hydrogen in emission.

Another group of stars which exhibits emission – especially at Hα – are the supergiants of type Ia, and less often the Ib supergiants. Usually the emission profiles are then of the P Cygni type.

Spectral peculiarities. We have mentioned some of these already. We have the stars with emission lines of various types, designated as Be, shell or B[e] (section 9.2). Some of the emission line stars belong to the group of Herbig Be–Ae stars (section 10.6).

Table 9.7. *Some lines characteristic of B-type stars.*

λ	Identification	Comments
1175	C II	Maximum is at B1 and the line disappears at B6 in the Lα wing.
1216	Lα	When not affected by interstellar or circumstellar effects, the line has a half width increasing monotonically from 10 Å at O9 to 100 Å at B8.
1265	Si II	Visible from B1 on, and increases towards B9.
1300		Blend of Si III and Si II; in later stars Si II predominates; the line grows between B2 to B8, where its usefulness disappears because of Lα
1336	C II	Maximum at B8.
1400		Blend of λ1349 and λ1403 of Si IV; has a maximum at B1 and disappears at about B8.
1465		Non-identified, except with C III (?), appears at about B2 and increases toward later types.
1548	C IV	Strong in O-type stars, disappears at B2 in dwarfs (see table 9.8).

Then we have the stars with anomalies in carbon, nitrogen and/or oxygen, called CNO stars (section 9.3); the stars with anomalies in helium (section 9.4) and the stars with anomalies of heavier elements, called Bp (section 9.5). Besides these there are the B-type subdwarfs (section 9.8) and the horizontal branch stars (section 10.4). The β Cep stars are looked at in section 9.10.

Rotation. Many B-type spectra have fuzzy lines because of high rotation. In fact, the highest stellar rotational velocities observed so far belong to B-type objects. An illustration of the effect of different speeds of rotation is provided in the atlas by Slettebak *et al.* (1975). The general distribution of rotational velocities is shown in table 9.9, taken from Uesugi and Fukuda (1981).

It is often found that Be stars have high values of $\overline{V \sin i}$, around 450 km/s; however, the converse that all Be stars have high rotation is not true. On the other hand B-type stars with low $\overline{V \sin i}$ values often have peculiar spectra. Close binaries tend to rotate more slowly than non-binary stars; this is attributed to the presence of a companion which forces a synchronization between the periods of rotation and revolution. This is illustrated in table 9.10, taken from Jaschek (1970).

Photometry. We have already discussed some of the photometry done on B-type stars in previous sections and also in the chapter on UBV photometry.

Table 9.8. *Lines showing luminosity effects in B-type stars.*

λ	Comments
1400	Blend of λ1394 and λ1403 of Si IV; this line has a pronounced positive luminosity effect and can be used for supergiants up to B5.
1548	C IV; has a pronounced positive luminosity effect; since in dwarfs it disappears at B2, its appearance in middle B-type stars indicates a supergiant.
1608 1622 1629 1640	Fe II, M.8; the strongest line is λ1708; all four lines are enhanced in supergiants in B-type stars and show little variation with temperature.
1723	Al II, M.6; positive luminosity effect (blend of three lines).
1855	Al III, M.1; blended with λ1862 Al II + Al III
1891	Fe III, M.52 + 53
1926	Fe III, M.57 + 34
1967	Fe III; blend of several lines

We shall present here simply in figure 9.6 the results of broad band photometry beyond UBV, as taken from Straizys (1977). As can be seen, a very close relation exists between spectral types and de-reddened multicolor photometry. Observe that all colors are taken with regard to A0 stars, which explains the behavior of the curves; the line for A0 should be strictly equal to zero, and is so except for the statistical error. This error exists because one averages over a large number of stars which may not be identically A0 dwarfs.

Narrow band photometry has been done on several Balmer lines, which we have already mentioned. A general compilation was published by Mermilliod and Mermilliod (1980). For lines other than hydrogen, there exists only the photometry by Nissen (1974) on He I λ4026. This photometry, taken together with the Strömgren and Hβ photometry, allows us to pin-

Table 9.9. *Rotation of B-type stars:* $\overline{V \sin i}$.

	V	III	I
B0	154	180	88
B2	146		
B5	184	57	40
B8	173	55	
A0	119	22	

Note: the bar over $V \sin i$ means 'average $V \sin i$. Values in km/s.

Table 9.10. *Distribution of* $V \sin i$ *for normal B-type stars.*

$V \sin i$ (km/s)	Spectroscopic binaries	Normal
0–100	53	91
100–200	39	39
200–300	8	18
300–400		8
400–500		4
Total number of stars		277

point with high precision anomalies in the strength of the helium and/or hydrogen lines.

Radial velocities. Most of the B-type stars have low radial velocity. If the solar component is taken out, we find a dispersion of ± 12 km/s. When a boundary of three times this value is taken, less than 5% of the stars have radial velocity larger than 36 km/s.

High velocities appear if the star is a runaway (see section 9.6) or if the star is an old object (see section 11.5). Spurious high velocities may appear when the star is a spectroscopic binary and only a part of the orbit has been observed.

In supergiants different radial velocities are often derived if different lines are used. In most cases the differences are modest (up to 10 km/s) in the classic region, except in the Balmer lines. In general the lower Balmer lines (α, β) have larger negative velocities than the higher lines. This variation of radial velocity with the quantum number is called the 'Balmer progression'.

Figure 9.6. Broad band UBV... L photometry for B-type stars. The positions of the mean wavelengths are indicated on the abscissa.

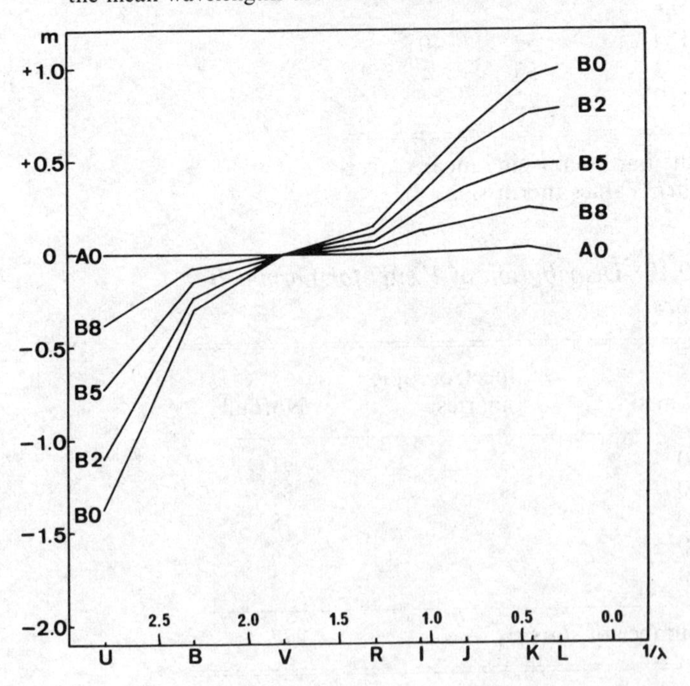

Frequency of binaries. The proportion of visual binaries having a primary with a B-type spectrum is normal (Abrams 1947), i.e. about 20%. The percentage of spectroscopic binaries has been discussed several times in the literature and seems to be now located at about 40%. A general discussion with complete bibliography can be found in Gieseking (1983).

Magnetic fields. No magnetic field greater than the possible observational errors has been measured in B-type stars with a non-peculiar spectrum (Landstreet 1982). In Ap and Bp stars on the other hand many average longitudinal fields of the order of several kilogauss have been measured. Probably it is more correct to say that all B-type stars with measurable magnetic fields have a peculiar spectrum.

Distribution in the galaxy. The B-type stars are concentrated in the galactic plane; the concentration becomes more marked when stars of tenth or eleventh apparent magnitudes are considered. Brighter stars, because of their nearness, are occasionally seen at high galactic latitudes. If we consider only the stars brighter than $6^{m}\!.5$, we find that they tend to concentrate along a plane tilted by some twenty degrees with regard to the galactic plane. This concentration is called 'Gould's belt', in honor of its discoverer. A recent study of Gould's belt was carried out by Stothers and Frogel (1974). It seems that the belt is a local deformation of the galactic structure.

Another characteristic of the distribution of B-type stars is the clustering of their positions on the sky. This was discovered long ago by Pannekoek (1929), and Ambartsumian in the late 1940s formulated an interpretation of the phenomenon which is best seen in early (B0–B3) stars. He showed that B-type stars, being very young, cannot be far away from their birth-places. In fact, an object having a velocity of 1 km/s displaces itself by one parsec in 10^6 years.

Concentrations of O-type and early B-type stars with surrounding nebulosity are called 'O–B associations'. When the nebulosity is of the reflecting type, it is called an 'R association'. When the group also contains somewhat later stars and is concentrated in a small volume, one speaks of an (open) cluster. An old cluster may have no B-type star left.

We should also mention the fundamental role played by O- and B-type stars in the ionization of the interstellar medium. Due to the abundance of energetic radiation, each early type star ionizes a volume of gas surrounding it. Since the ionization energy of hydrogen is 13.65 eV, which is equivalent to 912 Å, all radiation shorter than this limit ionizes interstellar hydrogen. In a medium of about constant hydrogen density, a sphere of hydrogen ionization is formed, which is called the 'Strömgren sphere' in honor of its discoverer (Strömgren 1939) or the 'H II region'. The size of the H II region depends sharply upon the temperature of the central star. For very early stars (O6) the size of this region reaches a hundred parsecs, whereas for late B-type stars (B5) the size shrinks to a few parsecs.

In the interior of an H II region, recombination of electrons and protons produces an emission line spectrum of hydrogen, which includes Hα. H II regions are therefore powerful Hα-emission regions.

Number of stars. Table 9.11 gives the number of objects within three limits. The very large difference between the two first lines of the table and the third is explained by the large absolute magnitudes of the B-type stars. These absolute magnitudes make them visible from great distances and in fact an enormous volume of space is surveyed when considering B-type stars of the ninth apparent magnitude. On the other hand when we consider a distance-limited volume, B-type stars are very rare.

Because of the difference in absolute magnitude among early and late B-type stars, the proportion of early to late B-type stars is also falsified when surveys are done to a magnitude limit. The corrected percentages are for B0–B5 stars 35% ($m < 9$) and 8% (constant distances) and for B5–B9 stars 65% and 92% respectively. Thus the early B-type stars are much less frequent than the late ones.

Absolute magnitude. Since most B-type stars are too far away to determine their distance by trigonometric parallax, luminosities are mostly determined from objects in clusters and associations. Luminosities can be calibrated on Balmer line photometry (Hα, Hβ or Hγ), or on color, or on spectral type. We have already mentioned (see table 9.5) the results of Hβ

Table 9.11. *Number of B-type stars.*

Limit	Number of stars	Percentage	Source
$m = 6.5$	1760	17	Hoffleit and Jaschek (1982)
$m = 9$	17000	8	Charlier (1926)
$r = 100$ pc	50	1	D.A. Allen (1973)

Note: 'Percentage' means percentage of B-type stars relative to stars of all spectral types.

photometry. In table 9.12 is given the calibration by spectral classification, taken from Schmidt-Kaler (1982).

For supergiants the uncertainties are of the order of at least $\pm 0^{\rm m}.5$.

Bibliography. The best monograph is A. Underhill's part of the book *B stars with and without emission lines*, edited by A. Underhill and V. Doazan (CNRS–NASA, 1982).

9.1 Be Stars

A Be star is defined to be a star of B-type which has shown emission in at least one of the Balmer lines at some time. As previously mentioned, emission in the Balmer lines follows the rule that the emission in Hα is stronger than that in Hβ, which in turn is stronger than that in Hγ, and so on.

Usually the restriction that the star should not be a supergiant is included in the definition, since these also show emissions in Hα. So the working definition is 'a non-supergiant B-type star whose spectrum has or had at some time one or more hydrogen lines in emission' (Jaschek, Slettebak and Jaschek 1981). Be stars are illustrated in figure 9.7.

With regard to the variability implicit in the definition, we must add that all Be phenomena are variable, although our knowledge of the time scales is rather meager. For many stars time scales of the order of tens of years have been found; periods of years are less frequent. Furthermore, a large literature, with sometimes conflicting evidence, exists on short time scales of seconds, minutes or hours (Harmanec 1984a, 1984b). Probably several scales can coexist in the same star – for instance a short cycle together with a long one. Statistically the only significant result is that Be stars of later spectral class have longer time scales than early type stars (Jaschek *et al.* 1980).

Although Be stars belong to those stars with particular spectra which were

Table 9.12. *Visual absolute magnitude calibration.*

Sp. type	V	III	Ib	Ia
B0	$-4^{\rm m}.0$	$-5^{\rm m}.1$	$-6^{\rm m}.1$	$-6^{\rm m}.9$
B1	$-3^{\rm m}.2$	$-4^{\rm m}.4$	$-5^{\rm m}.8$	$-6^{\rm m}.9$
B2	$-2^{\rm m}.4$	$-3^{\rm m}.9$	$-5^{\rm m}.7$	$-6^{\rm m}.9$
B5	$-1^{\rm m}.2$	$-2^{\rm m}.2$	$-5^{\rm m}.4$	$-7^{\rm m}.0$
B8	$-0^{\rm m}.2$	$-1^{\rm m}.2$	$-5^{\rm m}.2$	$-7^{\rm m}.1$

discovered by the pioneer spectroscopists (Secchi 1878), they were not studied systematically until the 1920s and 1930s by the Michigan astronomers, Curtiss and McLaughlin, and the Mt Wilson astronomers, Merrill and Burwell. Merrill initiated the first Hα star survey by observing the region of Hα on objective prism plates. He suppressed the rest of the spectrum to avoid overlapping the spectra of nearby stars; the drawback of this is that the

Figure 9.7. Be and shell stars. The figure shows a B2e star compared to a B2v star, and a B-type shell star (BD 52°3147) compared to a B-type dwarf and an A-type supergiant.

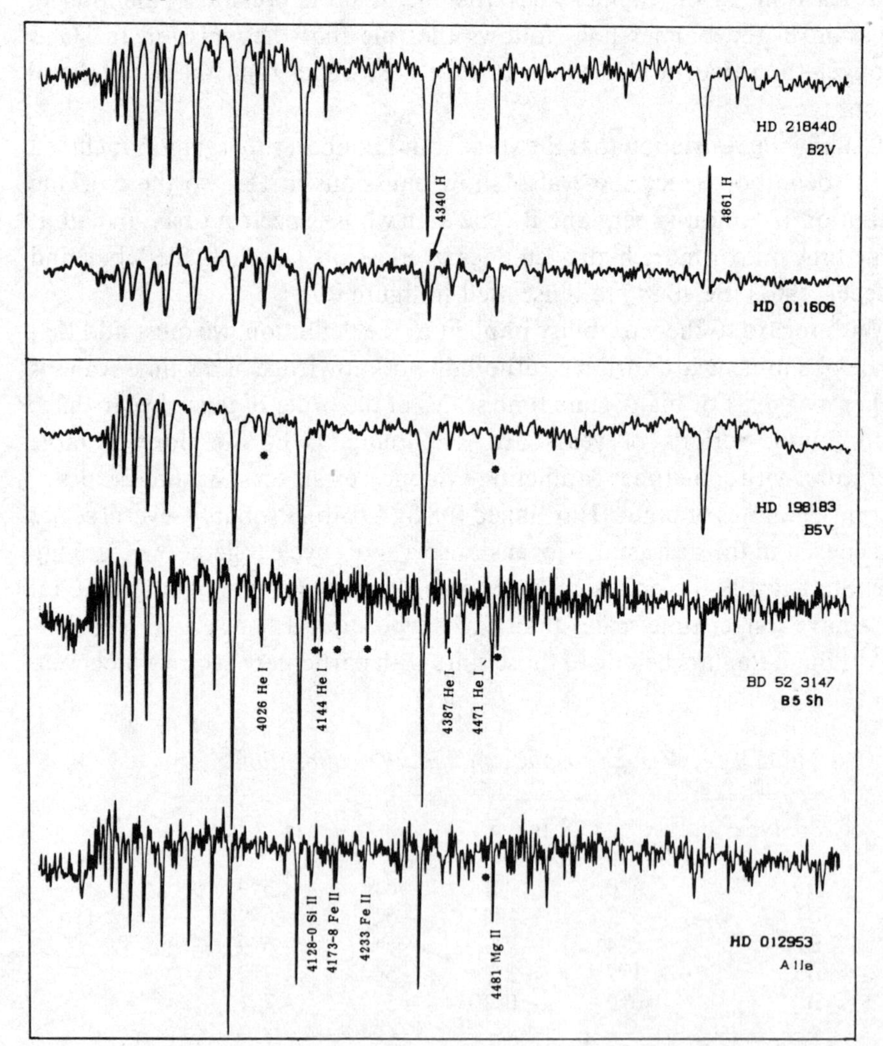

spectrum cannot then be used to judge if the object is truly of B-type. This is why supergiants, planetary nebulae and late type giants, all of which may show Hα in emission, are also found in surveys of this type.

Objective prism work is usually done at 200 or 300 Å/mm which enables us to pick out objects in which the emission is fairly strong. Stars with fainter emission can obviously go undetected unless a higher dispersion is used.

Spectral classification of Be stars is usually not easy. As we shall see, many stars have fast rotations, so that the lines are generally broad. Then there are emission features in the hydrogen lines, and sometimes in other lines, which vary over time. Because of this it would be obviously preferable to classify the star whenever the emissions are weak or absent but this imposes severe restrictions upon observers. Most of the bright Be stars were classified by Jaschek *et al.* (1980) and Slettebak (1982). The latter compares in detail the work of different classifiers.

Table 9.13 gives the proportion of Be stars found among bright stars ($m \leqslant 6.5$) (Jaschek and Jaschek 1983). The table shows that Be stars are most frequent at B2, declining sharply toward earlier types and somewhat slower toward later types. The Be stars form a small part (about 11%) of all B-type stars. We should add that the percentages of table 9.13 include all stars which were once Be. Thus at any given moment, the percentage is certainly smaller.

Table 9.13. *Percentage of Be stars among B-type stars.*

Sp. type	Percentage
O8–O9.5	6
B0–1	13
B1.5–2.5	18
B3	10
B4–5	15
B6–7	12
B8	7
B9	4
B9.5	3
A0–A1	1

Note: The percentage gives Be stars with regard to all B-type stars of luminosity classes III–V. The total number of emission stars is 169.

Let us describe briefly the different phenomena found in Be stars, at classification dispersion. The first fact is that the emission in the Balmer lines is closely related to the spectral type of the star. Figure 9.8 taken from Jaschek *et al.* (1980) shows that the quantum number of the highest Balmer line in which emission is seen, diminishes steadily toward later spectral types; i.e. statistically late Be stars only exhibit Hα in emission.

The emission features themselves have a great variety of profiles; figure 9.9 shows the most common types. Typical widths of the features are of the order of 50–200 km/s (5 Å at Hα) although the definition of the width constitutes a problem by itself. Moreover, the two emission peaks (violet = V, red = R)

Figure 9.8. Quantum number of the highest Balmer line seen in emission as a function of spectral type. From Jaschek *et al.* (1980).

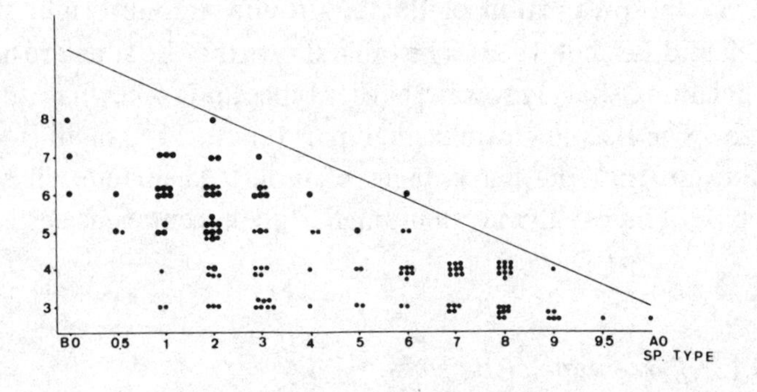

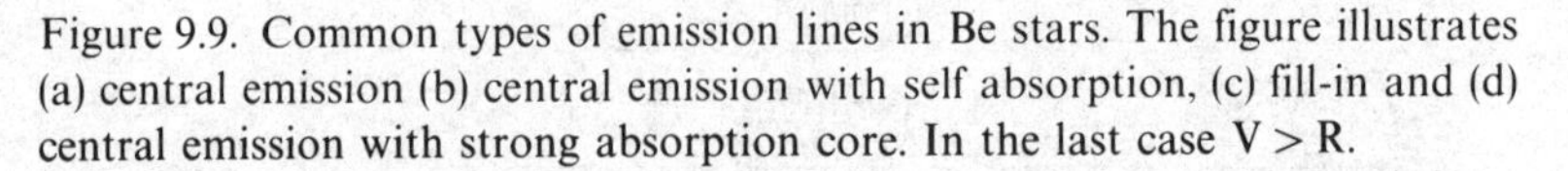
Figure 9.9. Common types of emission lines in Be stars. The figure illustrates (a) central emission (b) central emission with self absorption, (c) fill-in and (d) central emission with strong absorption core. In the last case V > R.

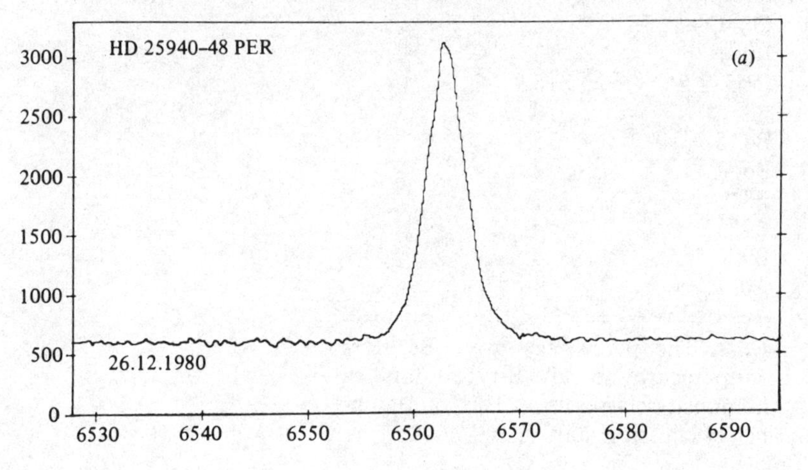

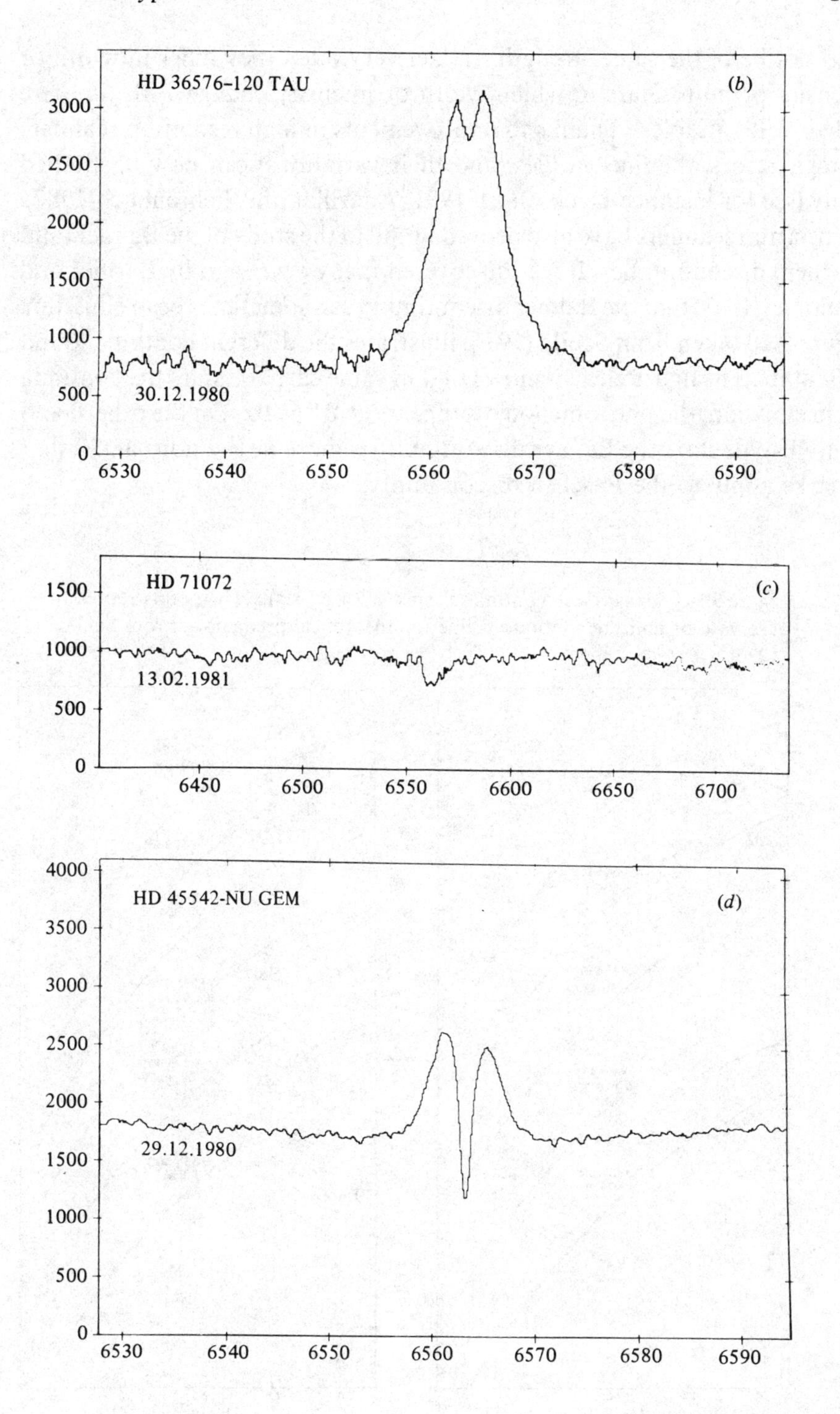
HD 36576–120 TAU
(b)
30.12.1980
HD 71072
(c)
13.02.1981
HD 45542-NU GEM
(d)
29.12.1980

need not be of the same strength. In fact very often they differ in width or intensity or both. Stars in which width or intensity change with time are called 'V/R variables'. Thanks to improvements in high resolution scanners or registrators, the line profiles (and their variations) can now be studied easily (see for instance Dachs *et al.* 1981; Andrillat and Fehrenbach 1982).

Spectrum scanners have also proved useful in the study of the Balmer (and Paschen) discontinuities. It was discovered many years ago by Barbier and Chalonge (1941) that the Balmer discontinuity can sometimes be in emission. Figure 9.10 taken from Schild (1976) illustrates the different continua found in Be stars. The figure clearly shows that in some early Be stars the continua are in emission; the phenomenon disappears at B1 or B2. On the other hand in late B-type stars the Balmer discontinuity is more or less normal. Similar remarks apply to the Paschen discontinuity.

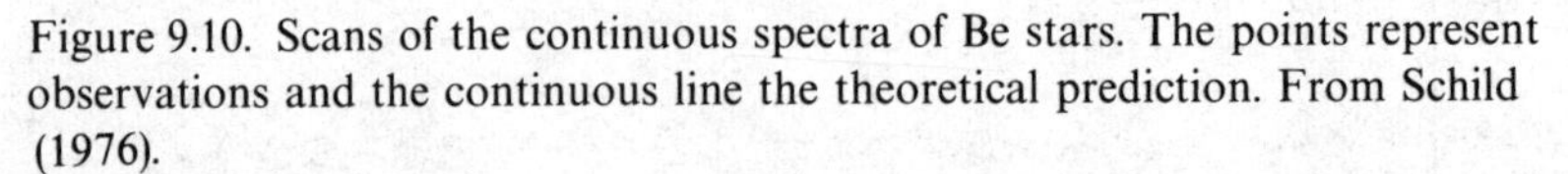
Figure 9.10. Scans of the continuous spectra of Be stars. The points represent observations and the continuous line the theoretical prediction. From Schild (1976).

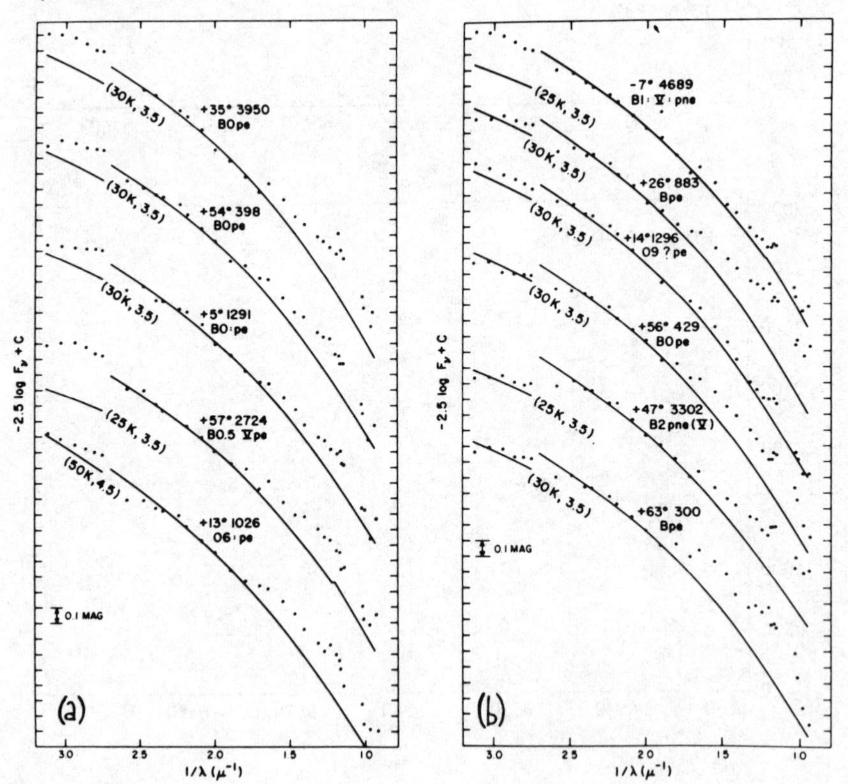

Besides the Balmer lines, helium and iron can also appear in emission. Neutral helium lines, especially $\lambda 5876$, may appear in the earliest Be stars. However, Fe II emission lines are more frequent, especially those in the red. These lines can be present in stars of spectral types between B0 and B5. If stars with Fe II in emission change their intensity, the change runs parallel to the strength of the Balmer emissions; both strengthen or weaken simultaneously.

A term often used in connection with Be stars is 'shell'. It was introduced by Struve in the 1930s and has become rather ill-defined. A 'B-type shell star' is a Be star whose spectrum is characterized by the simultaneous presence of (a) broad absorption lines and (b) sharp absorption lines which arise from ground states or metastable levels (Jaschek, Slettebak and Jaschek 1981). Usually strong and narrow absorption cores are observed in the Balmer lines, either in only the lower lines or in all lines. Simultaneously with these narrow and strong absorption profiles in H I, there are broad and ill-defined He I lines.

In general, if the spectrum of the star is later than B4–B5, narrow lines of other elements (Fe II, Ti II, ...) are also seen. The number of lines and the species present suggest the spectrum of a late B- or early A-type supergiant, although not all lines present in a supergiant spectrum are present with their normal intensity. Often quoted exceptions are Mg II $\lambda 4481$ and Si II $\lambda 4128$–30 which are weak or absent in shell spectra (see figure 9.7). However, the presence of a supergiant spectrum contradicts the spectral type derived from the presence and strength of the He I lines, which indicate an earlier type. The He I spectrum also often presents the peculiarity that the lines arising from metastable levels are very strong, for instance $\lambda 3889$.

The appearance of absorption cores in H I and/or metallic absorption lines qualifies a star as exhibiting a 'shell phenomenon' or passing through a 'shell episode'. If only the H I cores are present, one speaks of a 'hydrogen shell', and these can be present at all spectral subclasses.

It is unfortunate that 'shell' is also used in the geometrical sense of an extended atmosphere surrounding a star, instead of the spectroscopic sense. In many cases one has to find out what the author meant exactly. In the spectroscopic sense we have used here, no difference exists between a 'shell' and a 'Be' star. Depending upon the epoch and the dispersion used, probably all Be stars pass through a 'shell' phase.

Other terms used in connection with Be stars are 'Bex', 'pole-on' and 'B[e]'. The designation 'Bex' was introduced by Schild (1966) to denote extreme Be stars, characterized by hydrogen shells, quite weak He lines and

Fe II emission lines in the region $\lambda\lambda$3600–4800. The notation is used as B1 IIIex, denoting an extreme Be star of type B1 III.

The term 'pole-on' was introduced by Slettebak (1949) to denote stars in which the helium lines are very sharp and the hydrogen lines very broad. The term derives from the probable explanation that the lines have narrow profiles because the star is seen 'pole-on'; the hydrogen lines are very broad because gravity is stronger at the poles. Since stronger gravity also implies larger rotation flux, we should expect 'pole-on' stars to emit more ultraviolet flux, and that is what Kodaira and Hoekstra (1979) found.

For the term B[e] see section 9.2.

It should be stressed that the description of a Be star spectrum depends very much on the dispersion of the plate material used and the epoch of observation. Since all Be phenomena are variable in time, studies of these stars must be based upon long series of observations. This point is well documented in the atlas of Hubert-Delplace and Hubert (1979). The atlas shows convincingly that a spectrum can pass in a few years from that of a normal star to a Be spectrum, have a shell phase and reappear as a normal spectrum. Descriptions of isolated spectra taken at a random phase should therefore be considered with due caution.

The material of the atlas has been used to set up a scheme for further subdivision of the Be stars; the reader is referred for details to the paper by Jaschek *et al.* (1980).

A final note should be made with regard to notation. A Be star is denoted as B3 IVe for instance, the 'e' standing for emission. Some confusion can arise because if the emission is present only at Hα, a classifier working in the classic region will see a normal spectrum and will not attach an 'e' to the classification, despite the fact that the star is a Be object. Shell stars are sometimes, but not by all astronomers, denoted Bep.

Infrared. We have already given in section 9.0 a list of the lines visible in this region; the most important lines are those of the Paschen series. This series is usually seen in absorption, except in early type Be stars, where it may be in emission. If in early type stars the Balmer emission is intense, usually some Paschen lines are also in emission. Furthermore if Paschen lines are in emission, the star has an infrared excess (Briot 1981).

Of other interesting elements, there are lines from oxygen and calcium. The O I triplet at λ7774 (72, 74 and 75) is usually present in absorption. When it is seen in emission it usually accompanies the emission of Fe II λ7712. O I λ8446 is more frequently in emission in early type Be stars, and its emission strength correlates with that of the Hα emission (Kitchin and Meadows 1970). The Ca II triplet at λ8498, λ8542 and λ8662 is difficult to separate from Paschen lines, except at high dispersion. The triplet appears sometimes in emission (Andrillat 1979; Briot 1981) but the emission strength is apparently not related to the behavior of other lines, nor is it concentrated in the early type stars.

More studies in the infrared region were made after the discovery of the existence of infrared color excess in some Be stars; see section 9.2 on B[e] stars.

Rotation. We have already mentioned that the spectral lines are usually wide and shallow. This can be clearly seen in the statistics by Hardorp and Strittmatter (1970) based upon Slettebak's data. Figure 9.11 allows us to compare the statistics of both normal stars and Be stars. We notice first that

Figure 9.11. $V \sin i$ for normal and emission line stars. In the two upper graphs is given the distribution of $V \sin i$ for two groups of normal stars, and in the lower part the distribution for Be and shell stars. From Hardorp and Strittmatter (1970).

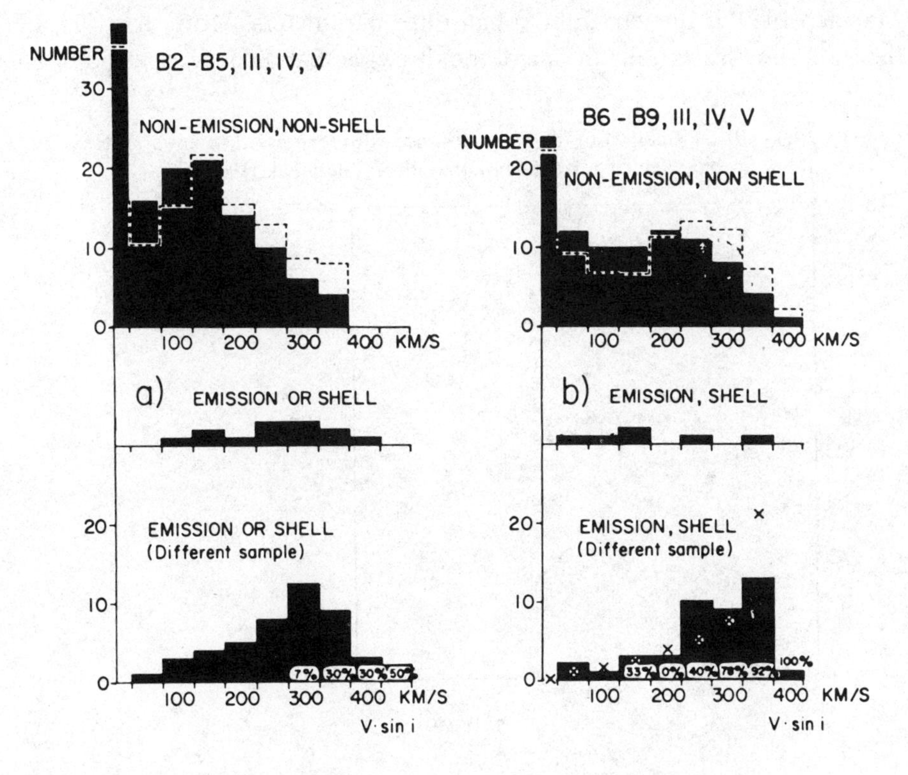

frequently the rotation of Be stars is fast and secondly that it differs from that of normal dwarfs. In fact the distribution of $V \sin i$ for Be stars is consistent with the hypothesis that all Be stars rotate at high speed ($\bar{V} \sim 375$ km/s) and that the observed distribution of $V \sin i$ is produced only by the random inclination of the rotation axis. The Be stars with low $V \sin i$ are what we have called 'pole-on' stars. There is thus a direct link between fast rotation and the fact that a B-type star has Be characteristics. Such a link is underlined by the fact that Be stars are most frequent at B2–B3, which is where stellar rotation comes closest to the 'critical' velocity.

A further link between rotation and Be characteristics exists in that the width of the Balmer emission is related to the value of $V \sin i$, as can be seen from figure 9.12, taken from Slettebak (1976). This relation is understandable if rotation is the reason behind the Be phenomenon.

Photometry. We start by observing that since the emission features in Be stars are rather narrow, their effect on wide band filter photometries is almost zero.

This is exactly what happens with UBV, in which Be stars are similar to B dwarfs. The only feature which comes out clearly in UBV (Feinstein and Marraco 1979) is the variability of all three parameters. More than 77% of the stars show variations in one of the three passbands over a period of 14

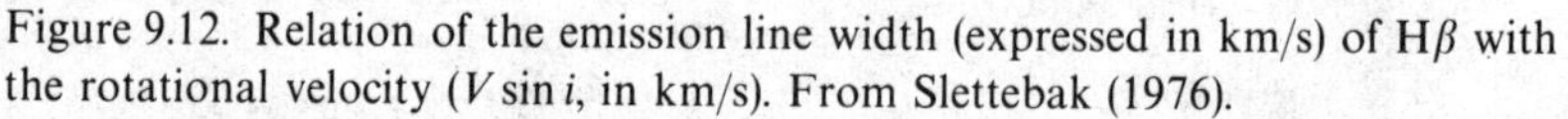

Figure 9.12. Relation of the emission line width (expressed in km/s) of Hβ with the rotational velocity ($V \sin i$, in km/s). From Slettebak (1976).

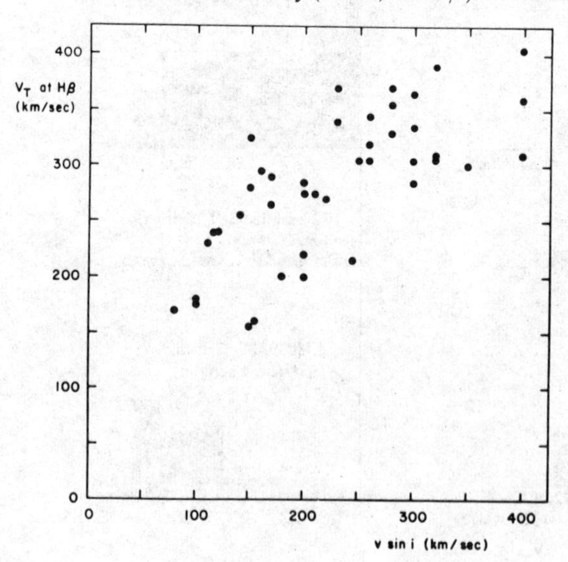

years; this means that over a longer period, probably *all* stars are variable. The practical insensibility of wide and intermediate passband photometry is also found in other photometric systems. For details see Golay (1974).

Clearly, it is better to work with narrow band filters centered upon Balmer lines and this has been done by several photometrists. The principle is to measure Hβ, for instance, and to derive the 'beta-index' related to the equivalent width (see section 5.2). The only difficulty is that if the emission is weak, and is superimposed upon a broad absorption, the photocell only perceives a weak Hβ line and attributes this to a high luminosity or an early type object. For this reason, it is safer to measure *two* Balmer lines, for instance α and β. Figure 9.13 gives the relation between both indices. We see that the emission line stars fall completely away from the normal relation indicated by the straight line in the upper right corner. It is thus possible to detect Be stars with a purely photometric technique, and to follow their behavior.

Photometry at longer wavelengths has been carried out by different astronomers. An important discovery was made by Johnson (1967) who found an excess in the infrared in many Be stars, when they are compared to stars of the same spectral type. This excess starts at 1 μm and increases

Figure 9.13. Relation between Hα and Hβ narrow band photometries. The continuous line corresponds to normal stars. Dots represent dwarfs, open circles giants and triangles supergiants.

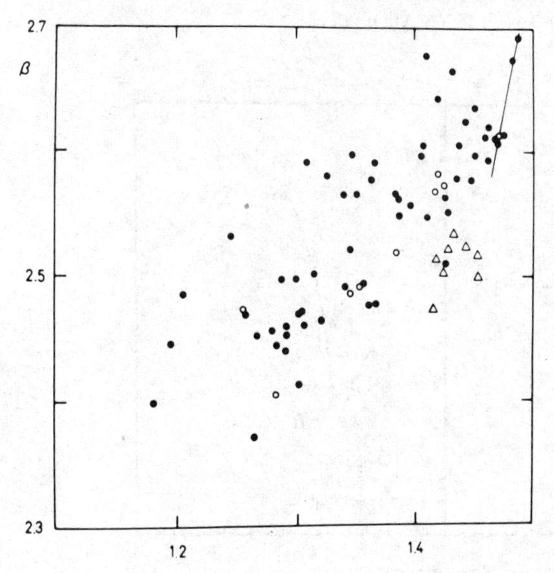

toward longer wavelengths. Figure 9.14 from C.W. Allen (1973) represents the measurements of the radiation in the interval 0.3–11 μm for two Be stars. In both cases the radiation does not fall off as predicted by the behavior of normal stars, but increases again to a broad maximum at long wavelengths. It is thought that a cool dust cloud in the vicinity of the stars is responsible for this. The dust cloud re-radiates, at longer wavelengths, the stellar flux.

It is possible to use this fact to discriminate stars with cool dust shells from normal stars. This is done most easily by constructing diagrams using measurements in the bands H (1.6 μm), K (2.2 μm) and L (3.5 μm). For normal B-type stars, H–K $\sim$ K–L within a few tenths of a magnitude. Stars for which H–K and K–L are both larger than $0^{m}.5$ are clearly stars with infrared excess. Among these are found predominantly B[e] stars, often (50%) associated with nebulosity (see figure 9.15).

Radio observations. Most of the radio observations carried out have been negative. Only a few emitters have been detected, and are all Be objects with associated nebulosity. These objects emit in the range 1–100 GHz. Furthermore, radio emission is only seen in objects with huge infrared excess (Purton *et al.* 1982).

Figure 9.14. Energy distribution in two Be stars. The points represent observations made at the mean wavelengths (at the top, broad band name; at bottom, λ in μm). The broken curve represents theoretical models (star + shell + free-free radiation). From C.W. Allen (1973).

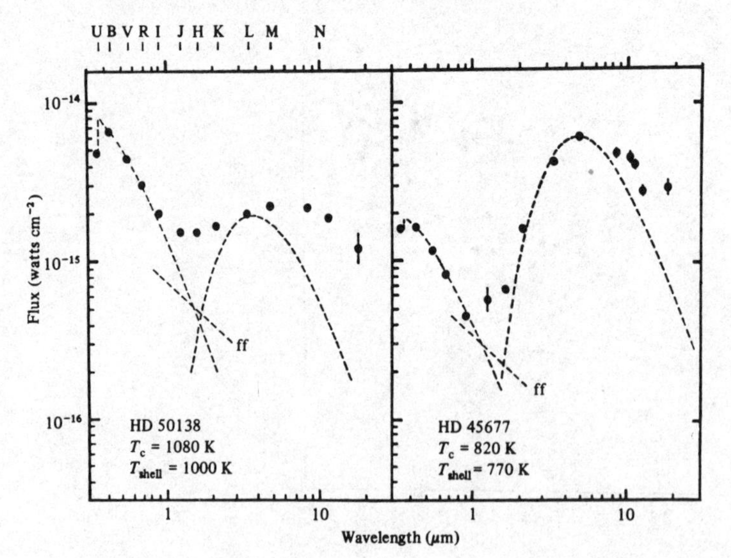

Be stars in binaries. As far as visual binaries are concerned, Be stars seem to have the same proportion of binaries as B-type dwarfs. With regard to spectroscopic binaries the problem is difficult because of the large rotational velocities, which seriously hinder radial velocity studies and the detection of their binary nature. Kriz and Harmanec (1975) suggest that all Be stars are interacting, i.e. close binaries; however up to 1986 a periodic radial velocity variation had been found in only a few cases. The interested reader is referred to the papers by Harmanec (1982) and Peters (1982) for more details.

Stars in clusters, frequency and absolute magnitude. Since many B stars occur in clusters, it is to be expected that Be stars also appear there, and this is confirmed observationally. Schild and Romanishin (1976) obtained, from a survey of 566 stars in 29 young clusters, a frequency of 7%. Moreover clusters allow us to obtain the absolute magnitude, which corresponds to stars on, or slightly above, the main sequence ($< 1^m$) (Mermilliod 1982a). This agrees well with the fact that most Be stars are classified spectroscopically as V or IV.

Distribution. The distribution does not differ from that of normal B-type stars. The number of Be stars with $m \leqslant 6^m5$ is 165.

Atlases and bibliography. Some Be stars are illustrated in all spectral atlases.

Figure 9.15. Location of Be stars in infrared photometric diagrams. The line *BB* represents the black body; in region *D* are found classical Be stars, whereas Be stars with associated dust clouds are found in region *C*. IR denotes the direction of interstellar reddening.

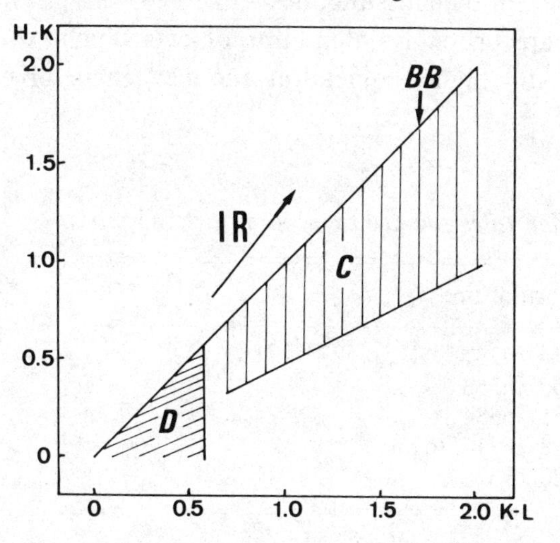

The first atlas devoted to Be stars was the above-mentioned Hubert-Delplace and Hubert (1979) which illustrates the behavior of 40 stars out of a sample of 120 Be stars followed over more than 20 years with the same instruments.

For further reading, we recommend

IAU Symposium N. 70 (1976) 'Be and shell stars'

IAU Symposium N. 98 (1981) 'Be stars'.

The bibliography on Be stars is updated regularly in the *Be star Newsletter*.

9.2 B[e] stars

A B[e] star is a Be star exhibiting forbidden lines in emission. The notation B[e] was apparently first used consistently by Wackerling (1970).

B[e] objects were known long ago, but for a long time their faint magnitudes ($m > 10$) prevented further studies. Only the discovery that many of these objects have associated large infrared excesses brought them back to attention. The most extensive survey was carried out by Allen and Swings (1976), who studied about 65 objects.

The most common forbidden emission lines found in these stars are listed in table 9.14; they correspond to ions with ionization potentials of the order of less than 25 eV. Not all of the lines listed appear in all objects; the most frequent are those of [Fe II] and [O I]. For an illustration, see figure 9.16. A typical object of the group is HD 45677, studied in detail by Swings (1973).

The group constitutes an extension of the Be stars with Fe II lines in emission, which are found in spectral types B0–B3. There is a smooth transition toward stars which occasionally show [O I] and then toward stars which show forbidden lines of low ionization species, as listed in table 9.14. Sometimes forbidden and permitted emission lines – like Fe II and [Fe II] – are found together. The groups probably extend into objects which exhibit forbidden lines of ions of still higher ionization energies; these objects

Table 9.14. *Forbidden lines found in emission in B[e] stars.*

Element	Important lines
[Fe II]	$\lambda 4244, \lambda 4287, \lambda 4415, \lambda 5273, \lambda 7155$
[O I]	$\lambda 6300, \lambda 6363$
[N II]	$\lambda 5755, \lambda 6584$
[S II]	$\lambda 4068, \lambda 6717, \lambda 6730$
[O II]	$\lambda 7320, \lambda 7330$

however grade smoothly into planetary nebulae which lie outside the scope of this book.

Many of the B[e] objects are also radio emitters (Purton *et al.* 1982).

9.3 CNO stars

A CNO star is a late O-type or an early B-type star (O8–B4) in whose spectrum the lines of some of the elements C, N and O are weaker or stronger than in the standard stars. The first stars of this group were found by Jaschek and Jaschek (1967); later the group was enlarged and described by Walborn (1970, 1971a).

Since the lines of C, N and O are used as classification criteria, the segregation of members depends on precise classification based upon spectra of higher dispersion than those used in the MK system. Typically 40–60 Å/mm are required, since at 100 Å/mm many of the lines of C, N and O are blended with other features.

Since all three elements show a positive luminosity effect, it is clear that it is easier to detect supergiants with weak CNO lines or dwarfs with strong CNO lines; the other two cases are more difficult to recognise. In order to ensure that luminosity-sensitive lines are not misused, the luminosity has to be derived from lines other than those of CNO. Usually lines of He and Si are used since the hydrogen lines are too broad for classification. Moreover,

Figure 9.16. A late B-type B[e] star. The star HD 190073 is compared to a B9.5 object. The Ca II lines of HD 190073 are at least partially interstellar. Notice the emissions λ4244 and λ4287 of [Fe II] and λ4233 and λ4583 of Fe II. Observe also the central emission in Hβ and Hγ.

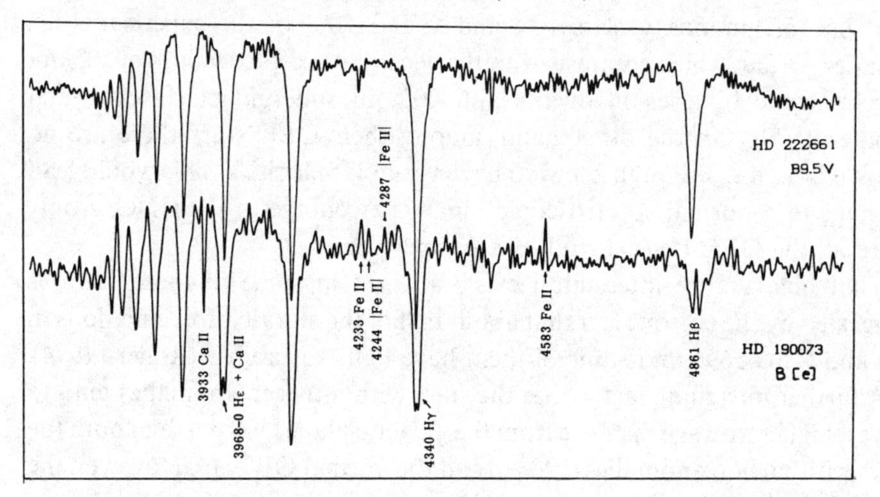

because the three elements C, N and O do not reach their maximum strength simultaneously, misclassifications in spectral type may occur: O II has its maximum at B1, N II at B2 and C III at B0.

A certain number of objects have been put into this group by various observers, with some disagreement over isolated cases. The disagreement is most often due to the lack of consistency of the spectral range. Walborn (1976) has defined a number of standards for the different types of anomalies at intermediate dispersion (40–60 Å/mm). He observes that taken at 100 Å/mm these anomalies are just barely perceptible and the stars can therefore be considered as 'normal stars'. Such a procedure is dangerous because stars known to be peculiar in one way or another should not be used as standard stars.

Walborn (1971b) suggests that these peculiar stars can be divided into two groups; one with enhanced nitrogen and weakened carbon and/or oxygen lines – which he calls OBN stars – and the other with deficient nitrogen and enhanced carbon lines which he calls OBC stars, the OB in both designations standing for the spectral types in which they can occur.

Jaschek and Jaschek (1974a) challenged this schema, arguing that it seems artificial to introduce it because O-rich stars are also found and because there might be spectrum variability in a large proportion of these stars. For the first argument, the term 'CNO stars' rather than 'OBC and OBN stars' is now used. The second point, namely the variability, has not yet been checked on suitable material.

Bolton and Rogers (1978) have analysed some of the stars for radial velocity variations. By using their own material and that of colleagues, they conclude that at least 50% of the OBN stars are spectroscopic binaries. All of the stars of luminosity classes IV and V and 50% of the supergiants are binaries – a fact which contrasts with the generally accepted values of 50% for spectroscopic binaries in dwarfs and 20% in supergiants (Jaschek and Gómez 1970). On the other hand, among their OBC stars there are no known binaries, although a few do have variable velocities. This would lead perhaps to a normal spectroscopic binary percentage in the CNO group, since all the OBC objects studied are supergiants.

That binary type interaction exists at least in some of these stars, is suggested by the curious fact that about half of the stars exhibit emissions in Hα and C III λ5696 and some of them have light variability (Rogers 1974).

A further intriguing fact is that the stars with nitrogen anomalies tend to have a higher average distance from the galactic plane (300 pc) than both the stars with carbon anomalies (120 pc) and the normal OB ones. However, the average radial velocities are low, so that they cannot be runaway objects.

Another possibility is that the absolute magnitude of these objects might be different.

Some of these objects belong to associations; Schild (1985) provides a list of the nitrogen anomalous stars in clusters. He notes that both nitrogen poor and nitrogen rich objects can be found in the same association.

In the Small Magellanic Cloud Dubois, Jaschek and Jaschek (1976) found that instead of an enhancement of one of the three elements C, N and O, all three are weakened in the B-type supergiants.

CNO stars on the whole are rare. About fifty are known, all brighter than $V = 10$. A conservative estimate puts their proportion among O8–B4 stars at the 5% level.

9.4 Helium abnormal stars

In some B-type stars the helium lines are stronger or weaker than in normal stars. These stars are called respectively He strong and He weak line stars; for brevity, helium strong and helium weak stars, and, when taken together, helium abnormal stars.

He strong stars are defined as stars having abnormally strong lines of He I; the He I lines can be comparable to or much stronger than the H I Balmer lines, and the Balmer lines may even be invisible at classification dispersion. When the Balmer lines are invisible, the star is called

H poor = H deficient = extreme He star

These stars will be discussed separately under 'H deficient stars' (section 9.12). When the He I and the H I lines are comparable in strength, they are called

He strong = intermediate He rich

The latter designation distinguishes them from 'extreme He stars'.

The first helium strong star was discovered by Popper (1942), but the group as such was defined by Bidelman (1952). The first systematic survey was made by McConnell, Frye and Bidelman (1970, 1972) in the southern sky and constitutes our only firm basis to evaluate the number of these objects. Almost all these stars correspond to B2 V except of course for the He I lines, which are much stronger. The spectrum of helium strong stars can be described as follows.

Hydrogen lines are present, but He I lines are of comparable strength, as can be seen by comparing He I λ4471 with H I λ4340. All series of He I are well developed far down into the ultraviolet. The singlet/triplet ratios in the

He I series are similar to those in normal dwarfs, for instance λ4387/λ4471. Balmer and Paschen discontinuities are present, but are weak. Usually lines of Si, C and N, and sometimes of S and Al, are present, and their intensities correspond to what might be expected in early type dwarf spectra (B0–B3). HD 64740 or HD 96446 are typical stars of the class.

The spectra of a number of He strong stars are illustrated by Kaufmann and Theil (1980), and Walborn (1983b) has given equivalent width measurements for many of them.

He weak stars are defined as stars having abnormally weak helium lines. They are found principally in middle and late B-type stars, where at classification dispersion there are few lines present belonging to elements other than hydrogen and helium. Because of the weak helium lines, such a star is classified as being of later spectral type than it really is. Discovery is not easy, except if either the color is measured or the star is observed at higher dispersion. If the color is taken into account (Jaschek, Jaschek and Arnal 1969) the spectral type inferred from the continuum is in disagreement with the strength of the helium lines, and the star is recognized as being peculiar. If medium dispersion is used (Garrison 1967) the hydrogen profiles may reveal inconsistencies with the helium line strength. Finally with high dispersion (for instance Bidelman 1960), fainter lines of other elements become visible; from their presence the helium line strength can be judged as incompatible with that in normal stars.

If spectra at classification dispersion are analyzed the definition can be made precise in the sense that He weak objects are those B-type stars in which the helium line strength does not correspond to the spectral type as given by the hydrogen lines (and the color), the helium lines being weaker than expected. Since the hydrogen line strengths are difficult to estimate, colors are usually preferred. Stars showing weak helium lines may further show enhanced lines of Si II or Mn–Hg, in which case they are called Ap stars (see section 9.5). It is clear that because of unavoidable errors in classification, some marginal Si or Hg–Mn stars will appear in the He weak group.

He weak stars occur in the range B2–B7, being more frequent at later types. If the types taken from the literature (i.e. according to the helium lines) are considered, they are systematically two or more spectral subtypes later.

The group was first defined by Garrison (1967) on 90 Å/mm plates, although isolated examples of the group had been known before. Sargent and Searle (1968) at 50 Å/mm define a group of Bw stars, which are essentially identical to the ones considered here.

A prototype of these stars is HD 191980, described by Keenan, Slettebak and Bottemiller (1969). In this star the Balmer lines correspond to B5, the C II lines to B3, and the ratio He I λ4471/Mg II λ4481 to B8.

Besides the He strong and the He weak stars, there are also *He variable stars* in which the strength of the He I lines varies periodically. The classic example of these HD 125823 (Jaschek, Jaschek and Kucewicz 1968). The helium lines vary enormously in strength, from those of a B2 to those of a B9 star. All other lines, however, remain constant, and photometric and radial velocity variation are at most marginal (Norris 1971). In some of the stars, Hα is seen in variable emission, as in σ Ori E (Walborn 1974). Landstreet and Borra (1978) suggest that the Hα emission exists only in rapid rotators.

Generally the He variable stars fall between the (early type) He rich and the (later type) He poor stars.

Further subdivisions of the He weak stars were proposed by Jaschek and Jaschek (1974b) and by Baschek (1975) but because of the variability of many He weak stars it seems premature to attach too much weight to them. Let us note simply that some strange objects with strongly enhanced Ga II and P II lines belong to this group.

Photometry. UBV photometry has been performed on the majority of the He stars. Figure 9.17 shows the rather neat subdivision of the stars into the two groups, with the variables falling toward the limit of the two groups. The stars exhibit resonable amounts of interstellar reddening. uvby photometry shows a similar behavior; Osmer and Peterson (1974) find the separation at [u–v] = + 0.40.

Narrow line photometry has been carried out on different features. For the Balmer lines we give in figure 9.18 a graph of β versus [u–b]. He strong stars fall below the main sequence as expected, because the H I lines (Hβ) are weaker than normal. As expected, He weak stars fall on the main sequence. What is unexpected is that a certain number of them fall below the main sequence. Feinstein (1978) found a similar effect. This author measured with narrowband filters Hα, Hβ and Hγ in He weak stars and found that whereas the UBV and the UBVRI indicate the same spectral type, the spectral type deduced from α and β line photometry is in several cases *earlier* than the one corresponding to UBVRI photometry.

Narrow line photometry on the He I line λ4026 was done by Nissen (1974, 1976), Pedersen (1976, 1979) and Pedersen and Thomsen (1977), in connection with uvby photometry. The main result is that He abnormal stars can be singled out easily from normal objects. In view of the difficulties of detecting He weak stars, the photoelectric method is an objective way of deciding difficult cases, except for stars later than about B8, where the λ4026 line becomes too weak to be usable in photometry.

Narrow band photometry on the λ5200 flux depression was done by Maitzen (1981). His main conclusion is that He weak stars show the same depression as Ap stars; however, it gets weaker toward earlier types and disappears for He strong stars.

As a byproduct of photometric observations, it became clear that a number of He abnormal stars are variables in light and/or in λ4026 strength.

Figure 9.17. UBV photometry of helium abnormal stars. Crosses, helium strong stars; points, helium weak stars; circles, variables. The straight line represents the main sequence and the arrow the reddening line. From Jaschek and Jaschek (1981).

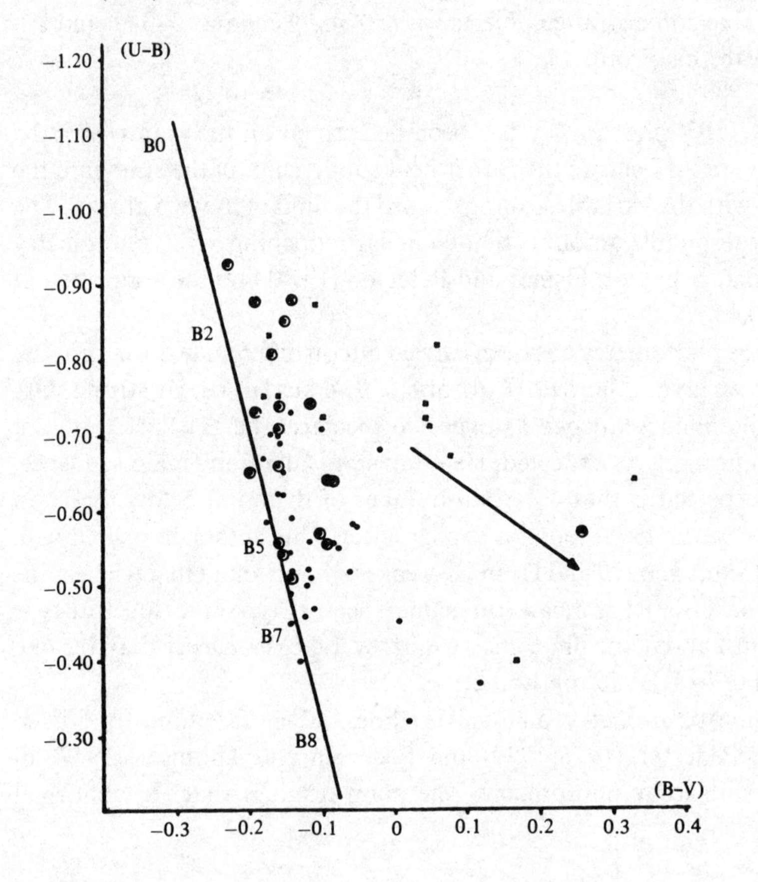

About one third of the stars are variables, as can be seen from the following values:

	Number	Variables	Percentage
He strong stars	19	7	37 ± 11
He weak stars	44	13	29 ± 7

Pedersen (1979) stresses the fact that helium line variability is usually combined with variable Hα line intensity, photometric variability and a variable magnetic field. Radial velocity variations are, however, only seen in some stars.

As far as the λ4026 variability is concerned, which is generally more marked than the variability in uvby, the light curves are in a few cases

Figure 9.18. Strömgren photometry of helium abnormal stars. Line index versus u–b color. The straight line represents the main sequence. Crosses, helium strong stars; points, helium weak stars. From Jaschek and Jaschek (1981).

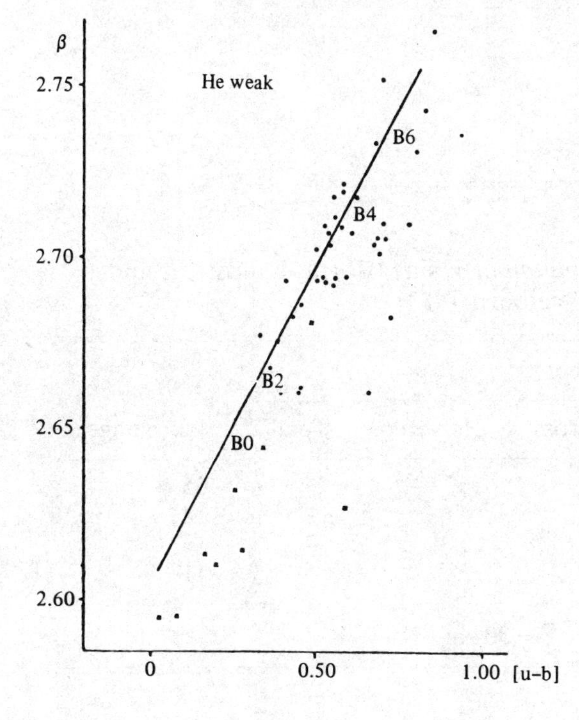

sinusoidal, but in general they show a complicated behavior. There is no clear-cut relation between λ4026 strength and the light curve.

Magnetic fields have been measured in several objects. Borra and Landstreet (1979) analysed several He strong stars and found strong longitudinal magnetic fields in two-thirds of them (six out of nine stars analysed). The fields are relatively high in σ Ori E, (+ 2800 to − 2200 G) and are high even in stars with large rotational velocity. Borra, Landstreet and Thompson (1983) have measured the magnetic fields in a large sample of He weak stars. They find that many of them (40%) have a magnetic field. The fields appear less strong than in He strong stars, but are still stronger than in Ap stars.

Rotational velocities are available for many He abnormal stars. The statistics are given in table 9.16. The distribution of $V \sin i$ is rather different from that of normal B-type stars, where $150 < \overline{V \sin i} < 200$ km/s, for He stars, $\overline{V \sin i} = 71$ km/s.

Table 9.15. *Distribution of periods of He variables.*

Log P (days)	Number
− 0.5 to 0.0	2
0.0 to 0.5	7
0.5 to 1.0	4
1.0 to 1.5	3
Modal value ~ 2 days	

Table 9.16. *Distribution of* $V \sin i$ (Borra, Landstreet and Thompson 1983; Walborn 1983).

$V \sin i$ range	Number			
	He strong	He weak	Total	Percentage
0–50	9	15	24	55
51–101	3	5	8	18
101–151	5	2	7	16
151–201	3	2	5	11
$\overline{V \sin i} = 71$ km/s	$\bar{V} = 90$ km/s			

A correlation exists between $V \sin i$ and the period of variability, as in classical Ap stars.

Radial velocities have been measured for a number of stars, but usually only to obtain an average velocity. At present all radial velocities known are small ($\leqslant 40$ km/s).

Binaries. Some He abnormal stars have companions. Jaschek and Jaschek (1981) list the known cases, but our knowledge is very limited. He abnormal stars are found in clusters and associations, as for instance in Sco-Cen (Garrison 1967). For other cases see Ciatti and Bernacca (1971), Bernacca and Ciatti (1972) and Abt (1979). Since all these associations are young, the He abnormal stars are obviously young objects. Let us note that the Ori OB1 association contains two helium weak stars (Abt 1979) and one helium rich object.

The frequency of the He abnormal stars is difficult to ascertain, but is probably not more than a few percent of the normal B-type stars. (See also Egret and Jaschek 1981.) On the other hand Nissen (1974) found in field stars and in Sco-Cen 6% of stars He abnormal, which is of the same order of magnitude.

9.5 Bp and Ap stars

A B-type or A-type star in which lines of one or several elements are abnormally enhanced is traditionally called an Ap star, although the majority of the stars are really of B-type. Therefore we prefer to speak more appropriately of Bp and Ap stars, and since they form a unique group, we describe them together here.

The first systematic study of the Bp and Ap stars was undertaken by Morgan (1933) although isolated members of the group were known earlier. Morgan used 30 Å/mm material in the $\lambda\lambda$3800–4800 region. He distinguished six types of stars, called respectively Mn, λ4200, Si, Cr, Eu and Sr stars, the order of the groups being a temperature order, except for the Si stars which do not have a definite place in the sequence. The lines of the elements which are most prominently enhanced in each group are given in table 9.17.

Later Bidelman (1961) and Jaschek and Jaschek (1962) showed that λ4200 is a line arising from a high excitation level of Si II, so this group is really a subgroup of the Si stars. Bidelman (1962) showed that a line at λ3984 usually found in Mn stars belongs to Hg II, so that the Mn stars are sometimes called Mn–Hg or Hg–Mn stars.

The first difficulty with the schema arises when more than one element is enhanced. Morgan himself called some stars Cr–Eu, but usually he assigned the group according to the most prominent element. However, different observers have different opinions on this point, because many stars exhibit spectral variability (see below). As a consequence the classifications given for a particular star frequently disagree. Furthermore, some spectroscopists provide a list of the most enhanced elements, a fact obviously dangerous since it depends on the plate factor. As a rule astronomers using lower plate factors provide descriptions, whereas objective prism observers use the classical groupings.

As a result, the classification schema becomes less and less practical. Osawa (1965), in a study of 244 stars at 60 Å/mm, used 13 groups, made up mostly from combinations of the classical ones but adding some new ones like 'Hg–λ4077'.

More recently, a tendency to simplify has set in, which is well illustrated in Jaschek and Jaschek (1974b) and Preston (1974). In the first paper, only three broad groups were retained, namely the Si, the Mn and the late type group. Since by now it seems that this schema is better suited for classification purposes, let us describe the groups briefly.

(*a*) *Silicon stars.* The enhanced lines are λλ4128–30, λλ3856–62–54 and, in the earliest stars, λ4200 and λ3954. In the red regions λ5041 and λ5056 can be added to this list. A fact pointed out by Deutsch (1956), and which is regularly rediscovered, is that there is discrepancy between the UBV color of the star and the spectral type, which is usually assigned on the basis of the He I line

Table 9.17. *Characteristic lines of elements defining different groups.*

Si-λ4200	Si II λ3856, λ3862, λ3853; λ4128, λ4130; λ4200, λ3954 (high excitation levels)
Mn	Mn II λ4206, λ4136
Cr	Cr II blend at λ4111. In addition λ4171 and λ4233 are also seen. Observe that these lines are not the strongest lines according to the laboratory work and that the latter are blended with Fe. λ4233 is not even the strongest line of the multiplet
Eu	Eu II λ4205, λ4129, λ3919, λ3930. At low dispersion λ4129 should be used with caution, since it may be due to the blend of Si II. In this case λλ3853–56–62 may be used as discriminants for Si II
Sr	Sr II λ4077, λ4215. The presence of λ4077 alone is insufficient evidence, since in Sr stars the line can be due to Cr II λ4076, λ4077

strength. Therefore helium is weak if the color is taken as correct, or the color is too blue if He I lines are taken as correct. However, in the UV region, the spectral types correspond well to the spectral type in the blue and not to the UBV colors (Cucchiaro *et al.* 1978).

The most pronounced Si-λ4200 stars can be singled out at 300 Å/mm (Honeycutt and McCuskey 1966), but for recognition of the less pronounced stars, plate factors of about 100 Å/mm are necessary.

To examine the behavior of other elements, plate factors of 20 Å/mm or less have to be used. The picture then becomes blurred because a variety of elements can be enhanced, and they are usually not similarly enhanced in all stars. Examples of this are Cl, Mg and the heavy elements. But Mn is never enhanced in Si stars, and Eu appears very sporadically.

The question of whether a definitive border exists between Si and normal stars has been examined by Durrant (1970). He concludes that this is not the case, although in general Si stars do not rotate rapidly ($V \sin i < 100$ km/s), a fact that distinguishes them from normal stars.

(*b*) *Manganese stars.* Since most of the manganese lines are rather faint, a plate factor of 40 Å/mm or less is needed to recognize members of this group, although extreme members can be discovered at 100 Å/mm. Sometimes the Hg II λ3984 line is used to detect Mn stars. An analysis of the behavior of other elements shows that He is abnormally weak when compared with standards in all stars, and that Ga II, Y II, Hg II and Pt II (Dworetsky 1969) are present in the majority of them. Si usually is normal, rare earths are absent, and the Balmer lines are narrow.

It should also be mentioned that because the lines are sharp, it is possible to measure in some heavy elements like Hg and Pt the minute wavelength shifts produced by anomalies in the proportion of isotopes. The reader is referred for more details to the papers in the 23rd Liège colloquium (1981).

(*c*) *Late Ap-type stars.* We place in this group all objects in which heavier elements like Cr, Sr and Eu are enhanced, in combination with others. This group makes up about half of all Ap stars, of which about one-third are known spectrum variables. Extreme stars of this type can be recognized at 300 Å/mm (Honeycutt and McCuskey 1966), although detection is reasonably complete only at about 100 Å/mm.

Rotation of the objects is slow ($V \sin i < 100$ km/s); the hydrogen profiles suggest main sequence objects.

Examination of these stars at high dispersion reveals the presence of absorption lines of many heavy elements and in fact in these stars all the rare earths and heavy elements have been found, up to the end of the periodic table (uranium was found by Jaschek and Malaroda 1970). Probably, but this has not been completely established, elements of nuclear fission are also present (Kuchowicz 1975). Finally, up to now no systematic pattern has been found which could lead to a further subdivision of this group (Cowley 1979). Figure 9.19 illustrates some peculiar stars.

Before closing this section let us give a very definite warning. The phenomena displayed in Bp and Ap stars are very complex and it can be said that even at 20 Å/mm no two stars really look alike. Therefore any classification scheme is open to criticism, in the sense that exceptions can always be found. There are freaks like HD 101 065 (Pryzybylski 1963) where the strongest lines belong to holmium and where all other elements lighter than barium are absent, except H and Ca. Such a star cannot be accommodated in any classification scheme! The object is certainly unique, but it is far from being the only object which is hard to accommodate.

Photometry. Broad band photometry has been carried out for many Bp and Ap stars. A summary of the main results is given in table 9.18 and figures 9.20 and 9.21 taken from Cowley, Jaschek and Jaschek (1970).

As can be seen from figure 9.20, the stars tend to fall on or very close to the main sequence, except the Hg–Mn and the Sr–Cr–Eu stars. Figure 9.21 shows that the colors of the different groups overlap considerably, so that the correlation of table 9.18 (color and groups) is only a statistical one. The interesting point that Sr–Cr–Eu stars deviate from the relation for dwarfs can be undertsood easily through increased blanketing by the large number of (enhanced) lines present in the spectrum.

Intermediate photometries, like uvby for instance, can use two features, namely the increased blanketing in late type Ap stars and the anomalous Balmer discontinuity; the latter was discovered by Glagolevski (1966). A larger m_1 and a smaller c_1 index is then expected. This is what Cameron (1967) found; he was able to separate Ap- and A-type normal stars by uvby photometry alone. The method fails only for Hg–Mn stars. For a more detailed study of the photometric behavior of Ap stars, see Adelman (1981) and Hauck (1975).

Adelman (1975) and Maitzen (1975) discovered that the continuous spectrum of the Ap stars presents depressions at λ4100, λ5200 and λ6700. Of these the most easily measurable is λ5200 which is a 300–400 Å wide feature, whose equivalent width can be up to 20–30 Å (figure 9.22). This feature can

Figure 9.19. Spectra of Bp and Ap stars. The tracings illustrate a Si-type object (HD 112413) compared with a B8 v star, and a Sr–Cr–Eu object compared with an A0 v star. For explanation see table 9.17.

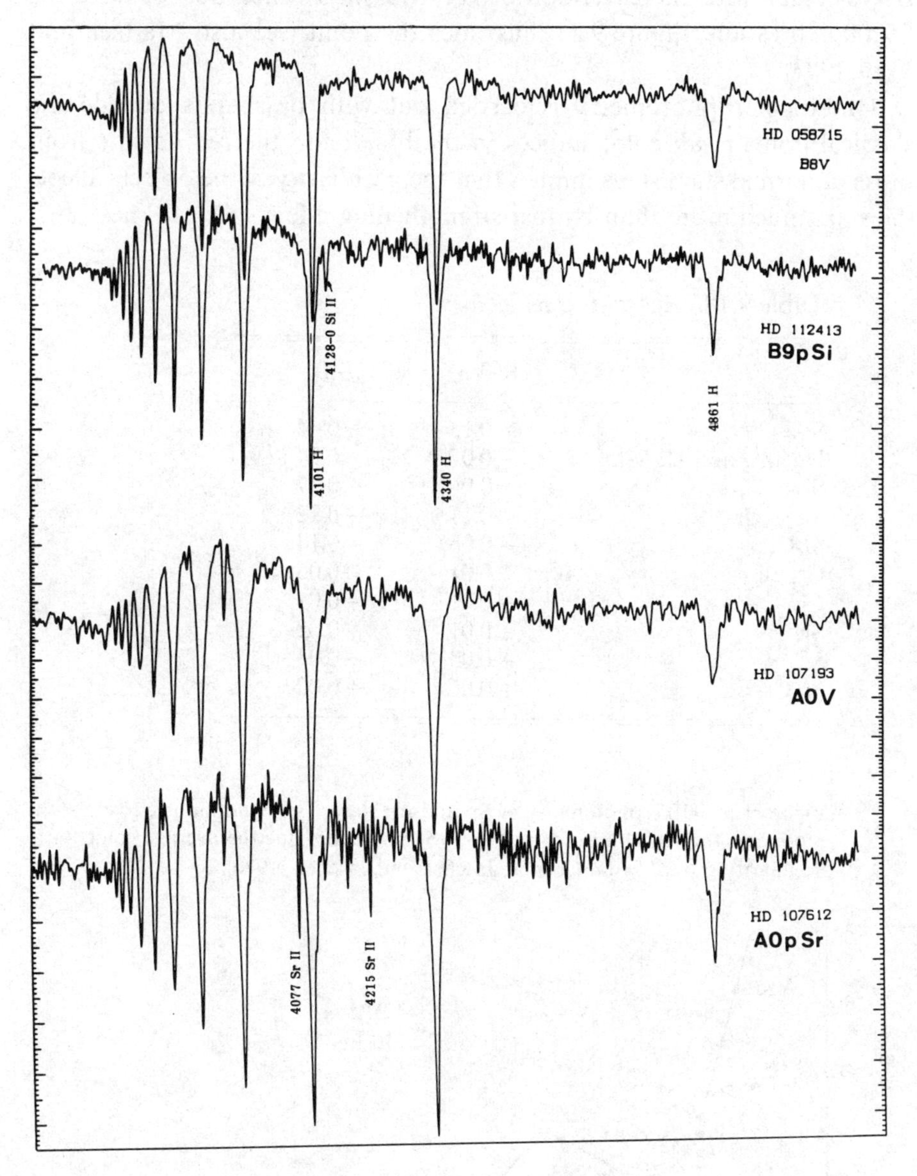

be measured with interference filters at $\lambda 5000$, $\lambda 5200$ and $\lambda 5400$, and an *a* index can be defined. This *a* index permits the separation of Ap and normal A-type stars, with the exception of Hg–Mn stars which do not have the $\lambda 5200$ depression. Figure 9.23 illustrates the point (see also Maitzen and Vogt 1983).

If multicolor photometry is carried out with passbands outside the classical domain, all color indices in the blue or in the red deviate from those of normal stars. This implies that the peculiarity of the objects affects the stars much more than by just strengthening a few spectral lines.

Table 9.18. *Average UBV colors.*

	B–V	U–B
Si-λ4200	−0.13	−0.44
Hg, Mn and HgMn	−0.07	−0.34
Si	−0.06	−0.27
Si, Sr, Cr,	−0.11	−0.22
Si, Cr	−0.05	−0.14
Cr, Eu	−0.01	−0.06
Cr	−0.01	−0.01
Sr, Cr	+0.01	+0.02
Sr	+0.03	+0.03
Sr, Cr, Eu	+0.06	+0.02

Figure 9.20. UBV photometry of Bp and Ap stars. The continuous curve represents the relation for dwarfs and the points provide the average color for a peculiarity group. From Cowley, Jaschek and Jaschek (1970).

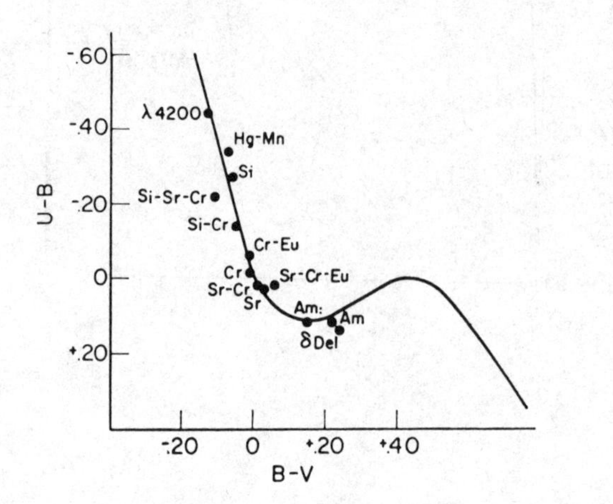

Infrared. No outstanding features have been found in the infrared up to now. Usable lines include N I ($\lambda\lambda$8680–86) and O I ($\lambda\lambda$7774 and 8446). N I is usually very weak or absent. The O I triplet was first studied by Keenan and Hynek (1950) and found to be weak or absent in the late Ap stars. Mendoza (1977) used narrow band photoelectric filters and confirmed this result for the 'late-group'. Hg–Mn stars are again an exception. For more details see the proceedings of the workshop *Ap stars in the infrared* (1978).

Ultraviolet. Bp and Ap stars are easy to detect in the region below λ3000, through very heavy line blanketing (Cucchiaro *et al.* 1977) or specific anomalous features. As an example of the latter let us quote the Ga II line λ1414 characteristic of Mn stars (Takada and Jugaku 1981).

Rotation. We have mentioned that Bp and Ap stars rotate slowly, as shown

Figure 9.21. UBV photometry for Bp and Ap stars. Frequency of the different (U–B) values for different subgroups of peculiarity. From Cowley, Jaschek and Jaschek (1970).

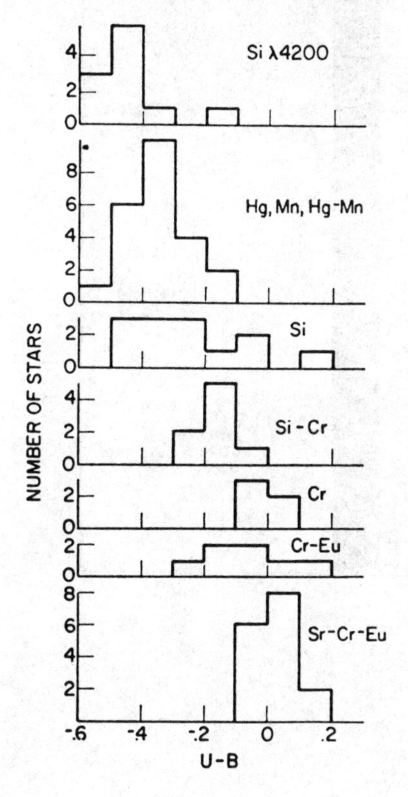

Figure 9.22. The depression of the continuum at λ5200 in Bp and Ap stars. The figure represents scans of some peculiar stars. The black areas represent the position of the narrow band filters used to measure the λ5200 depression. The vertical scale is given by the segment indicating 5% of the intensity. From Maitzen (1975).

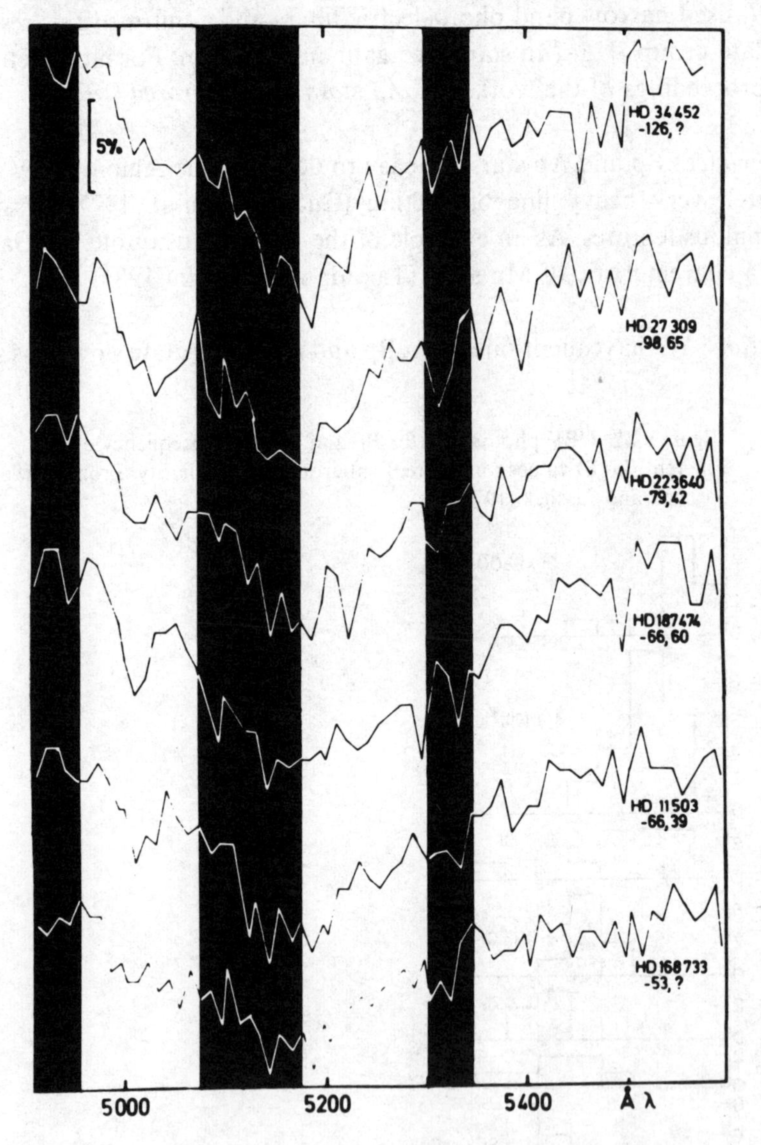

by Slettebak (1954); table 9.19 gives the results for a sample of nearby Ap stars, taken from Jaschek and Jaschek (1974b).

Abt, Chaffee and Suffolk (1972) conclude from the study of a large sample of Ap stars that they rotate on average one-quarter times as fast as normal A stars, and that slow rotation is a necessary but not a sufficient criterion to produce an Ap star.

Table 9.19. *Rotation characteristics for Ap stars.*

Group	$V \sin i$ (km/s)	Range (km/s)
Si	46	16–105
Hg-Mn	29	5–100
Cr-Eu-Sr	30	5–85

Figure 9.23. Photometric separation of normal and peculiar stars with the λ5200 index. Open circles, peculiar stars; filled circles, normal stars. The broken line represents the average value for dwarfs in which there exists no depression. From Joncas and Borra (1981).

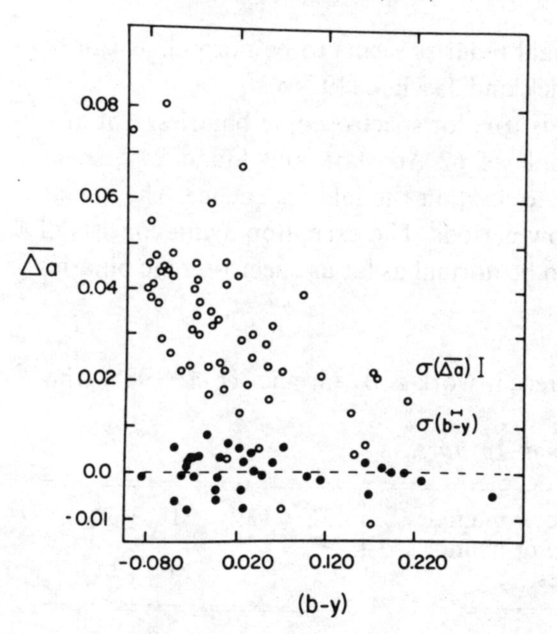

Magnetic fields. The existence of strong magnetic fields in Ap stars was discovered by Babcock (1947) through the Zeeman effect. As we have seen in section 2.3.3 the Zeeman effect is generally a small fraction of one angstrom. This means that magnetic fields can only be measured in stars with low rotational velocities, because high rotation completely blurs the lines. Due to the observing errors, the lowest field measurable is of the order of 3×10^2 G and the largest field so far measured is of 4×10^4 G. Unfortunately no close association exists between the size of the field and spectral peculiarities; stars with very large fields do not exhibit outstanding line enhancements. So far about five hundred magnetic fields have been measured, and we can conclude that *all* Ap stars have a magnetic field, except the Hg–Mn stars, where the field is very weak or non-existent.

Photoelectric scanning methods have become available only recently. These measure the circular polarization in the violet and red wings of a strong line like Hβ, permitting the intensity of the average longitudinal magnetic field to be obtained (Borra and Landstreet 1980). Provided $V \sin i < 300$ km/s, the measures are unaffected by stellar rotation. Some evidence was found that rapid rotators might have surface magnetic fields smaller than those of the slow rotators.

All magnetic fields observed so far are variable (see also the section on spectrum variables). The magnetic field measurements have been collected in a catalog by Didelon (1983).

Binaries. The proportion of visual binaries seems to be normal, as can be seen from table 9.20 from Jaschek and Jaschek (1974b).

We do not know if the same is true for spectroscopic binaries. Abt and Snowden (1973) studied a sample of 62 Ap stars and found 24% spectroscopic binaries as compared to 45% on the main sequence. The deficit appears mostly in stars with slow periods. The exception again consists of the Hg–Mn stars, which seem to be normal as far as spectroscopic binaries are concerned (Aikman 1976).

Absolute magnitude. The most recent work is by Grenier *et al.* (1985) who

Table 9.20. *Visual binaries among Ap stars.*

Ap	14 stars	6 double or multiples	43% ± 13%	31% ± 5%
	56 others	25 double or multiples	29% ± 5%	
Normal	100	29 doubles		29% ± 5%

discussed all available information. The main result is that Ap stars have the same luminosities as dwarfs with the same UBV photometric indices.

Frequency. The frequency of Ap stars has been investigated several times; we will quote here the results of Hartoog (1976). He computed the frequency of Ap and marginal Ap stars with respect to the number of dwarfs in the range B5–A5 and found 8%.

Other authors have found similar frequencies, usually about 10%, although as explained above, Mn stars are more difficult to detect and therefore the known percentage may be a lower limit to the true frequency. As for the relative frequency of different subgroups, we quote the results from Cowley, Jaschek and Jaschek (1970) that the Si and the Mn–Hg groups each contain about 25% of the stars while the late type group has 50%.

Clusters and associations. Ap stars occur in clusters and associations and from this fact the age of the stars can be estimated. The largest effort in this direction was made by Abt (1979), who analyzed 455 stars in 12 open clusters and associations. He observed Ap stars in clusters with ages between 10^6 and 5×10^8 years.

The percentage of Ap stars in clusters does not differ from that of field stars and there is an age effect in the sense that more Ap stars appear in older than in younger clusters. Abt and Cardona (1983) found a similar effect in visual doubles, which can be regarded a mini-clusters.

Because of their presence in young open clusters and associations, and their low radial velocity, the Ap stars must belong to population I. It is therefore surprising that in the past few years a few objects have been found which have Ap characteristics and a large space velocity (Jaschek *et al.* 1983).

Number, distribution on the sky. Ap stars are common objects. Down to $m = 6^m5$, 250 are known, and down to $m = 9$, about three thousand. Because of possible bias against the recognition of Hg–Mn and marginal peculiarities, the latter number is likely to be a lower limit to the true number. The distribution on the sky of the Ap stars follows that of the normal late B- and early A-type dwarfs.

Spectrum variables. We have already mentioned the fact that many Ap stars are spectrum variables. The only notable exceptions are the Mn stars in which no spectral variability has ever been found. The classical technique (Guthnick 1931; Deutsch 1947) is to take a series of plates of the star and to make estimates by eye of the intensity ratio of neighboring lines that are seen to vary over the series, for instance Fe II λ4233/Eu II λ4206. The resulting values of the ratio are then plotted as a function of time, and if enough plates are available, the period can be determined. Other authors prefer to work on microphotometric tracings of the spectrum rather than visual estimates.

Usually a number of elements display intensity variations with the same period, but with differences in phase (figure 9.24). So for instance in some stars the iron group elements Cr, Ti and Fe vary together, but with a different phase from that of the more rare earth elements. Other elements do not vary at all. In other stars the grouping of the elements is different and no clear general rule has emerged as yet.

As for periods, there are more than two hundred known, with values of between half a day and several thousand days, and a modal value of a few days. The variations are explained by means of a selective grouping of

Figure 9.24. Visual estimates of line strength (ordinates) versus phase in the 20-day cycle of 73 Dra. From Preston (1967).

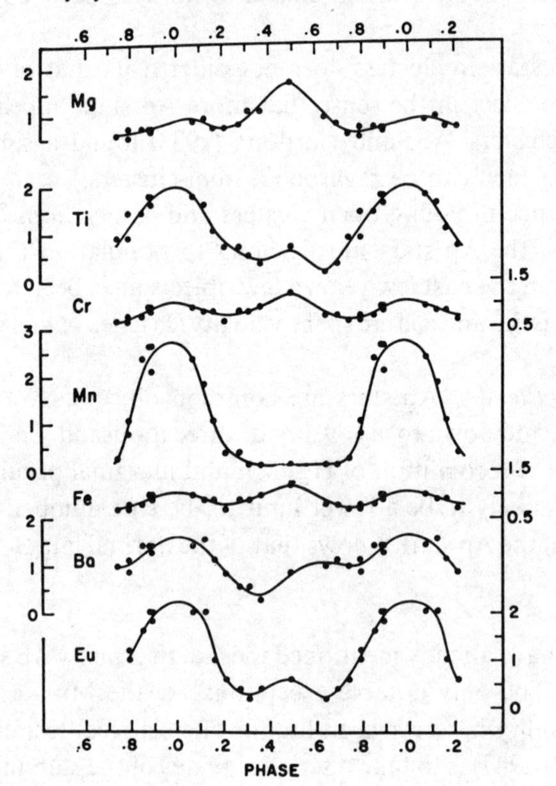

certain elements at certain places on the stellar surface. The rotation of a star carries the spots across the visible disk, producing changes in intensity. The rotation period is thus identical with the spectral variation period and with the variation in radial velocity of each element (figure 9.25). Observe that zero radial velocity is always expected for the element whose spot is moving tangentially with respect to the observer, i.e. zero radial velocity occurs when the line intensity is a maximum.

Since we are dealing with rotating stars, we should expect a relation between the equatorial velocity V and the period P (Deutsch 1947). If P is large, V has to be small, and vice versa. Since V is not directly accessible, we use $V \sin i$. A plot of $V \sin i$ versus P is given in figure 9.26, where we see that practically all the points fall below an upper envelope.

The spectrum variables are also photometric variables. Typically the variations in both color and magnitude are of the order of a few hundredths of a magnitude and have the same period as the spectral variations. Since it is relatively easy to measure variations of this order, the period determinations are made nowadays almost exclusively by photometry. Attention

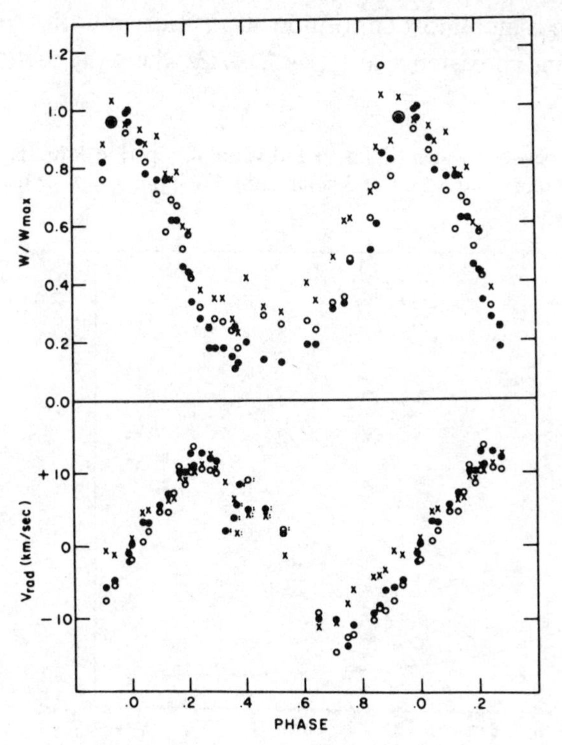

Figure 9.25. Equivalent widths and radial velocities for Eu II lines (filled circles), Ge II lines (crosses) and Dy II lines (open circles). From Pyper (1969).

should be paid to the fact that light curves in different colors vary both in amplitude and in phase.

This phenomenon is especially visible in the ultraviolet, where the change in amplitude and phase can be followed for different wavelengths (see figure 9.27). Usually a certain 'zero' wavelength exists, at which the star does not exhibit any variation at all. Besides this long term variability, there are variations of the order of a few hours or minutes, in the cooler Ap stars (Kurtz 1983).

The cause of the accumulation of certain elements at certain regions on the stellar surface is almost certainly related to the presence of magnetic fields. Since the measured magnetic fields also vary over the cycle, both in strength and often also in polarity, we are led to assume that the rotation axis does not coincide with the magnetic axis. At different phases different parts of the stars are seen and an – apparently – variable magnetic field perceived, which varies over the same period as the rotation period. The question of how constant over time the magnetic field is in reality has not been tackled.

Relation to other groups. The most obvious links to other groups are with Am stars on the cooler side and with He stars on the hotter side.

With the Am stars there is an almost continuous transition in both UBV and spectral type, with some coexistence at types A1–A3, where the earliest

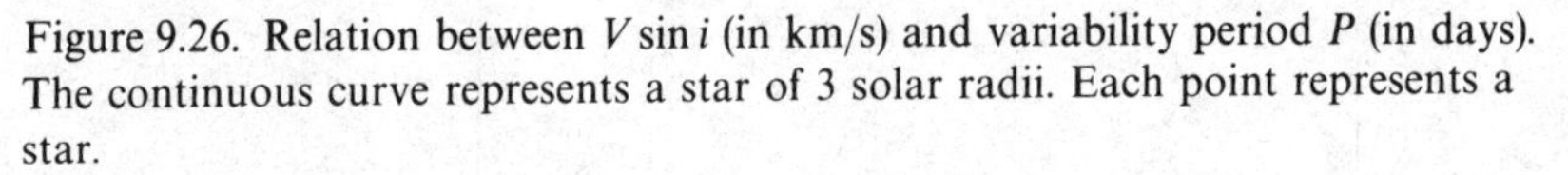

Figure 9.26. Relation between $V \sin i$ (in km/s) and variability period P (in days). The continuous curve represents a star of 3 solar radii. Each point represents a star.

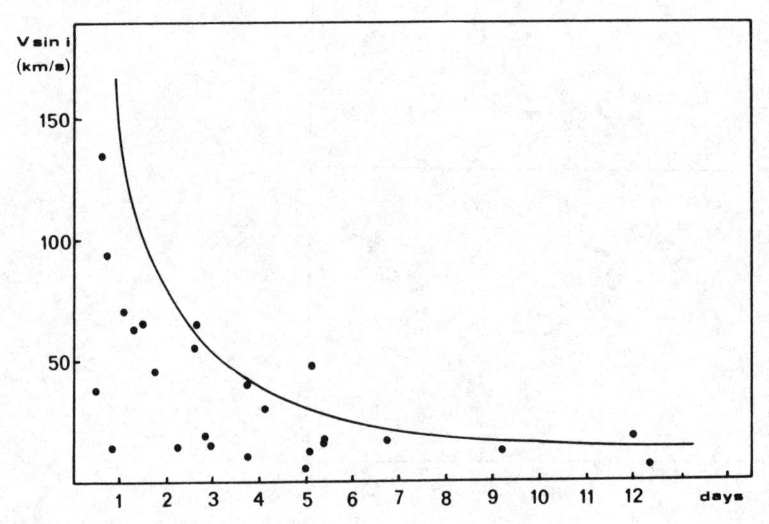

Am stars appear. Curiously there are not many Ap stars in the region where Am stars are most frequent, but they are found again at F0.

Several authors (e.g. Wolff and Wolff 1975) have stressed the similarity of Am stars with the Hg–Mn stars, suggesting that they constitute a single family. Fundamentally both have weak or non-existent magnetic fields, no spectral variability, no superposition in the two-color diagram, low rotation and are not deficient in spectroscopic binaries.

On the early side, Ap stars are next to the He weak objects, which themselves continue towards the (intermediate) He strong objects. The

Figure 9.27. Ultraviolet light curves for α^2C Vn. Each light curve represents the variation at the wavelength (in Å) indicated on the right side. From Molnar (1973).

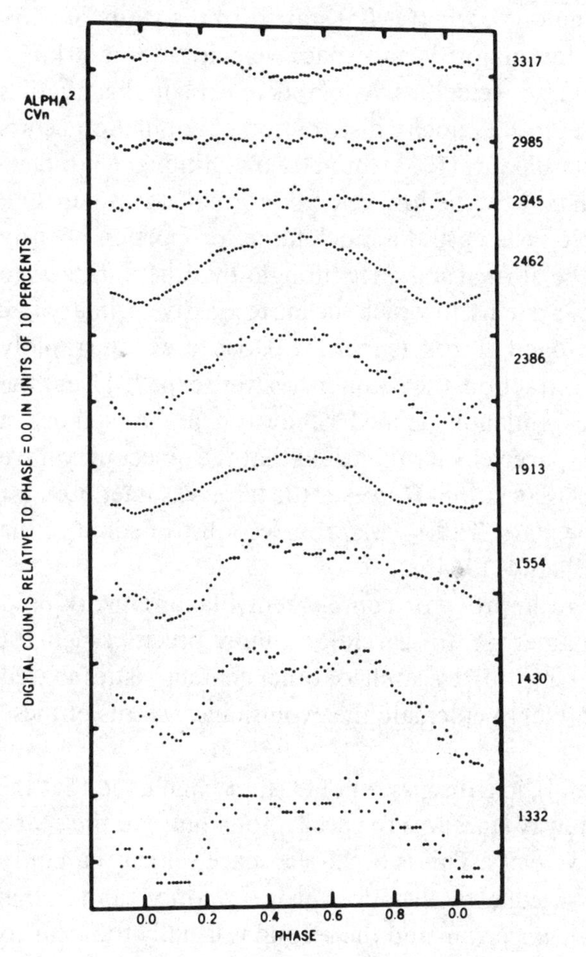

bluest Si stars share the helium weakness with the He weak stars and their places are contiguous in the two-color (U–B, B–V) diagram. Moreover spectrum variables and magnetic fields exist in both groups; probably the arguments are sufficient to create a single group with all these stars. This was done by Preston (1974) who grouped the Bp and Ap stars with related groups and introduced the division into four chemically peculiar (CP) stars: (1) the Am stars; (2) the magnetic Ap stars; (3) the Hg–Mn stars; and (4) the He weak stars. An objection we can make against such a scheme is that it joins groups which spectroscopically look rather different. For an observer, 'He weak' probably means more than CP4.

9.6 **Runaway stars**

Blaauw (1956) defined a group of runaway stars on the basis of kinematic characteristics among early (O–B5) stars. From a study of early type stars he found that the majority had space velocities $|v| < 30$ km/s, except a group which has larger velocities. Attempts to explain this fraction as the 'tail' of a 'reasonable' space velocity distribution of population I stars were unsuccessful. Later on Blaauw (1964) found two additional characteristics of these stars, not shared by the low velocity normal stars, namely a lack of double and multiple systems and a much larger proportion of early type objects than among the normal stars. He thought that these stars were once components of binary systems, in which the more massive primary had shed most of its mass in a violent process (supernova). Because of the rapidly diminishing gravitational attraction, the secondaries were expelled from the system at high space velocity. Blaauw defined 'runaway' stars to be those in which the direction of the space velocity indicated that the object may have originated in a known OB association. However the term was later used for all high velocity early type stars. Table 9.21 provides a list of some of the better known members (Blaauw 1964).

Subsequent work tried to improve or complement Blaauw's work, and went in three directions, namely (a) to search for a more precise kinematic definition of group membership, (b) to search for other characteristics shared by the group members, and (c) to elucidate the evolutionary status of these objects.

Cruz-Gonzalez *et al.* (1974), in a discussion of O stars, concluded that the best criterion to select runaway stars was to specify not a limit on the space velocity, but on the radial velocity. This is because a space velocity can only be obtained through knowledge of distance, proper motion and radial velocity, of which the first is uncertain and the second is usually too near to its possible error to be trustworthy. These authors provided a list of 72 runaway stars, out of a total of 664 O-type stars. The percentage of runaways is thus of the order of 10%, which goes up to 30% if certain statistical corrections are applied. They also found that a considerable fraction of the runaway stars belong to double or multiple systems, contrary to Blaauw's findings. However Conti, Leep and Lorre (1977) found only 7% of 'runaway' stars, although their criterion is slightly different from the one used by Cruz-Gonzalez *et al.* Stone (1978) gets a larger fraction – 20% – and states that almost all stars whose distance from the galactic plane $|z| > 80$ pc are 'runaways'.

Carrasco and Creze (1978) analyzed the absolute magnitude of these objects with the help of statistical parallaxes, and suggested that 'runaways' are underluminous by two or three magnitudes, which would put them between the dwarf and subdwarf sequences. Such a feature should certainly show up in the spectrum, and in the recent literature a number of papers have dealt with this problem. Carrasco *et al.* (1979, 1980) proposed that the 'runaway' objects are low mass, underluminous old-disc population objects, similar to the UV bright stars (section 8.5). This proposal ran into several objections. Walborn (1983a) found that the suggested candidates for the group, on the basis of galactic latitude and color excess alone, seem *not* to be nearby objects. Tobin and Kilkenny (1981) compared samples of stars of low and high radial velocity and concluded that except for a few subdwarfs, no spectroscopic difference can be found at 50 Å/mm. Keenan, Dufton and McKeith (1982), Keenan and Dufton (1983) and Tobin and Kaufmann (1984) arrived at the same conclusion from atmospheric analysis of about thirty

Table 9.21. *Runaway stars.*

Name	HD	MK type	Sp. velocity (km/s)	Associated with	$\|z\|$(pc)
λ Cep	210 839	O6 f	64	I Cep	34
ξ Per	24 912	O7	50	II Per	90
	152 408	O7–O8 fp	> 109	I Sco	10
68 Cyg	203 064	O8	49	I Cep	67
AE Aur	34 708	09.5 v	106	I Ori	7
ζ Oph	149 757	O9.5 v	39	II Sco	64
μ Col	38 666	B0v	123	I Ori	250
53 Ari	19 374	B2 v	59	I Ori	190
72 Col	41 534	B3 v	191	I Sco	98
	201 910	B5 v	58	I Lac	45

Note: $|z|$ = distance from galactic plane.

stars. In conclusion, at the moment the evidence is against a spectroscopic difference between low and high velocity stars, and the designation refers to a kinematic group.

9.7 Blue stragglers

A blue straggler is a globular cluster star which in the HR diagram lies above the horizontal branch, at or close to the zero-age main sequence. If the blue straggler belongs to an open cluster, it can be defined as a star lying above and to the left of the turn-up point in the color magnitude diagram. In both cases the separation between the turn off and the blue straggler has to be large, although it is hard to formulate this precisely. Blue stragglers are found in a variety of clusters, from the globular cluster M3, to old galactic clusters like NGC 752, and younger ones like NGC 2281. The term 'blue straggler' seems to have been introduced by Strom and Strom (1970).

A crucial point in the definition is the question of cluster membership, which must be ascertained on kinematic grounds (proper motions and/or radial velocity), since neither photometry nor position in the cluster alone can prevent the appearance of fictitious blue stragglers.

The definition is clearly not a spectroscopic one and therefore there are no a-priori reasons why an examination of the spectra of the group members should reveal common characteristics. This is exactly what happened. Just as an example of the kind of stars we find, let us quote the case of Barnard 29, a member of M13, which lies 3^m above the HB of this globular cluster. It was studied by Stoeckly and Greenstein (1968) and reanalyzed by Auer and Norris (1974). The spectrum shows hydrogen and sharp helium lines, with helium definitely not weak. The Balmer jump is small and lines up to $n = 16$ are observed. Of other elements N II λ3995 and the Si III triplet at λ4552 are visible. O II lines are not seen. This would suggest a B star (because of Si III), no dwarf (because of λ3995) and abnormal (because of the absence of O II). The UBV photometry gives $U-B = -0.84$, $B-V = 0.14$ indicating again an early B-type is too red for the U–B. Since the radial velocity is variable, with a range of 20 km/s, the authors suggest that the star has an (unobservable) subgiant K companion. This is however challenged by Auer and Norris (1974).

Mermilliod (1982b) has summarized the available information on the stragglers in open clusters. He finds that whereas few young clusters have a blue straggler, most of the clusters older than the Hyades have at least one. Secondly, the spectra of half of the blue stragglers are normal, whereas the spectra of the other half are peculiar – of the type Am, Ap, Bp or He abnormal.

Note also that the blue stragglers appear to share the peculiarity in their respective color domain – that is, if a blue straggler falls in a position on the two-color diagram where Si stars lie and if the blue straggler has a peculiar spectrum, it is of the Si-type.

Some (but not all) blue stragglers are binaries and some (but not all) have low $V \sin i$ (Mermilliod 1982b; Wheeler 1979).

We have therefore a rather heterogeneous collection of objects, from a spectroscopic point of view. Also kinematically the group is heterogeneous, because blue stragglers of globular clusters have high velocity, whereas those of open clusters belong clearly to the low velocity population. This remark is important because some authors have proposed to find 'field blue stragglers'. The reason for this is that most globular cluster blue stragglers are very faint, so that brighter field analogs would be most welcome. Bond and McConnell (1971) therefore searched for blue high velocity stars with large proper motion, since with high radial velocity alone the far away blue supergiants cannot be eliminated. Their list of candidates was then observed in uvby photometry to discriminate against other types of objects. HB stars can be discriminated by their large Balmer discontinuity (see figure 9.30 in section 9.8). As a result they found a few candidates (BD – 12°2669, HD 100 363, 224 927 and perhaps 214 539) which according to uvby photometry are metal weak.

Similar work was done by Olsen (1978, 1979) on photometric grounds and by Carney and Peterson (1981) who analyzed spectroscopically BD + 25°1981, BD – 12°2669 and HD 100 363. The three objects have F-type dwarf spectra with weak metallic lines, which are characteristics of halo stars.

Clearly such searches only produce analogs of globular (or old open) cluster blue stragglers. If the blue stragglers of younger open clusters should constitute a single family with the former, we would expect them to be metal weak, a fact which is not borne out by the many existing spectral classifications of the (bright) blue stragglers in open clusters.

It is therefore clear that the 'blue stragglers' are not a spectroscopically homogeneous group

9.8 B-type subdwarfs

Subdwarf B (sdB) stars exhibit broad and shallow Balmer lines, which can be seen up to $n = 10$–12. (In normal stars $n \sim 15$.) Furthermore, no

λ4686 line should be seen, since this line is characteristic of sdO stars. Members of this group can be detected at dispersions as low as 180 Å/mm if the Balmer discontinuity is visible. (For an illustration see figure 9.28.)

The sdB stars were defined first by Greenstein and Münch around 1954. Sargent and Searle (1968) define them in a slightly different way: 'a star which has colors corresponding to those of a B star and in which the Balmer lines are abnormally broad for the color, as compared to population I main-sequence stars. Such stars may be recognized by the early confluence of the Balmer series which is only seen up to $n \sim 12$. Some, but not all, have He I lines that are weak for their color'. As can be seen, the emphasis is placed upon the appearance of the hydrogen line spectrum. When examined in more detail, helium lines behave abnormally. In general He I lines are weak, as remarked by Sargent and Searle (1968), and no star has normal line strength. Furthermore singlet lines (for instance λ4387) are much weaker with respect to the triplet lines (for instance λ4471) than in normal dwarfs.

Of the other elements (C II, Mg II, O II) visible occasionally in sdB stars, we can only say that they are usually weaker than in dwarfs (Greenstein and Sargent 1974) and the same is true in the ultraviolet. Only nitrogen and silicon seem to behave normally (Heber *et al.* 1984).

Illustrations of some spectra are provided by Greenstein (1960) and Graham and Slettebak (1973).

Figure 9.28. Spectrum of a B subdwarf star. The spectrum of sdB HD 76431 is compared to HD 4727, a B5 V star. Notice the short Balmer series, and abnormal intensities of the different He I series. See text.

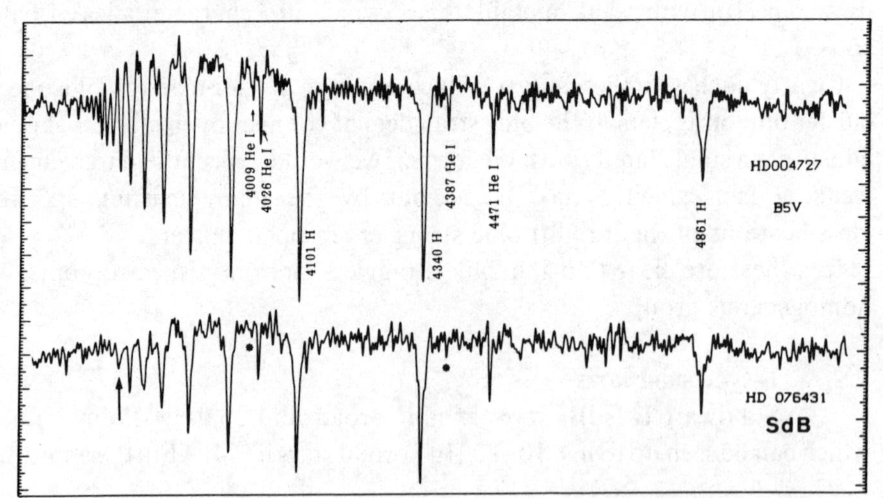

Photometry. Photometrically, sdB stars are difficult to segregate from normal stars. It can be done by uvby photometry, using the fact that both the Balmer lines and metallic lines are weak. The conditions are $b-y < -0.10$; $0.10 < m_1 < 0.15$; $c_1 < 0$ (Graham 1970; Kilkenny and Hill 1975). The place of these stars in the different diagrams are shown in figures 9.29 and 9.30 taken from Kilkenny and Hill (1975).

Radial velocities. The radial velocities are in general high. Greenstein and Sargent (1974) quoted $\bar{\rho} = -30\,\text{km/s}$, $\sigma = 49\,\text{km/s}$. Baschek and Norris (1975) have used this fact (large velocities, but not extreme) in conjunction with proper motions to suggest that these objects do not belong to the halo, but to the older disc population.

Binaries. Since one of the existing theories of the origin of sdB stars suggests that they are binaries, the few available data have been compiled by different authors (Baschek and Norris 1975; Mengel and Norris 1976). They found a few (three) cases of possible radial velocity variations and two cases (perhaps even four) of stars in which the observed colors are such as to indicate clearly a composite object.

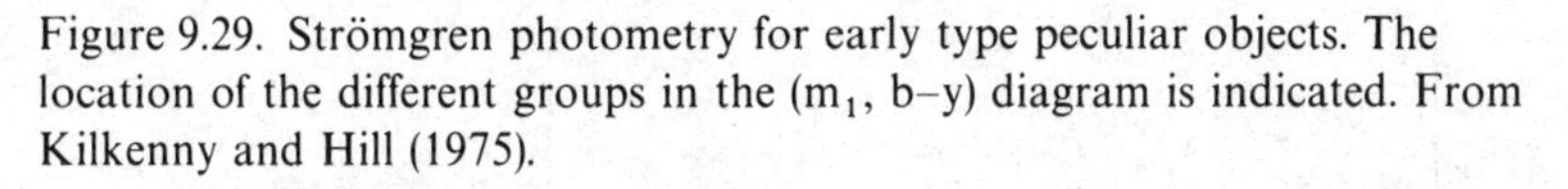
Figure 9.29. Strömgren photometry for early type peculiar objects. The location of the different groups in the (m_1, b–y) diagram is indicated. From Kilkenny and Hill (1975).

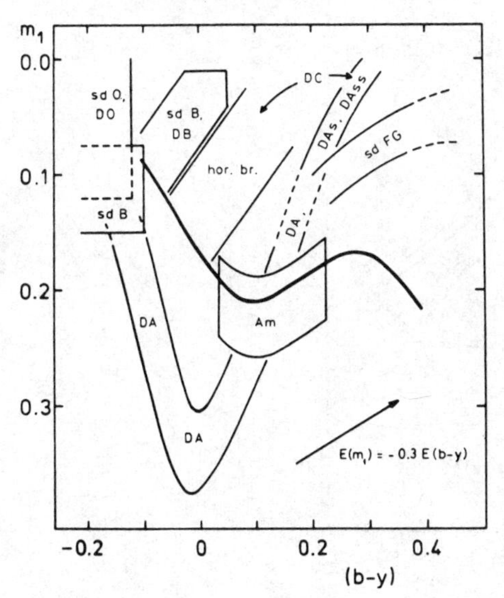

Absolute magnitude. From a discussion of all the available data, Greenstein and Sargent (1974) conclude an average of $M_v \sim +2$, with a tendency of earlier stars to have fainter magnitudes. This value makes them fall several magnitudes below the main sequence – from 6 magnitudes at early B to 1–2 at later B-types.

Relation to other groups. Clearly sdB stars are related to the sdO; they probably constitute a single group, sharing the hydrogen and helium line anomalies. Toward later types, the situation is less clear. Greenstein advocates a single family of stars embracing the sdO, sdB and HB stars, which he calls 'extended horizontal branch'. (It should be cautioned that the branch is only horizontal in M_{bol}, not in M_v, where the bluest part slopes considerably toward fainter M_v.) The arguments are essentially that the absolute magnitudes fit, the He anomalies are the same and the colors show a continuity. The 'extended horizontal branch' crosses the main sequence at around A0 and creates a region of confusion there. It seems that unless much

Figure 9.30. Strömgren photometry for early type peculiar objects. The location of the different groups in the (c_1, b–y) diagram is indicated. From Kilkenny and Hill (1975).

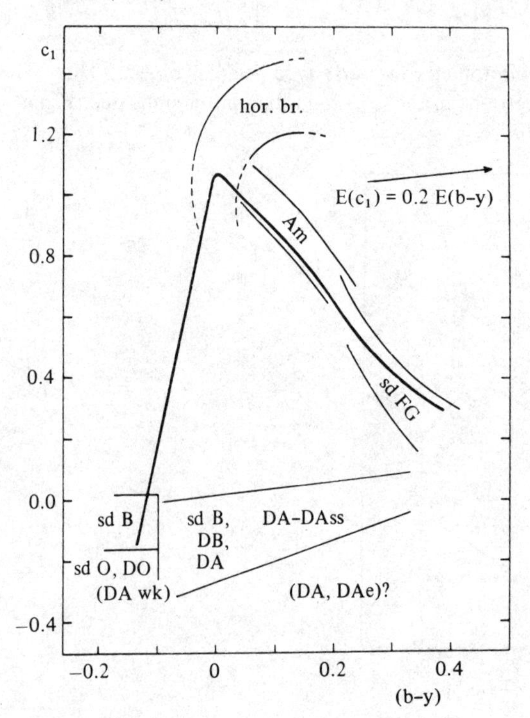

more is known about the individual stars, the Greenstein hypothesis is a very convenient one.

Number. The number of sdB stars identified spectroscopically as such is of the order of 10^2. Carnochan and Wilson (1983) think that the number of subdwarfs is much larger than previously thought; they find roughly comparable numbers of main sequence stars and underluminous objects. Another result pointing toward a large number of subluminous B-type stars is that by Green (1980), who found in a survey of 1434 square degrees above galactic latitude $+37°$, 27 B subdwarfs, as compared to 89 white dwarfs and 8 O-type subdwarfs.

Further reading. A summary of the knowledge of these stars is given by Baschek (1981).

9.9 'sn' stars

The 'sn' designation was introduced by Abt and Levato (1977) when classifying stars in the Orion OB1 association at 39 Å/mm. These stars range in type from B2 to B7 and in luminosity class from V to III. Their spectra have both sharp (s) and nebulous (n) lines, suggesting shell stars. The authors wrote that the stars have both sharp and broad He I lines.

Later Abt and Levato (1978), while classifying the brightest stars in Pleiades, again labelled some stars 'sn' and added 'The sharp lines are generally due to Ca II, Si II, Fe II and C II; the broad lines are mostly due to He I. These stars show mild examples of the effect exhibited by extreme shell stars such as Pleione and 48 Librae.' They called them explicitly mild shell stars. A list of members of this group (all discovered in clusters and associations) is given by Abt (1979). It is a very heterogeneous group with stars with a variety of peculiarities, as the authors remarked. A summary of other observational data is given by Mermilliod (1983). At the moment it is not clear if this is really an independent group.

9.10 β Cep stars = β CMa stars

A β Cep star is an early B-type object which presents short period light and radial velocity variations. The first object of this group to be discovered was β Cep (Frost 1902) and the second was β CMa. The group as such was described first by Henroteau (1928) and is designated indifferently as β CMa or β Cep.

The definition put forward is usually sharpened by excluding periods longer than 7^h (Underhill 1966), radial velocity variations larger than 150 km/s and light variations greater than $0^m.1$. The limitations are imposed to eliminate close binary nature as a cause of variability; the minimum period for a contact binary in early B-type stars is about twelve hours.

Further investigation shows that spectroscopically the β CMa variables are indistinguishable from normal stars in the $\lambda\lambda$3800–4800 region, and are even used as MK standards. So far all known β CMa variables fall in the interval B0.5 to B2, luminosity classes II, III and IV (Lesh and Aizenman 1973).

It is thus clear that the group cannot be spectroscopically defined and we omit further description. We refer the reader to Lesh (1982) and G.E.V.O.N. and Sterken (1982) for further information.

9.11 P Cyg stars

The name of this group comes from P Cyg, a nova which flared up in 1600. The star has a complicated photometric behavior, with irregular fades below the limit of visibility. Nowadays it is a 4^m9 B1 supergiant.

The most outstanding spectroscopic characteristic of the star is that most if not all lines show a 'P Cyg' type profile: an emission component to the red side of the absorption line. When examined at low plate factors, the radial velocities of the absorption lines are negative and vary with the excitation energy. (See figure 9.31.)

Because of the fact that at low plate factors only a few strong lines show

Figure 9.31. The spectrum of P Cyg. Notice the P Cyg profiles in the Balmer series and in He II lines. The comparison star is HD 167264, B0 Ia.

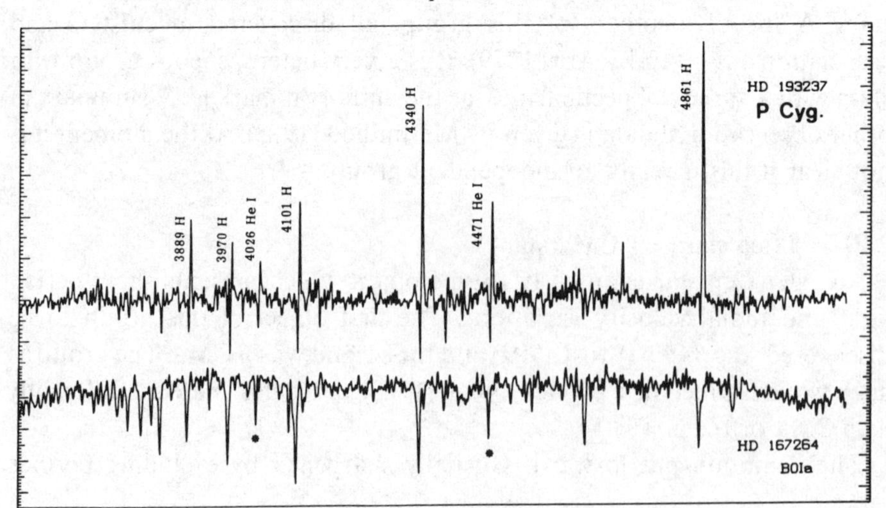

clearly the P Cyg type profile, astronomers have compiled lists of P Cyg type objects, comprising stars which show *some* lines with a P Cyg type profile. The oldest of these lists is the one by Beals (1950), but a close examination (de Groot 1982) reveals that no stars in the list are really similar to P Cyg, because most of them show these profiles only in some Balmer lines. The only remaining candidates are AG Car and R 81 in the Large Magellanic Cloud.

For more details see the workshop on P Cyg (de Groot and Lamers 1983) and Lamers, de Groot and Cassatella (1983).

9.12 **Hydrogen deficient early type stars**

We call hydrogen deficient early type stars a group of early stars (types O, B and A) in which the hydrogen lines are very weak or absent. The term 'hydrogen deficient' (Hd) refers to the purely observational fact that the 'hydrogen lines are very weak or absent', but it does *not* mean that hydrogen is underabundant, since this can only be determined from interpretation. The group has also been called 'H poor'.

The O- and B-type stars of this group are also called 'extreme He stars'. They differ from the helium stars dealt with in section 9.4 in that in the 'extreme He stars' no hydrogen lines are seen, whereas in the helium stars the strengths of the helium and hydrogen lines are comparable.

The first example of hydrogen deficient star, HD 124 448, was described by Popper (1942), and this star can still be considered as a prototype. The group was defined as such by Bidelman (1950).

The major spectral feature is that hydrogen lines are very weak or absent. In consequence both the Paschen and the Balmer discontinuities are also very weak or absent. If the star is of type O or B, helium lines are very strong. In O-type stars, He II lines replace the Balmer lines. To ascertain whether the lines seen correspond to H I or to He II, λ4200 is examined; this line is similar in strength to λ4101 and λ4340 only if the lines are due to He II.

If the star is of type B, He I is prominent and all series are easily visible and complete. Furthermore singlets are stronger with respect to triplets, as can be seen for instance in the ratio $\lambda 4387/\lambda 4471$.

Of other lines observable in B-type stars one finds those which normally exist but the elements present and their intensity do not seem to follow strict rules. Usually O seems to be weak and C and N are strong, when compared to a suitable standard; Mg and Si seem to follow the C, N trend. There is however a B-type Hd star with strong O (Drilling 1978).

If the star is of type A, strong Ca II lines and numerous lines of metals are observed, which would suggest a supergiant type spectrum. He I lines can also be observed – they are weak, but are definitely present.

A number of O- and B-type Hd stars are illustrated in the atlas by Kaufmann and Theil (1980).

Work done in the UV region shows that strong C lines are present in Hd stars; for the hotter stars we find strong C IV λ1550 and for B-type stars C II λ1335 (Drilling *et al.* 1984). These authors also illustrate a number of spectra in the region $\lambda\lambda$1300–2850.

Photometry. The lack of Balmer lines should be perceptible in UBV photometry by a shift of the Hd stars toward the black body curve and away from normal stars. Figure 9.32 shows this to be true. Broad band photometry at longer wavelengths shows no infrared excesses, except for stars which are binaries, like υ Sgr.

In Strömgren photometry there are similar effects due to the lack of H. Here the c_1 index, indicative of the Balmer discontinuity, is displaced as expected toward smaller values. The c_1 index measures essentially (in B-type stars) the He I discontinuities at λ3725 and λ3450 (Walker and Kilkenny 1980). These authors also found that many, if not all, Hd stars vary in luminosity, a fact already suspected by Hill (1969), with small amplitudes ($\Delta m \lesssim 0^{m}\!.1$).

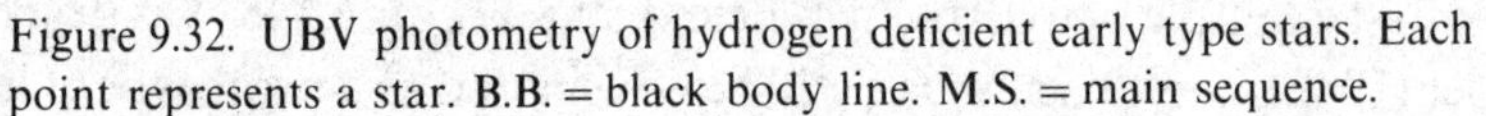
Figure 9.32. UBV photometry of hydrogen deficient early type stars. Each point represents a star. B.B. = black body line. M.S. = main sequence.

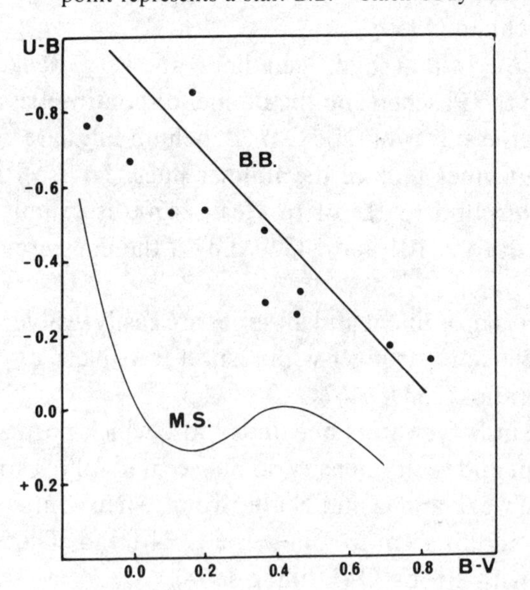

Radial velocity. The radial velocity is known for only a few Hd stars. It is in general large, with $|\bar{V}| \sim 130$ km/s, implying that the objects belong to an old population. Four stars are known to be spectroscopic binaries.

Absolute magnitude. The available evidence is scanty and does not contradict the assumption that these stars are near to (or on) the main sequence.

If this is assumed, Hd stars obey galactic rotation, although it should be noted that for stars out of the galactic plane this is not very strong evidence in either sense.

Number and distribution. The group is not numerous, since fewer than 30 objects of this type are known. Only 5 of these are in the HD ($m \lesssim 9^{m}$). They are not very concentrated toward the galactic plane; $|\bar{\beta}| \sim 26°$.

Further reading. The best summary is still from Hunger (1975).

References

Abrams J.V. (1947) *PASP* **59**, 78
Abt H.A. (1979) *Ap. J.* **230**, 485
Abt H.A. and Cardona O. (1983) *Ap. J.* **272**, 182
Abt H.A. and Levato H. (1977) *PASP* **89**, 797
Abt H.A. and Levato H. (1978) *PASP* **90**, 201
Abt H.A. and Snowden M. (1973) *Ap. J. Suppl.* **25**, 137
Abt H.A., Chaffee F.H. and Suffolk G. (1972) *Ap. J.* **175**, 779
Adelman S.J. (1975) *Ap. J.* **195**, 397
Adelman S.J. (1981) *Liège 23rd Colloquium*, p. 13
Aikman G.C.L. (1976) *Publ. D.A.O. Victoria* **14**, 379
Allen C.W. (1973) *Astrophysical quantities*, The Athlone Press
Allen D.A. (1973) *MNRAS* **161**, 145
Allen D.A. and Swings J.P. (1976) *AA* **47**, 293
Andrews P.J. (1978) *Mem. RAS* **72**, 35
Andrillat Y. (1979) *Publ. Obs. Strasbourg* **6**, 70
Andrillat Y. and Fehrenbach C. (1982) *AA Suppl.* **48**, 93
Andrillat Y. and Houziaux L. (1967) *AA* **50**, 107
Auer L.H. and Norris J. (1974) *Ap. J.* **194**, 87
Babcock (1947) *Ap. J.* **105**, 105
Balona L.A. and Crampton D. (1974) *MNRAS* **166**, 203
Bappu M.K.V., Chandra S., Sanvol N.B., and Sinval S.D. (1962) *MNRAS* **123**, 521
Barbier D. and Chalonge D. (1941) *Ann. d'Astroph.* **4**, 293
Baschek B. (1975) in *Problems in stellar atmospheres and envelopes*, Baschek *et al.* (ed), Springer Verlag

Baschek B. (1981) *Troisième Journee de Strasbourg*, Strasbourg Observatory, p. 3
Baschek B. and Norris J. (1975) *Ap. J.* **199**, 694
Beals C.S. (1950) *Publ. Dom. Obs. Victoria* **9**, 1
Bernacca P.L. and Ciatti F. (1972) *AA* **19**, 482
Bidelman W.P. (1950) *Ap. J.* **111**, 333
Bidelman W.P. (1952) *A. J.* **116**, 227
Bidelman W.P. (1960) *PASP* **72**, 24
Bidelman W.P. (1961) *Ap. J.* **135**, 651
Bidelman W.P. (1962) *Sky and telescope* **23**, 140
Blaauw A. (1956) *Bull. Astr. Inst. Netherlands* **15**, 265
Blaauw A. (1964) *Ann. Rev. AA* **2**, 213
Bolton C.T. and Rogers G.L. (1978) *Ap. J.* **222**, 234
Bond H.E. and McConnell D.J. (1971) *Ap. J.* **165**, 51
Borra E.F. and Landstreet J.D. (1979) *A. J.* **228**, 809
Borra E.F. and Landstreet J.D. (1980) *Ap. J. Suppl.* **42**, 421
Borra E.F., Landstreet J.D. and Thompson I. (1983) *Ap. J. Suppl.* **53**, 151
Briot D. (1981) *AA* **103**, 5
Cameron R.C. (1967) in *The magnetic and related stars*, Mono Book Corp. Baltimore, p. 471
Carney B.W. and Peterson R.C. (1981) *Ap. J.* **251**, 190
Carnochan D.J. and Wilson R. (1983) *MNRAS* **202**, 317
Carrasco L. and Creze M. (1978) *AA* **65**, 279
Carrasco L., Bisiacchi G.F., Costero R. and Firmani C. (1979) *IAU Symp.* **83**, 299 Conti P.S. and de Loore C.W.H. (ed.), Reidel Publ. Co.
Carrasco L., Bisiacchi G.F., Cruz-Gonzalez C., Firmani C. and Costero R. (1980) *AA* **92**, 253
Charlier C.V.L. (1926) *Medd Lund* **36**
Chiatti F. and Bernacca P.L. (1971) *AA* **11**, 485
Conti P.S., Leep E.M. and Lorre J.J. (1977) *Ap. J.* **214**, 759
Cowley A.P., Jaschek M. and Jaschek C. (1970) *A. J.* **75**, 939
Cowley C.R. (1979) *Ap. J.* **233**, 633
Crawford D.L. (1958) *Ap. J.* **128**, 190
Crawford D.L. (1978) *A. J.* **83**, 48
Cruz-Gonzalez C., Recillas-Cruz E., Costero R., Peimbert M. and Torres-Irubert S. (1974) *Rev. Mexicana AA* **1**, 211
Cucchiaro A., Macau-Hercot D., Jaschek M. and Jaschek C. (1977) *AA Suppl.* **30**, 71
Cucchiaro A., Macau-Hercot D., Jaschek M. and Jaschek C. (1978) *AA Suppl.* **33**, 15
Dachs J., Eichendorf W., Schleicher H., Schmidt-Kaler T., Stift M. and Tug H. (1981) *AA Suppl.* **43**, 427
de Groot M. (1982) *Irish A. J.* **15**, 216
de Groot M. and Lamers H.J.G.L. (1983) *Irish A. J.* **16**, 162
Deutsch A.J. (1947) *Ap. J.* **105**, 283

Deutsch A.J. (1956) *PASP* **68**, 92
Didelon P. (1982) *AA Suppl.* **50**, 199
Didelon P. (1983) *AA Suppl.* **53**, 119
Drilling J.S. (1978) *Ap. J.* **223**, L29
Drilling J.S., Schonberner D., Heber U. and Lynas-Gray A.E. (1984) *Ap. J.* **278**, 224
Dubois P., Jaschek M. and Jaschek C. (1976) *IAU Symp.* **72**, 149, Hauck B. and Keenan P.C. (ed), D. Reidel Publ. Co., Dordrecht, p. 149
Durrant C.J. (1970) *MNRAS* **147**, 75
Dworetsky M.M. (1969) *Ap. J.* **156**, L101
Egret D. and Jaschek M. (1981) *23 Liège Astroph. Coll.*, p. 495
Fehrenbach C. (1958) *Handbuch der Physik*, Flugge S. (ed.) vol. L, p. 1. Springer
Feinstein A. (1978) *Rev. Mex. AA* **2**, 331
Feinstein A. and Marraco H.G. (1979) *A. J.* **84**, 1713
Frost E.B. (1902) *Ap. J.* **15**, 340
Garrison R.F. (1967) *Ap. J.* **147**, 1003
G.E.V.O.N. (Groupement Étoiles Variables Observatoire Nice) and Sterken C. (1982) *Proceedings of the workshop on pulsating B stars*, Nice Observatory
Gieseking W. (1983) *Compt. rendus Journees Strasbourg*, 5 eme reunion, p. 4
Glagolevski J.W. (1966) *A. J. URSS* **43**, 73
Golay M. (1974) *Introduction to astronomical photometry*, Reidel D. Publ. Co
Graham J.A. (1970) *PASP* **82**, 1305
Graham J.A. and Slettebak A. (1973) in *Spectral classification and multicolor photometry*, Fehrenbach C. and Westerlund B.E. (ed.), Reidel D. Publ. Co., p. 245
Green R.F. (1980) *Ap. J.* **238**, 685
Greenstein J.L. (1960) in *Stellar atmospheres*, Greestein J.L. (ed.), Univ. Chicago Press.
Greenstein J.L. and Sargent A. (1974) *Ap. J. Suppl.* **28**, 157
Grenier S., Gomez a., Jaschek C., Jaschek M. and Heck A. (1985) *AA* **145**, 33
Guthnick P. (1931) *Sitzungsber Preuss. Ak. Berlin N.* **27**
Hack M. (1953) *AA* **16**, 417
Hardorp J. and Strittmatter P.A. (1970) *Stellar rotation*, Reidel D. Publ. Co., p. 48
Harmanec P. (1982) *IAU Symp.* **98**, 279, Jaschek and Groth H.G. (ed.), Reidel D. Publ. Co
Harmanec P. (1948a) 'Rapid variability of early type stars', *Hvar Observatory Bulletin* **7**, 55
Harmanec P. (1984b) *Be Newsletter* **10**, 48
Hartoog M.R. (1976) *Ap. J.* **205**, 807
Hauck B. (1975) *IAU Coll. 32, Physics of Ap stars* Weiss W.W., Jenker H. and Wood H.J. (ed.), p. 365
Heck A., Egret D., Jaschek M. and Jaschek C. (1984) *IUE low dispersion spectra reference atlas*, ESA SP-1052
Henroteau F.C. (1928) *Handbuch der Astrophysik* **6**, 436, Springer Verlag

Heber U., Hunger K., Jonas G. and Kudritzki R.P. (1984) *AA* **130**, 119
Hill P.W. (1969) *Inf. Bull. Var. Stars* **357**
Hoffleit D. and Jaschek C. (1982) *Bright star catalogue*, 4th edition, Yale Univ. Obs.
Honeycutt R.K. and McCuskey S.W. (1966) *PASP* **78**, 289
Hubert-Delplace A.M. and Hubert H. (1979) *An atlas of Be stars*, Paris Meudon Observatory
Hunger K. (1975) in *Problems in stellar atmospheres and envelopes*, Springer Verlag.
Jacoby G.H., Hunter D.A. and Christian C.A. (1984) *Ap. J. Suppl.* **56**, 257
Jaschek C. (1970) *IAU Coll., Stellar rotation*, Slettebak A. (ed.) Reidel D. Publ. Co., p. 219
Jaschek C. and Gomez A. (1970) *PASP* **82**, 809
Jaschek M. and Jaschek C. (1962) *PASP* **74**, 151
Jaschek M. and Jaschek C. (1967) *Ap. J.* **150**, 355
Jaschek M. and Jaschek C. (1974a) *AA* **36**, 401
Jaschek M. and Jaschek C. (1974b) *Vistas in Astronomy* **16**, 131, Beer A. (ed.), Pergamon Press
Jaschek C. and Jaschek M. (1981) *23 Liège Astroph. Coll.*, p. 417
Jaschek C. and Jaschek M. (1983) *AA* **117**, 357
Jaschek M. and Jaschek C. (1984) *The MK process and stellar classification*, Garrison R. (ed.), Toronto p. 290
Jaschek M. and Malaroda S. (1970) *Nature* **225**, 246
Jaschek M., Jaschek C. and Arnal M. (1969) *PASP* **81**, 650
Jaschek M., Jaschek C. and Kucewicz B. (1968) *Nature* **225**, 246
Jaschek M., Slettebak A. and Jaschek C. (1981) *Be Newsletter* **4**, 9
Jaschek M., Hubert-Delplace A.M., Hubert H. and Jaschek C. (1980) *AA Suppl.* **42**, 103
Jaschek C., Jaschek M., Gomez A. and Grenier S. (1983) *AA* **127**, 1
Johnson H.L. (1967) *Ap. J.* **150**, L39
Joncas G. and Borra E.F. (1981) *AA* **94**, 134
Kaufmann J.P. and Theil U. (1980) *AA Suppl.* **41**, 271
Keenan F.P. and Dufton P.L. (1983) *MNRAS* **205**, 435
Keenan F.P., Dufton P.L. and McKeith C.D. (1982) *MNRAS* **200**, 673
Keenan F.P. and Dufton P.L. (1983) *MNRAS* **205**, 435
Keenan P.C. and Hynek A. (1950) *Ap. J.* **111**, 1
Keenan P.C., Slettebak A. and Bottemiller R.L. (1969) *Astroph. Letters* **3**, 35
Kilkenny D. and Hill P.W. (1975) *MNRAS* **173**, 625
Kitchin C.R. and Meadows A.J. (1970) *Astroph. Space Sc.* **8**, 463
Kodaira K. and Hoekstra R. (1979) *AA* **78**, 292
Kriz S. and Harmanec P. (1975) *Bull. A. I. Czechoslovakia* **26**, 65
Kuchowicz B. (1975) *IAU Coll. 32, Physics of Ap stars*, Weiss W.W., Jenker H. and Wood H.J. (ed.), p. 169
Kurtz D.W. (1983) *MNRAS* **205**, 3
Lamers H., de Groot M. and Cassatella A. (1983) *AA* **123**, L8

Landstreet J.D. (1980) *A. J.* **85**, 611
Landstreet J. D. (1982) *Ap. J.* **258**, 639
Landstreet J. D. and Borra E.F. (1978) *A. J.* **224**, L5
Lesh J.R. (1982) *B stars with and without emission lines*, CNRS-NASA monographs
Lesh J.R. and Azienman M.L. (1973) *AA* **22**, 229
Liège 23 Colloquium (1981) Institut d'Astrophysique, Liège
Maitzen H.M. (1975) *IAU Coll. 32, Physics of Ap stars*, p. 233
Maitzen H.M. (1981) *AA* **95**, 213
Maitzen H.M. and Vogt N. (1983) *AA* **123**, 48
McConnell D.J., Frye R.L. and Bidelman W.P. (1970) *PASP* **82**, 730
McConnell D.J., Frye R.L. and Bidelman W.P. (1972) *PASP* **84**, 388
Mendoza E.V. (1977) in *HR diagram, IAU Symp.* **80**, 289, Philip Davis A.G. and Hayes D.S. (ed.), Reidel D. Publ. Co
Mengel J.G. and Norris J. (1976) *Ap. J.* **204**, 488
Mermilliod J.C. (1982a) in *IAU Symp.* **98**, 23, Jaschek M. and Groth H.G. (ed.), Reidel D. Publ. Co
Mermilliod J.C. (1982b) *AA* **109**, 37
Mermilliod J.C. (1983) *AA* **128**, 362
Mermilliod J.C. and Mermilliod M. (1980) *BICDS* **19**, 65
Molnar M.R. (1973) *Ap. J.* **179**, 527
Morgan W.W. (1933) *Ap. J.* **77**, 330
Nissen P.E. (1974) *AA* **36**, 57
Nissen P.E. (1976) *AA* **50**, 343
Norris J. (1971) *Ap. J. Suppl.* **23**, 235
Olsen E.H. (1978) in *Astronomical papers dedicated to R. Strömgren*, Reiz A. and Anderson T. (ed.), Copenhagen University Observatory
Olsen E.H. (1979) *AA Suppl.* **37**, 367
Osawa K. (1965) *Ann. Tokyo Astron. Obs.* (2) **9**, 123
Osmer P.S. and Peterson D.M. (1974) *A. J.* **187**, 117
Pannekoek A. (1929) *Publ. Amsterdam N.* **2**
Pedersen H. (1976) *AA* **49**, 217
Pedersen H. (1979) *AA Suppl.* **35**, 313
Pedersen H. and Thomsen B. (1977) *AA Suppl.* **30**, 11
Peters G. (1982) in *IAU Symp.* **98**, 311
Petrie R.M. (1952) *Publ. Dom. Obs. Victoria* **9**, 251
Popper D.M. (1942) *PASP* **54**, 160
Preston G.W. (1967) *Ap. J.* **150**, 871
Preston G.W. (1974) in *Annual review of astronomy and astrophysics*, vol. 12, Palo Alto, Cal., Ann. Rev. Inc.
Pryzybylski A. (1963) *Acta Astr.* **13**, 217
Purton C.R., Feldman P.A., Marsh K.A., Allen D.A. and Wright A.E. (1982) *MNRAS* **198**, 321
Pyper D. (1969) *Ap. J. Suppl.* **18**, 347
Rogers G. (1974) unpublished, Thesis

Sargent W.L.W. and Searle L. (1968) *A. J.* **152**, 443
Schalen C. (1926) *Medd. Uppsala N.* **10**
Schild H. (1985) *AA* **146**, 113
Schild R.E. (1966) *Ap. J.* **146**, 142
Schild R.E. (1976) in *IAU Symp.* **70**, 106, Slettebak A. (ed.), Reidel D. Publ. Co
Schild R.E. and Romanishin W. (1976) *Ap. J.* **204**, 493
Schmidt-Kaler Th. (1982) in Landolt–Börnstein, group VI, vol. 2b, p. 1.
Secchi A. (1878) *Die Sterne*, Brockhaus
Slettebak A. (1949) *Ap. J.* **110**, 587
Slettebak A. (1976) *IAU Symp.* **70**, Slettebak A. (ed.), p. 123
Slettebak A. (1982) *Ap. J. Suppl.* **50**, 55
Slettebak A. and Stock J. (1957) *Zeitschr. f. Astroph.* **42**, 67
Slettebak A., Collins G.W., Boyce P.B., White N.M. and Parkinson T.D. (1975) *Ap. J. Suppl.* **29**, 137
Stoeckly R. and Greenstein J. (1968) *Ap. J.* **154**, 900
Stone R.C. (1978) *A. J.* **83**, 393
Stothers R. and Frogel J.A. (1974) *A. J.* **79**, 456
Straizys V. (1977) *Multicolor stellar photometry*, Mokslas Publishers, Vilnius
Strom K.H. and Strom S.E. (1970) *Ap. J.* **162**, 523
Strömgren B. (1939) *Ap. J.* **89**, 526
Swings J.P. (1973) *AA* **26**, 443
Takada M. and Jugaku J. (1981) *23 Coll. Liège*, p. 163
Tobin W. and Kaufmann J.P. (1984) *MNRAS* **207**, 369
Tobin W. and Kilkenny D. (1981) *MNRAS* **194**, 937
Underhill A. (1966) *The early type stars*, Reidel D. Publ. Co
Underhill A. and Doazan V. (1982) *B stars with and without emission lines*, CNRS NASA SP-456
Uesugi A. and Fukuda I. (1981) *Proc. 7 Codata Confer.*, Pergamon Press
Wackerling L.R. (1970) *Mem. RAS* **73**, 153
Walborn N.R. (1970) *Ap. J.* **161**, L149
Walborn N.R. (1971a) *Ap. J.* **164**, L67
Walborn N.R. (1971b) *Ap. J. Suppl.* **23**, 257
Walborn N.R. (1974) *Ap. J.* **191**, L95
Walborn N.R. (1976) *Ap. J.* **205**, 419
Walborn N.R. (1983a) *Ap. J.* **267**, L59
Walborn N.R. (1983b) *Ap. J.* **268**, 195
Walker H.J. and Kilkenny D. (1980) *MNRAS* **190**, 299
Wheeler J.C. (1979) *Ap. J.* **234**, 569
Wolff S.C. and Wolff R.J. (1975) *IAU Coll. 32, Physics of Ap stars*, p. 503
Ap stars in the infrared (1978) Workshop, Vienna Observatory

10

A-type stars

10.0 Normal stars

According to the Harvard system, an A-type star is an object in which strong Balmer lines are accompanied by many faint to moderately strong lines. These metallic lines increase gradually in strength from A0 to A9. A-type stars differ from B-type stars in that in the former there is no He I line. The difference between A- and F-type stars is subtler; in the latter the metallic lines are more numerous and stronger.

The behavior of metallic lines can be illustrated by that of the Ca II lines H (λ3968) and K (λ3933). On 100 Å/mm plates these lines are very weak at A0. Since H is close to Hε (λ3970), only a faint K-line and a broad H + Hε line are seen. At A5 approximately, K is half as strong as H + Hε, and at F0, $I(K) \simeq I(H + H\varepsilon)$. Table 10.1 provides the equivalent widths of some typical lines.

Table 10.1 shows that hydrogen has its maximum at A2, decreasing from thereon; Ca II and metals increase in strength toward later types. In principle therefore the hydrogen lines and the Ca II lines, representing the metals, can be used for quantitative spectral type assignment. For visual classification of A-type stars, the lines are however too strong and fainter features have to be used.

The principal elements seen in A-type stars are lines of Fe I and Fe II, Cr I and II, and Ti I and II, which account for about two-thirds of all lines.

The number of lines visible grows rapidly toward F-type stars and therefore all lines become more or less blended. It is thus important to specify the plate factor. A line which at 2 Å/mm is isolated from its neighbors, at 20 Å/mm is a blend of several lines and at 100 Å/mm a blend of still more lines. For instance in a typical A0 star there are 13 lines stronger than 0.1 Å equivalent width between λ3900 and λ4500. At F2, there about 87 lines of this intensity. So in fact on average in A0 there is one of these lines for each 46 Å, whereas in F2 there is one line for each 7 Å.

A consequence of this crowding is that the continuum between lines becomes more and more chopped up and can be reached only at low plate factors. If the equivalent widths of all lines occurring in bands of 25 Å are integrated, the blocking fraction, i.e. the fraction of energy substracted by all lines, is obtained. Usually this quantity is called η. Table 10.2 provides the values of $\eta(\lambda)$ as taken from Ardeberg and Virdefors (1975).

As can be seen from the table, the fraction increases steadily toward both later types and shorter wavelengths. This is clearly visible on the spectra where the number and strength of the metallic lines increase smoothly towards F0.

The MK system subdivides the A-type stars into ten subclasses of spectral type. As a rule all criteria depend critically upon resolution. A feature visible at 125 Å/mm resolves habitually at 75 Å/mm into several lines; this implies that the feature useful at 125 Å/mm is partially or entirely useless at 75 Å/mm. Table 10.3 provides some criteria used by Morgan, Abt and Tapscott (1978), Yamashita, Nariai and Norimato (1977) and Landi, Jaschek and Jaschek (1977) at three different plate factors.

Besides spectral types, stars can be classified by luminosity. Ratios between intensities of singly ionized and neutral lines of either the same or

Table 10.1. *Equivalent widths (in Å) of some strong lines in A-type dwarfs.*

	Hα	Hβ	Hγ	K	Fe λ4045	Sr II
A0	9.0	12.2	13.6	0.3	0.1	0.1
A3	9.3	14.1	17	2.1		
A5	8.5	13.0	15.5	3.5	0.2	
A7	6.6	10.9	13	4.5		
F0	5.5	7.0	8	6.5	0.3	0.2

Table 10.2. *Values of the blocking function $\eta(\lambda)$ for A-type dwarfs.*

	λ3300	λ3509	λ4036	λ4255	λ4465	λ5000	λ6050
A0	0.03	0.02	0.01	0.01	0.01	0.01	0.00
A3	0.10	0.08	0.06	0.05	0.05	0.03	0.00
A5	0.15	0.14	0.08	0.08	0.07	0.04	0.01
A7	0.18	0.16	0.10	0.09	0.08	0.05	0.01
F0	0.23	0.22	0.13	0.12	0.10	0.06	0.02

similar elements are used almost exclusively as luminosity indicators. Typical line ratio factors used at 80–100 Å/mm plate factors are given in table 10.4.

In both luminosity and spectral type criteria care must be taken with stars

Table 10.3. *Spectral type criteria in A-type stars.*

Source	Plate factor	Criteria
Morgan, Abt and Tapscott (1978)	125 Å/mm	The most useful blends between B9.5 and A5 are Mn I λ4030–34 and Fe I λ4271. Also the feature at λ4300 (blend of Ti II and Fe I), and Ca I λ4226 can be used.
Yamashita, Nariai and Norimato (1977)	75 Å/mm	Ca I λ4226/Mg II λ4481 and Fe I λ4045/Fe II λ4173 for A3–F0. Fe II λ4233 has a maximum at A2.
Landi, Jaschek and Jaschek (1977)	40 Å/mm	Mg II λ4481/Fe I λ4385 can be used between A0 and A7. At A3 in dwarfs, Ca I λ4226 = Fe II λ4233. At A5–A7, Mg II λ4481/Fe II λ4416.

Table 10.4. *Luminosity criteria in A-type stars.*

Source	Plate factor	Criteria
Morgan, Abt and Tapscott (1978)	125 Å/mm	A0: the profiles of the hydrogen lines A2: λ4383–85 (Fe I, II)/λ4481 (Mg II) A5: λ4417 (Fe II, Ti II)/λ4481 (Mg II) (A2, A5): The blend is stronger at high luminosity
Yamashita, Nariai and Norimato (1977)	73 Å/mm	A2–A3: λ4417 (Fe II, Ti II)/Mg II λ4481 larger at high luminosity A5 : the same A7–F0: Sr II λ4215/Ca I λ4226 larger at high luminosity.
Landi, Jaschek and Jaschek (1977)	42 Å/mm	A0–A2: enhancement of Fe II λ4173–78, λ4233 at high luminosity A3 : Fe II λ4416/Mg II λ4481 larger at high luminosity A5 and A7: Fe II λ4351/Mg II λ4481 larger at high luminosity A7–F0: Sr II λ4215/Ca I λ4226 larger at high luminosity

with peculiar spectra. For instance, there are stars in which the K line is weak (see section 10.1, Am stars), and stars in which some elements are enhanced (see section 9.5, Ap stars). Ionized strontium is enhanced in many Ap stars and as a result it should never be used as the sole luminosity indicator.

Because the hydrogen lines are still by far the strongest lines in the spectrum, it is tempting to use their profile for luminosity classification. Although this is true in general (in the sense that supergiants do have less marked wings than dwarfs) one should keep in mind that in dwarfs low rotational velocity also produces sharp hydrogen lines. One must thus use luminosity criteria based upon line ratios. Spectral types and luminosities of some A stars are illustrated in figures 10.1 and 10.2.

Figure 10.1. Spectra of A-type stars. For explanation see text.

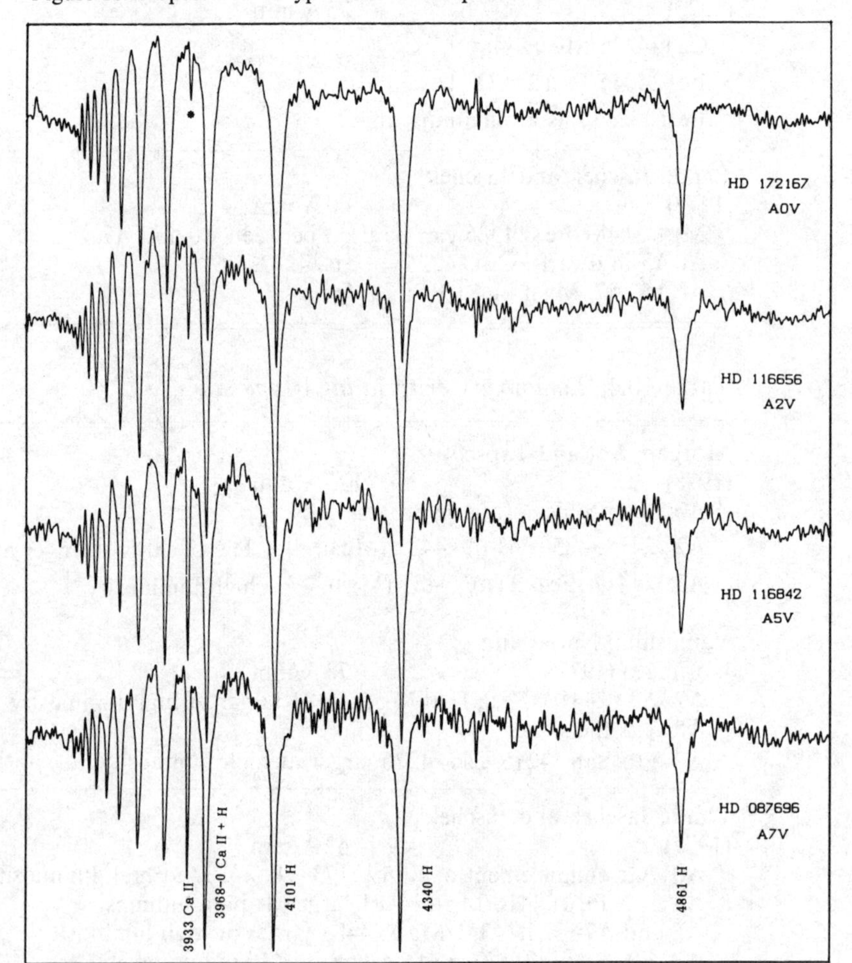

Ultraviolet. The region below λ3000 presents a continuum which decreases abruptly toward shorter wavelengths. We present in figure 10.3 the slopes as a function of spectral type, taken from Cucchiaro, Jaschek and Jaschek (1978). The fluxes were observed by the TD1 satellite and have a low resolution of 37 Å.

When observed at improved resolution, a number of strong lines (listed in table 10.5) become visible and can be used easily.

Curiously most features are sensitive to temperature, whereas features

Figure 10.2. Luminosity effects in A-type stars. For explanation see text.

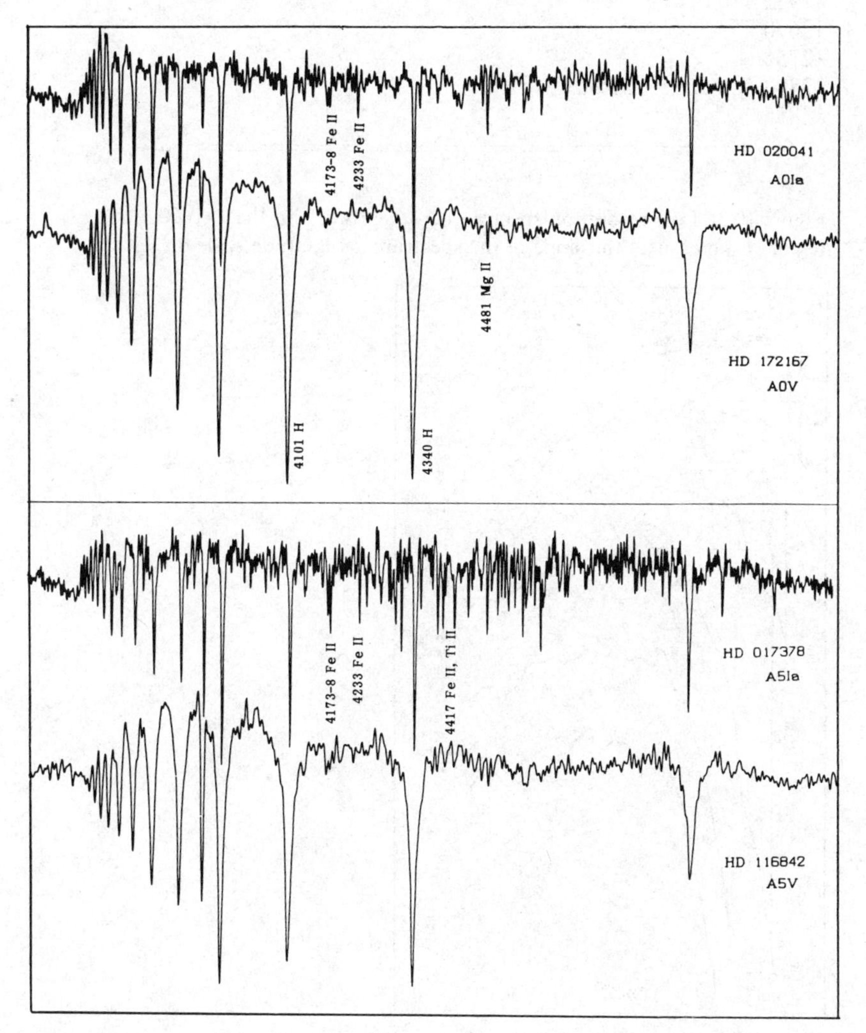

sensitive to luminosity are hard to find (Heck *et al.* 1984). On the other hand stars with weak metallic lines are easy to detect.

Infrared. In the near infrared the most conspicuous features are the Paschen series, which reaches its greatest strength in early A-type stars, the O I lines at λλ7771–5 and λλ8446, and the Ca I triplet at λλ8498, 8542 and 8662 (Merrill 1934). The O I lines exhibit a strong positive luminosity effect, as can be seen

Table 10.5. *Some strong features in A-type spectra.*

λ1850	This feature is sensitive to luminosity
λ1933	
λ2670	
λ2755	
λ2800	2795 + λ2803 Mg II
λ2855	

Figure 10.3. Low resolution spectra of A-type dwarfs in the λλ1400–2600 region. C_P measures the slope of the spectrum. Magnitude scale on left side.

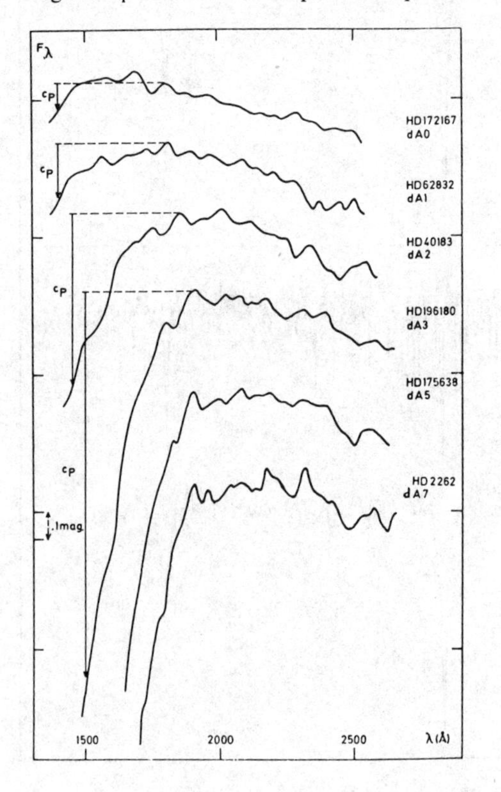

in figure 10.4 taken from Keenan and Hynek (1950). The measurement of these lines constitutes a powerful means for detecting supergiants, even at small dispersions (Parsons 1964).

Emission lines. In general few A-type stars show emissions. Among the rare stars which exhibit them there are some supergiants with emission in Hα, some dwarfs called Ae stars (see section 10.5) and some very isolated cases with emission in Ca II (K line).

Rotation. On the whole A-type stars rotate rapidly, but less than B-type stars and more than F-type stars. Table 10.6 provides a distribution of the average rotational velocities, from Uesugi and Fukuda (1979).

The higher value for A5 and A7 is due to the fact that Am stars (see

Table 10.6. *Rotation of A-type stars ($\overline{V \sin i}$ in km/s).*

	V	III	I	Ap	Am
A0	119	22	40		
A3	123	73			
A5	130	96	28	40	43
A7	118	110			
F0	71				

Figure 10.4. Luminosity effects in the O I λ7774 line. From Keenan and Hynek (1950).

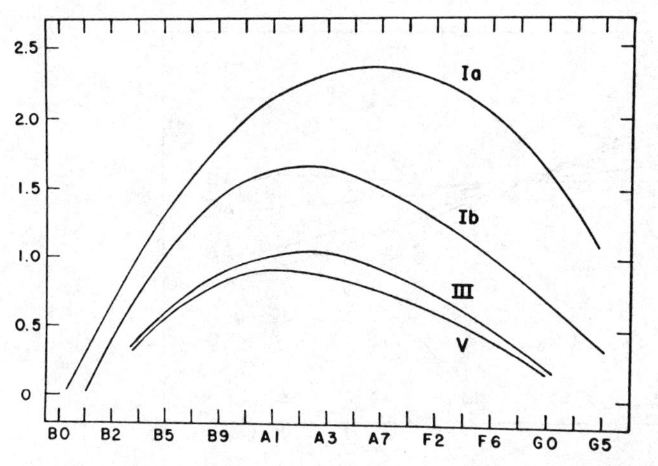

section 10.1), which constitute a large proportion of all A stars, have been omitted; had they been included, the average rotation would drop smoothly from A3 to F0.

Am and Ap stars always rotate slowly – no good case is known with $V \sin i > 100$ km/s.

Magnetic fields. Measurable magnetic fields have only been found in A-type stars with peculiar spectra (Ap stars), where the average longitudinal field is of the order of 10^3–10^4 G.

Photometry. We have already discussed some photometry on A-type stars in the previous sections, and in the chapter on UBV photometry. We shall present here simply in figure 10.5 the results of broad band photometry as taken from Straizys (1977). As can be seen, a very smooth relation exists between spectral type and photometry. Colors are as usual taken with regard to average A0 stars – therefore the color for A0 should be zero at all indices and this is approximately so. The remaining differences come from the fact that the average color index of a sample of various A0 stars is not strictly zero.

In figure 10.6 we have plotted the UBV indices for dwarfs and supergiants. As can be seen a difference is present only in early type stars; here the Balmer discontinuity is smaller for supergiants than for dwarfs, at the same spectral type. In the range A3–F0 the difference between luminosity classes I and V is too small for practical use.

Figure 10.5. Broad band photometry of A-type stars: U, B, . . . , L colors for dwarfs. The position of the passbands is given on the abscissa. V = 0.

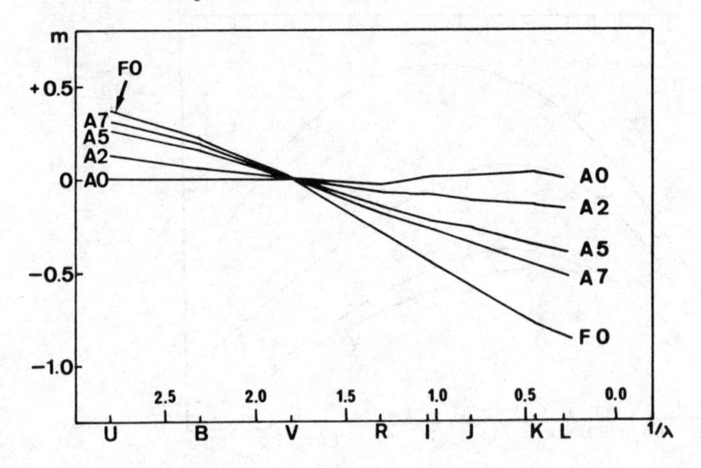

If other photometries are considered, uvbyβ for instance, the same facts appear in a slightly different form. Since the Balmer line strength and the Balmer discontinuity reach their maximum around A2, a graph using β versus [b–y] or [c] versus [b–y] should be two-valued, i.e. have two values of [b–y] for each c or β value. The effects described above in UBV concerning the sensitivity to luminosity should appear as a spread outside the sequence occupied by dwarfs. This is shown in figure 10.7.

Figure 10.6. Luminosity effects in UVB photometry for A-type stars.

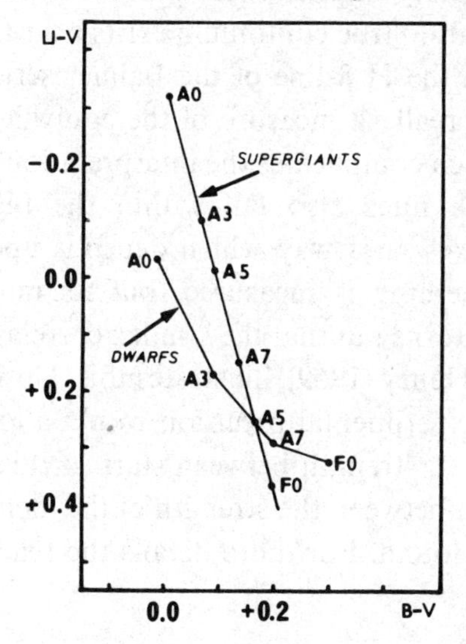

Figure 10.7. Strömgren photometry for A-type stars. [b–y] versus β diagram. The hatched area shows the location of giants and supergiants.

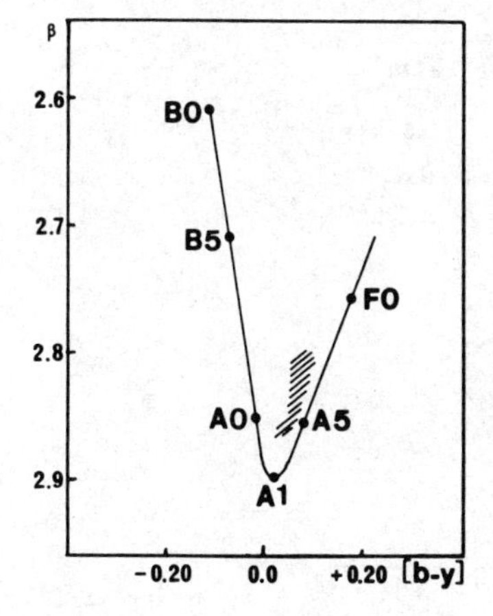

The m_1 parameter can also be used, which measures blanketing so that stars with more or with fewer lines than normal dwarfs are singled out. Application of this can be seen in the sections on Am (10.1) and λ Boo (10.3) stars.

Narrow band filters have been used specifically for hydrogen lines and the K line. We have already mentioned the β measures relating to Hβ, but the Hα line can also be used, as was done for instance by Mendoza (1978). The K line has been used by Lockwood (1968) and Henry (1969). The K line is measured through a filter 8.5 Å wide, and the continuum is measured through a similar filter centered on λ3915. A major difficulty appears with the continuum, which can be summarized by saying that no true continuum exists near the line. λ3915 lies in the extreme wing of the H 8 line of the Balmer series; therefore the ratio $\lambda 3933/\lambda 3915$ is not really a measure of the equivalent width of the K line. A second factor which complicates the interpretation of the measures is that numerous weak lines also fall within the filter passbands. These lines depress the fluxes, in a way which depends upon the spectral type. In conclusion, something is measured, but its interpretation is not obvious. Despite this it turns out that the k index correlates well with b–y. Figure 10.8, taken from Henry (1969), illustrates this. Notice that the dispersion is larger than the experimental precision would allow; Henry attributes this to variation in Ca II strength between stars. If this is true, we should expect to find a relation between the strength of the metals and the k index; indeed this is what is found. For more details the reader is referred to Henry (1978).

Figure 10.8. The narrow band Ca II K line index (k) as a function of b–y color. The bars indicate the limits of observational error.

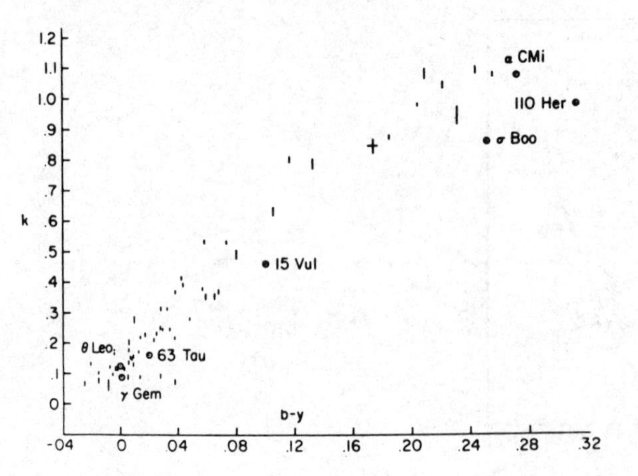

Another line extensively used for narrow band photometry is the O I feature at $\lambda\lambda$7771–4, which as we have mentioned is luminosity sensitive. Several photoelectric measures have been performed, for instance by Osmer (1972) and Mendoza (1971).

Binaries. The proportion of visual binaries seems to be normal, whereas the proportion of spectroscopic binaries is a somewhat controversial subject. Abt (1965) found the following percentages: subgiants, 30%; Ap 30%; Am, 80–100%. The average for A-type stars is about 50% (Jaschek and Gomez 1970). Furthermore Abt finds as a rule no spectroscopic binaries with normal spectra and periods less than a hundred days. If we consider *all* A-type stars together (i.e. A + Ap + Am) a normal percentage of spectroscopic binaries results. Such a procedure is not entirely unjustified, since the rotation statistics behave similarly.

There are few spectroscopic binaries among very rapid rotators, and this again is in line with the other facts. Am stars are slower rotators than normal stars, so that their proportion of binaries should be larger than in normal stars.

Absolute magnitude. A-type dwarfs and giants are on average still too rare and too far away as to be within the reach of trigonometric parallax measurements. Thus their absolute magnitude has to be established by statistical parallax, whereas for supergiants cluster stars must be used. In table 10.7 the results are summarized. They are taken from Schmidt-Kaler (1982) for luminosity classes I and II and from Grenier *et al.* (1985) for classes III and V.

It should be noted that, whenever possible, M should be derived through photometry, and not through spectral classification, since each MK

Table 10.7. *Visual absolute magnitude of A-type stars.*

	V	III	II	Ib	Ia
A0	0.65	0.0	− 3.0	− 5.2	− 7.2
A2	1.3	0.3	− 2.9	− 5.2	− 7.2
A5	1.95	0.7	− 2.8	− 5.1	− 7.2
F0	2.7	1.5	− 2.5	− 5.1	− 7.2

Note: r.m.s. errors ~ $\pm 0^m5$

luminosity class encompasses stars within a rather large absolute magnitude interval; for dwarfs, the range is 2^m (Jaschek and Mermilliod 1984).

Number, distribution in the galaxy. A-type stars are common objects in any survey complete down to a certain magnitude limit, as can be seen from table 10.8; they constitute between one-quarter and one-fifth of all stars visible. This proportion changes drastically when a distance-limited sample is considered; it falls then to less than 1%.

It is interesting to consider in more detail the distribution of the different groups of A stars. This was done on a sample of bright ($m \leq 6^m\!.5$) A stars classified by Cowley *et al.* (1969) (see table 10.9). From the table it is clear that stars other than dwarfs are less frequent than dwarfs, and that the proportion of Ap + Am stars is comparable to the number of all non-dwarfs. Most A-type dwarfs tend to be of early type; there are practically no stars classified as late dwarfs. This imbalance is slightly (but not entirely) reversed if Am stars are considered to be part of the normal late A-type stars. We have seen already that such a consideration is also suggested by the statistics of rotation and binaries. In contrast to dwarfs, giants are more evenly distributed over spectral type.

The A-type stars in the galaxy are concentrated toward the galactic plane. The average absolute galactic latitude is 18° and the average absolute

Table 10.8. *Number of A-type stars.*

Limit	Number of stars	Percentage of all stars
$m = 6^m\!.5$	1 500	24
$m = 9^m$	16 800	22
$r = 20$ pc	7	0.5

Table 10.9. *Distribution of A-type stars.*

Class	Percentage	Notes
V	63	Of these stars 85% are A0-A4
IV, III	17	Of these stars 54% are A0-A4
II, I	3	
Ap	5	
Am	12	

x-distance is of the order of 120 pc. Within the galactic plane the distribution seems to be rather uniform, with fluctuations of up to about 360 pc.

A-type stars are found in OB associations, but they have rarely been studied in detail because of their faintness even in nearby associations; for instance in the Sco-Cen association we know only the earliest A-type stars. In open clusters the situation is slightly better. From the data collected by Mermilliod (1980), there are 14 clusters in which the A9 stars brighter than $M_v = 10^m$ have been observed. However, not even all of these stars have been classified.

10.1 Am stars

An Am star is an A- or (early) F-type object to which no unique spectral type can be assigned. In particular the intensity of the K line of Ca II does not correspond to the strength of the Balmer lines and the metallic lines (see figure 10.9). By comparison with suitable standards a spectral type [Sp(K)] can be attributed to the K line, another type [Sp(H)] to the hydrogen lines and another [Sp(m)] to the metallic lines. We then find that Am stars obey the rule:

$$\mathrm{Sp(K)} \leqq \mathrm{Sp(H)} \leqq \mathrm{Sp(m)}$$

$$\mathrm{A1} < \mathrm{Sp(K)} < \mathrm{A6} \qquad \mathrm{A5} < \mathrm{Sp(H)} < \mathrm{F2} \qquad \mathrm{A5} < \mathrm{Sp(m)} < \mathrm{F6}$$

The only case excluded is Sp(K) = Sp(H) = Sp(m) which is obviously a normal star. Of these three types, Sp(K) and Sp(m) are easy to assign, whereas Sp(H) is difficult and has a large uncertainty attached.

The Am stars or metallic line stars were discovered by Titus and Morgan (1940) in the Hyades and the definition quoted above comes from Roman, Morgan and Eggen (1947); it can be applied to spectra having plate factors between 40 and 120 Å/m. Stars in which Sp(K) ≪ Sp(m) can be picked out even at 300 Å/m (Honeycutt and McCuskey 1966). Because of the difficulty of classifying Sp(H), not all authors quote it and usually the original paper has to be read to ascertain precisely what an author has done when classifying an Am star. Hauck (1986) has grouped all available classifications of Am stars in a catalog which greatly facilitates the task.

For several years nothing was added to these classification criteria until Bidelman in the fifties discovered that the Sc II lines (mostly λ4246, but also λ4320) are extremely weak in Am stars. The lines become visible only at 40 Å/m or better and offer a convenient way to characterize Am stars at such dispersions, since the Ca II and H I lines are too strong to be usable as classification criteria. Conti (1965) used the ratio Sc II λ4246/Sr II λ4215 to separate Am stars, because Sr II is enhanced and Sc II disappears in Am stars. We should however notice that both lines are luminosity sensitive, so that the Sc/Sr ratio should be handled carefully. A further difficulty arises from the fact that not all Sc II weak stars are Am. In fact in many late Ap stars of the Cr-Eu-Sr type, Sc II is also weak; on the other hand in some Am stars Sc II is *not* weak (Jaschek and Jaschek 1960).

The usefulness of the Sc II criterion is greatest in the early A-type stars, as Conti and Strom (1968a,b) showed. By means of stellar atmospheres analysis they were able to show that a group of stars around A0 characterized by low

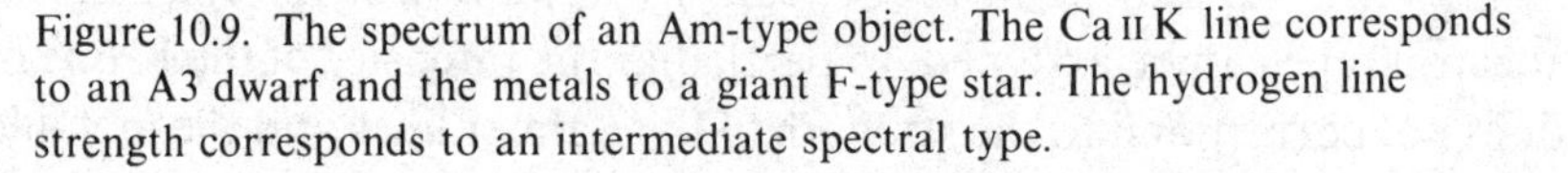

Figure 10.9. The spectrum of an Am-type object. The Ca II K line corresponds to an A3 dwarf and the metals to a giant F-type star. The hydrogen line strength corresponds to an intermediate spectral type.

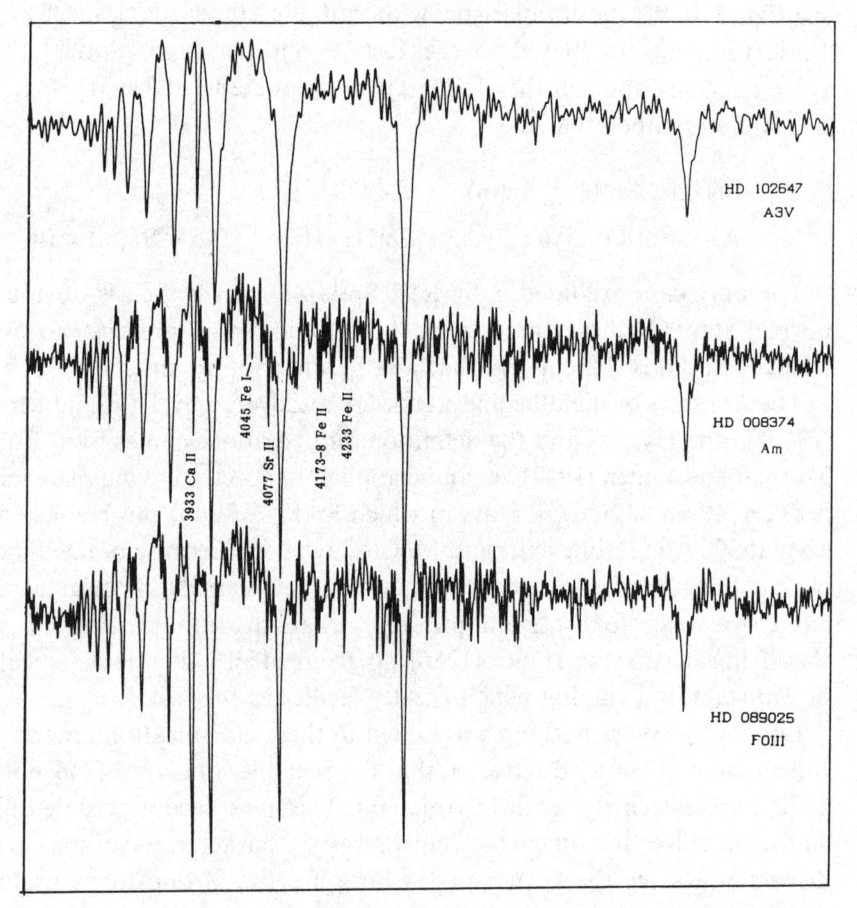

$\lambda 4246/\lambda 4215$ ratios had abundance anomalies similar to those of the typical Am and could be considered to be the early part of this group.

Other characteristics of Am stars are that metallic lines are sharp and that some lines usually strengthened in giants (Sr II $\lambda 4077$, Fe II $\lambda\lambda 4173$–8) are also enhanced in them. This pseudo-luminosity effect means that the metallic line spectrum is often classified as type III. However, there are differences between an Am and a giant spectrum; e.g. in Am stars $\lambda 4417 < \lambda 4481$ (Cowley *et al.* 1969), whereas $\lambda 4417 \sim \lambda 4481$ in giants.

Abt and Morgan (1976) made one further step in this direction (on 125 Å/mm plates) using two regions to provide luminosity criteria, namely $\lambda\lambda 3850$–4078 (called '39' for short) and $\lambda\lambda 4260$–4340 (called '43'). They found that according to the blue criteria, the star is a dwarf, whereas the violet criteria correspond to a giant or even a supergiant spectrum. They characterize Am stars by three types, namely the K line type, the spectral type in the '39' region and the type in the '43' region. For instance

HR 1376 = 63 Tau A2 – F3 III – F2 V

Objections have been raised by theoreticians of stellar atmospheres who do not find different physical parameters corresponding to the two regions. Obviously more work is required to clarify this point, but it is clear that when assigning metallic line spectral types it is vital to state what was done and how.

In their recent atlas, Morgan, Abt and Tapscott (1978) divided Am stars into two groups – the 'classical' and the 'proto'. They say 'we label classical Am stars' as those which have a difference between K line type and metallic line type greater than 0.5 spectral class. The 'proto Am stars' are those having a smaller difference. They add that the 'normal' metallic line type for the classical Am stars is determined over the spectral range $\lambda\lambda 3850$–4200, whereas for the proto Am stars it is determined over the spectral range $\lambda\lambda 3970$–4340. They use the notation kA4, mA9 (for instance) to denote Sp(K) = A4, Sp(m) = A9.

The division into 'classical' and 'proto' Am stars is not a new one because 'marginal' Am stars (i.e. stars with less definite Am characteristics) have been known for a long time. They were called 'marginal', 'mild', or 'fringe-metallic line stars' (Strömgren 1966) by different authors. In what follows we shall call 'mild' Am a star which does not possess definite Am characteristics, but whose Sp(K) differs little from Sp(H) and Sp(m).

A term also used in connection with Am stars is 'metallicity'. Jaschek and Jaschek (1959) used it in the sense Sp(m) – Sp(H) and we shall continue to use it in this sense. Hack (1959) used metallicity for Sp(m) – Sp(K) which has the advantage of being easier to determine, but is less significant.

On the basis of the different criteria given, it is not surprising that disagreements exist over the assignment of a given star to the class of Am stars. Misclassifications of Am stars can arise in different ways, but special care is needed when using low plate factors. Since Am stars usually have sharp lines, standards of similar rotational velocity must be used for comparison, since otherwise sharp lined late A-type stars are likely to be misclassified as Am stars.

Another difficulty arises when dealing with very early and very late Am stars. Conti (1965) showed for instance that when seen at low plate factors, Sirius (A1 v) exhibits weak Ca and Sc and strong lines of heavier elements. This pattern is typical of Am-type stars, and Sirius is thus, at this dispersion, an early Am object. At classification plate factors it is not detected as such, because the Sc II λ4246/Sr II λ4215 criterion is not usable. Similar difficulties arise with late Am stars (see section 11.1).

The interested reader is referred for a complete discussion of the classification problem to Jaschek and Jaschek (1974).

When low plate factors are used to analyse the behavior of the elements in Am stars, there is a rather large scatter between the results of different authors (see for instance Conti 1970). To characterize the general trend, we quote the results of Jaschek and Jaschek (1960) who estimated visually the intensities of 220 lines belonging to different elements in 26 Am and 7 normal stars. They found that those elements observed in the spectrum which lie before the iron peak are weaker than in normal stars, whereas elements after the iron peak tend to be stronger. So for instance Ca and Sc are weak and Mn, Y, Sr and Ba are enhanced. Rare earths may also appear, but are never very strong. A search for correlations among elements gave rather meager results, and in particular no families were found among the Am stars; a result confirmed by more recent work by Cowley and Henry (1979). This is surprising because no two Am stars are identical.

Photometry. There are more metallic lines present in an Am spectrum than there should be according to the Sp(H), which provides a fruitful basis for singling out Am stars photometrically. Any index measuring the crowding of lines will do the job. Furthermore we have already seen that the line blocking increases toward shorter wavelengths. As a result, in the UBV system, Am stars fall below the main sequence in the (U–B, B–V) plot. This is shown in figure 10.10 where the Am stars fall below the main curve by amounts between $0^m_.1$ and $0^m_.2$.

If blocking decreases with increasing λ, it can be assumed that the longer wavelength regions (colors R and I, for instance) are essentially unaffected by blanketing (Ferrer, Jaschek and Jaschek 1970). Normal stars and Am stars should thus have identical R–I colors. We can then mark in the (U–B, B–V) diagram the line uniting the observed colors of an Am star with the colors of a normal dwarf with the same R–I index. The result is shown in figure 10.11(a).

This procedure enables us to decide which of the three classifications (Sp(H), Sp(m) and Sp(K)) is the most significant: it should be the one which has the tightest correlation with R–I. It turns out that this is the hydrogen line type. Incidentally this type correlates very well, as it should, with the β narrow band index.

Returning to figure 10.11(a), observe that the moduli of the vectors uniting the observed colors with those of the 'de-blocked' dwarfs varies from star to star. It is tempting to see if these lengths are related to the metallicity m = Sp(m) – Sp(H); figure 10.11(b), taken from Ferrer, Jaschek and Jaschek (1970), shows this to be true.

Essentially similar results can be obtained with other photometric systems, provided they contain a temperature index and a metallicity index.

Narrow band photometry has been done in the K line and in the O I feature $\lambda\lambda$7772–5. The K line photometry has been discussed in section 10.0; we recall that the K line index varies smoothly with any temperature index

Figure 10.10. Position of Am-type stars in UBV photometry. The continuous curve corresponds to the dwarf sequence.

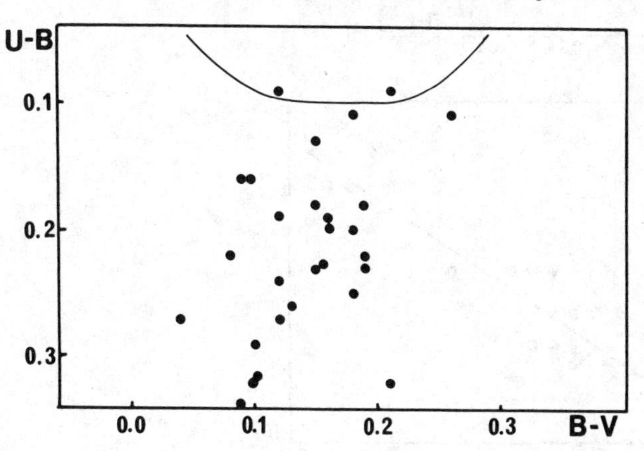

over the whole range of A-type stars. Since Am stars are characterized by Sp(K) < Sp(H), they should separate clearly from dwarfs. Figure 10.12 reproduces data from Henry and Hesser (1971); as can be seen Am stars segregate from dwarfs, although no sharp boundary exists. Henry (1978) has also found several cases of variable K line strengths, a fact which might explain part of the scatter.

Mendoza (1978) measured the O I $\lambda\lambda$7772–5 feature through narrow band photometry. By combining O I and Hα measures, he is able to segregate Am stars from normal dwarfs because Am stars have weak oxygen lines.

Figure 10.11(a). Position of Am-type stars in UBV photometry; Segments show the 'de-blocking'; for explanation see text. From Ferrer, Jaschek and Jaschek (1970). (b) Metallicity versus blanketing vector modulus. The length of the segments of (a) are indicated on the abscissa, the metallicity (see text) on the ordinate.

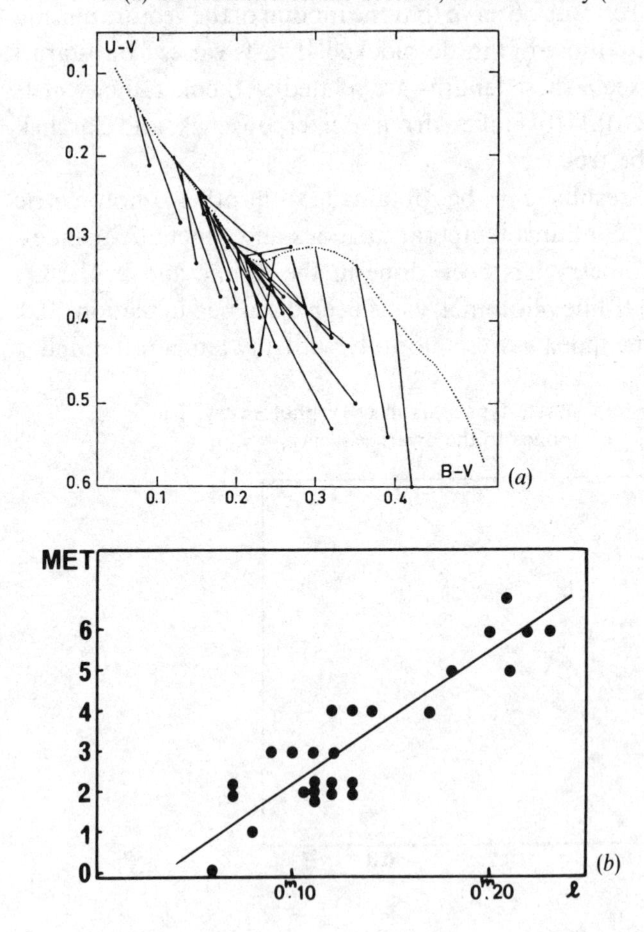

Rotation. In general Am stars have sharp lines, i.e. they rotate slowly.

Abt and Moyd (1973) have studied the question in detail, comparing a sample of A5–A9 luminosity classes IV and V stars (the 'normal stars') with a sample of Am stars. Table 10.10, taken from their work, summarizes the results. It is clear that Am stars rotate far less rapidly than A-type stars. The fact however that both distributions overlap shows that rotation alone cannot be the sole cause of 'metallicity'. Some authors have tried to link the rotation with the amount of metallicity (defined through a photometric index) in the hope of showing that a slower rotation produces a more pronounced Am star. The fact that different authors (Burkhart 1979; Hauck and Curchod 1980) have reached opposite conclusions on this point shows that at the least no clear-cut relation exists.

Table 10.10. *Rotational velocities of normal A and Am stars.*

$V \sin i$	Percentage of A stars	Percentage of Am stars
0–49	5%	67%
50–99	23%	23%
100–149	37%	
150–199	14%	
200–249	15%	
250–299	5%	
Number of stars	109	65

Figure 10.12. Narrow band Ca II K line photometry as a function of [b–y]. The continuous line represents the sequence for dwarfs, and the points the Am stars.

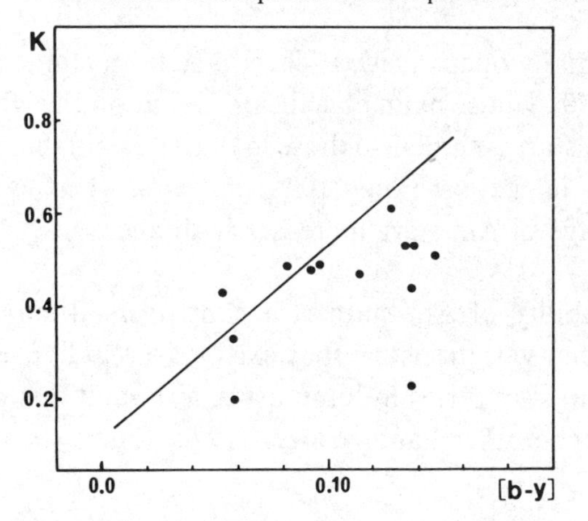

The marginal Am stars exhibit slightly larger rotational velocities than the classic Am stars, but again it is not the amount of rotation which decides if a star belongs to one or the other group.

Magnetic fields. All magnetic fields in Am stars measured up to now are near the possible observing errors (Borra and Landstreet 1980) so that we can suppose that the fields, if they exist, are substantially smaller than those in Ap stars.

Binaries. Abt (1961) found that all Am stars are spectroscopic binaries, mostly with periods $P < 100^d$; and according to him the few Am stars that are not SB can be regarded as stars seen pole-on. From a spectral classification of all SBs in the range A2–F3, Abt and Bidelman (1969) conclude that all SBs with periods between $2^d\!.5$ and 100^d have Am primaries. Ginestet *et al.* (1982) found some exceptions.

Visual binaries have the same frequency as in normal A stars (Abt 1965), implying essentially that wide binaries are unaffected by metallicity. However some curious rules emerge from a detailed analysis of both visual and spectroscopic binaries. If one of the two components of a binary is Am, generally (but not always) the other object is also Am or Ap. However, even if both components of the SB are Am, the elements which have anomalous line strengths are not the same (Stickland 1973). Furthermore if one normal and one Am star are present in a visual binary, usually the Am is the primary (Jaschek and Jaschek 1967). A curious fact is that among the many Am stars known (all of which are binaries) there should be many eclipsing binaries, but surprisingly very few cases are known (Popper 1980).

Clusters. Am stars exist in many open clusters. They have been studied by Hartoog (1976) and Abt (1979), whose main conclusion is that on the whole the overall percentage of Am stars is similar to those in the field, and that Am stars are present in clusters in the age range 10^6–10^8 years. There is the possibility that the percentage of Am stars increases with age.

Variability. The light variability of Am stars is a controversial subject, although all authors agree that variations – if they exist – are small. Breger (1970) concluded that no Am star varies in luminosity, although this may be a little too extreme. On the other hand Winzer (1974) found six stars out of a sample of 12 Am stars to be variable by amounts of $0^{m}_{.}01$ over a few days. These observations, as well as those of stars like 32 and 28 Vir, and which sometimes show variations up to $0^{m}_{.}05$, leave the question open. Eggen (1976) wrote that this might be just a problem of terminology because the stars which Breger found to be variable in this area are called δ Scu (see section 10.2).

Further important evidence was added by Böhm-Vitense and Johnson (1978) who found that some Am stars may be variable on a long time scale, as judged from a comparison of spectrum scans.

Absolute magnitude. According to recent work of Gomez *et al.* (1981), who used statistical parallaxes, membership in visual binaries and in open clusters, $M_v = 1.2 \pm 0.3$ at B–V = 0.16. This positions the stars one magnitude above the main sequence. At the moment the material is insufficient to decide if the absolute magnitude depends upon the metallicity and if marginal Am stars lie closer to the main sequence.

Proportion of Am stars. We have already said that the spectral type of the hydrogen lines of the Am stars falls between A5 and F2. Because of the difficulty of classifying Sp(H) it seems best to provide the statistics of the numerical incidence of Am stars by means of a photometric index, for instance B–V. Table 10.11 shows that Am stars constitute a sizable fraction of all A stars, especially around B–V $\sim$ 0.25. In fact few normal dwarfs are known of types A7 and A8.

Number of Am stars. The number of known Am stars with $V \leqslant 6^{m}_{.}5$ is 270 and with $V \leqslant 9^{m}$ about 1700. They are thus fairly common objects. Their distribution follows that of the normal dwarfs of the same Sp(H).

Table 10.11. *Proportion of Am stars.*

B–V interval	Am/AV + Am
0.10–0.19	0.10
0.20–0.29	0.36
0.30–0.39	0.12

Note: The B–V indices are not corrected for blanketing. Data from Jaschek

10.2 δ Scu and δ Del stars

δ Scu stars are defined as pulsating variables of spectral type A or F, with a period shorter than $0^{d}.3$. (Breger 1975, 1979). It should be mentioned that different authors have used slightly different definitions – some use different limits of the period and some introduce restrictions on the light amplitudes.

The light curves generally have a small amplitude of a few hundredths of a magnitude. The shape of the light curve is variable over time and variability may even fall below detection level at times, a fact which produces some difficulties for the assignment of group membership.

The group has also been called dwarf Cepheids, ultra short period variables or AI Vel type objects (Breger 1979).

δ Scuti, the prototype, was discovered as variable by Fath (1935), but the group was defined by Eggen (1956). Since in this book we do not deal with variable stars, the discussion could stop here, were it not for the fact that a few δ Scu stars have a spectroscopic peculiarity. These are the so-called 'δ Del' stars, whose history is rather complex.

Bidelman (1951) called attention to the spectrum of the star δ Pup in which 'the Ca II lines are rather weak for the given type' (F6 II). This characteristic was later used for a group, the so-called 'δ Del' stars (because δ Del was the brightest start of the new group). These stars may or may not be photometrically variable. Morgan and Abt (1972) classified the spectra of 14 δ Scu stars. They observe that 'the Ca II lines show a great range of intensity among objects having similar spectral types and luminosity classes'. They denote as 'p' only the most extreme cases of Ca II weakening (which are 4 out of the 14 spectra) and this is what other observers denote as 'δ Del' characteristics.

We may thus define δ Del stars as a group of late A to early F giants of luminosity classes II–IV, with weak Ca II lines.

The weakness of Ca II also characterizes the Am stars (section 10.1), so that both groups are probably related. Because of this Houk (1978) calls the stars 'Fm δ Del'. However in late A-type δ Del stars the hydrogen line spectrum closely matches the metallic line spectrum, a fact which does not always happen in Am stars.

At present no systematic work has been done on these stars. From the fragmentary data they do not seem to behave any differently than other stars of the same type. Many of the stars are variables of the δ Scu type, but we do not known if all δ Del stars are δ Scu stars or what fraction they form of the total. On the other hand, we know that *not* all δ Scu stars are δ Del stars.

About 25 δ Del stars have $V \leqslant 6^{\mathrm{m}}5$ and about 180 are known among HD stars.

10.3 λ Boo stars

These objects are defined as A-type stars with weakened metallic lines (see figure 10.13).

The prototype of the group, λ Boo, was described and illustrated by Morgan, Keenan and Kellman in their atlas (1943), on the basis of spectrograms having a plate factor of 110 Å/mm. When observed at higher resolution (40 Å/mm), λ Boo stars have low rotational velocity, with $V \sin i \leqslant 100$ km/s. When radial velocities are measured, the results are typically less than 20 km/s.

We can thus complete the preceding definition, saying 'λ Boo stars are A-type stars with metallic lines which are too weak for their spectral types, when the latter are determined from the ratio of their K line to Balmer line strengths. They have $V \sin i \geqslant 100$ km/s and low radial velocity.' Some authors use 'low space velocity' instead of 'low radial velocity'. We prefer the latter because it obviates the need to know the absolute magnitude of the star.

The definition makes clear that we are dealing with a subgroup of stars of

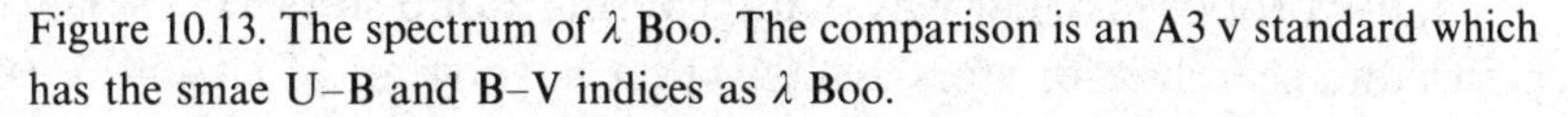

Figure 10.13. The spectrum of λ Boo. The comparison is an A3 v standard which has the smae U–B and B–V indices as λ Boo.

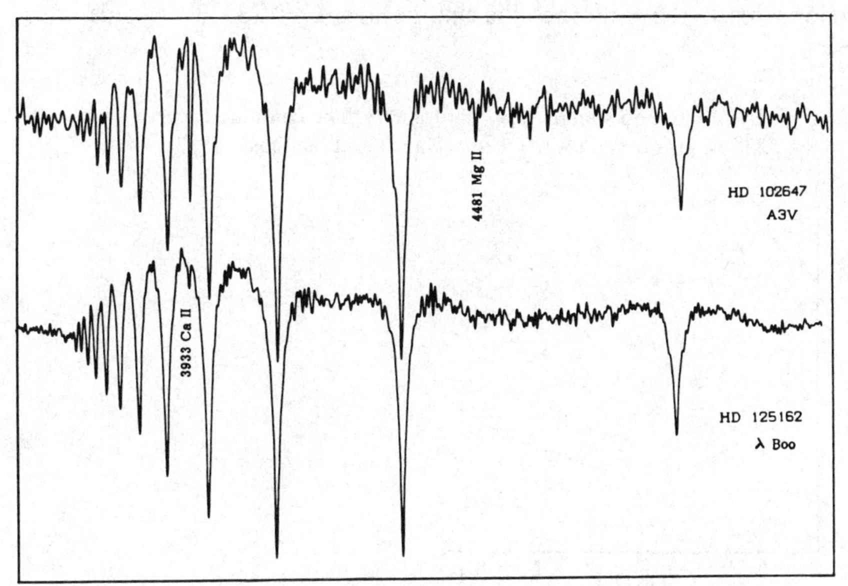

population I; the latter is based upon the low radial velocity. Hauck and Slettebak (1983) also include F-type stars in this group. We prefer to call these latter 'weak lined stars'.

In the ultraviolet ($\lambda < 3000$ Å) λ Boo stars are characterized by the presence of a strong C I line (λ1657), strong flux depressions around λ1600 and λ3040 and weakness of other metallic lines (Baschek *et al.* 1984).

The weakness of the metallic lines should be easy to measure photometrically, and it should thus be possible to segregate group members. Hauck and Slettebak (1983) have used both the Geneva photometry and the uvby system to carry this out; figure 10.14 illustrates the latter case.

However, care needs to be taken with this photometric method, because it segregates all metal weak stars, not just the λ Boo stars. This is because the conditions relating to the rotational velocity and the radial velocity need to be established using other techniques.

Until recently the number of known λ Boo objects was small (about ten), but the fact that some of them are very bright (for instance λ Boo, $V = 4^m18$) suggests that they should be numerous. On the other hand, Slettebak, Wright and Graham (1968) found, in a study of population I A-type stars in the direction of the galactic pole, three λ Boo stars among 77 A-type stars. The resulting percentage ($4 \pm 2\%$) shows that λ Boo stars are frequent, a fact which implies that many λ Boo stars have gone unnoticed. Recently Abt (1984) has proposed a new way of discovering λ Boo stars. He uses 40 Å/mm plates and searches for stars near A0 that have weak Mg II λ4481. (At this dispersion, this is the only metallic line besides Ca II λ3933.) Abt denotes

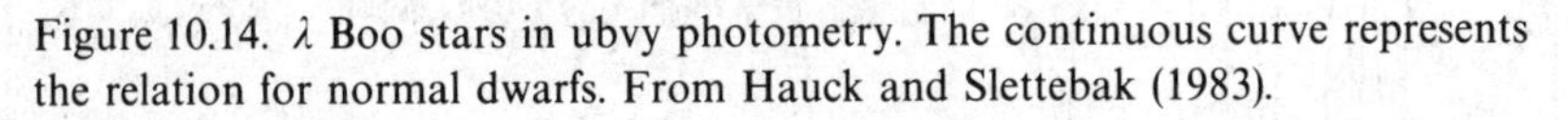

Figure 10.14. λ Boo stars in ubvy photometry. The continuous curve represents the relation for normal dwarfs. From Hauck and Slettebak (1983).

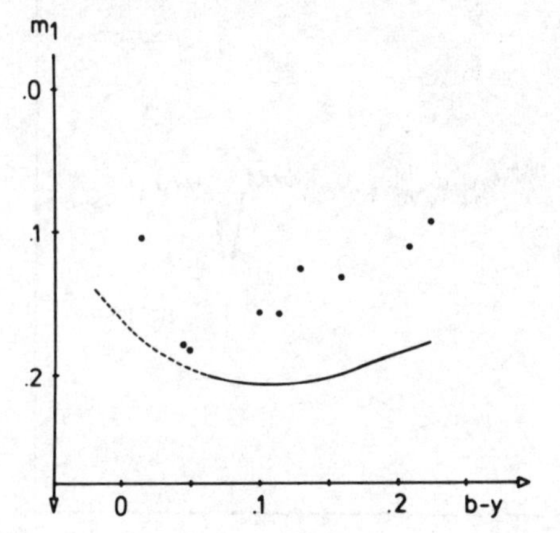

these stars as 'Ap (Mg weak)' or 'Ap (Ca, Mg weak)'. Most of his stars lie in the region A0–A3, luminosity classes III–V. In this range he finds 6% λ Boo stars, a percentage which agrees well with that of Slettebak *et al.* (1968).

10.4 Horizontal branch stars

Horizontal branch (HB) stars are defined by their position in the color–magnitude diagram of globular clusters. In figure 10.15 (Arp 1962) there are two branches – the red horizontal branch (RHB) and the blue horizontal branch (BHB) to the right and to the left of the position of the RR Lyrae stars. If the diagram is drawn in bolometric magnitude (m_{bol}) rather than in visual magnitude both branches fall along a horizontal line – hence the name 'horizontal branch'. The population of the horizontal branch varies significantly from one globular cluster to another. Clusters with stars in which metallic lines are weak have in general more stars on the HB than clusters in which there is no line weakening.

Observations of HB stars are difficult because in only a few southern clusters such as NGC 6347 do they have $m \leqslant 13$. Therefore a search was started to find brighter field HB stars. Candidates for HB stars are found

Figure 10.15. Observed HR diagram of the globular cluster M5. From Arp (1962). For terminology see figure 10.16.

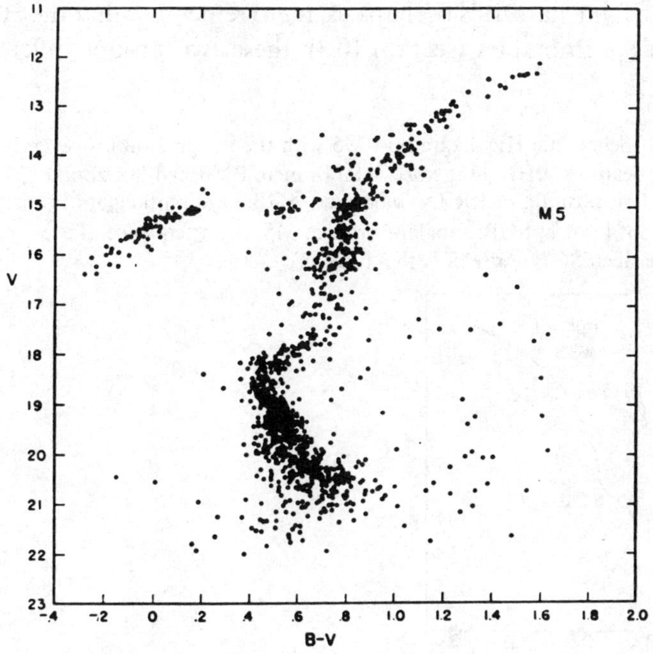

in the same samples as halo B-type stars, namely faint blue stars at high galactic latitude. This technique was first used by Humason and Zwicky (1947); a historical coverage is given in the '*First conference on faint blue stars*', Luyten (1965). In such a sample many different kinds of stars can be found, forcing us to define more stringent criteria to isolate HB field stars.

Spectroscopically HB stars are found to exhibit sharp and deep Balmer lines, a large Balmer discontinuity and very weak lines of other elements (Greenstein and Sargent 1974). So if the star is of type A, Ca II and Mg II are weak, and if of B-type, the He I lines are weakened. Such characteristics can be seen at low plate factors – 580 Å/mm to 180 Å/mm – but obviously weak metallic lines are not discovered at this dispersion.

This definition is similar to the one of Sargent and Searle (1968) who, from a study at 48 Å/mm, describe two groups, namely the Bw and Aw stars. 'Bw are B stars on the basis of their colors, but in whose spectra He I is absent or abnormally weak. These stars do have normal main sequence profiles with the Balmer series visible up to $n = 15$ or 16. This serves to distinguish them from helium weak sdB stars ... Aw ... is used for stars which have the colors of normal A stars but in whose spectra the metallic lines commonly encountered in A stars (e.g. the Ca II K line and Mg II λ4481) are abnormally weak.' It is clear that Greenstein and Sargent's definition is to be preferred since it is based upon both a larger sample and the higher plate factors needed for work on faint stars. It is perhaps useful to remember that the HB-type stars resemble λ Boo stars (section 10.3); these two groups differ in

Figure 10.16. Schematic HR diagram of M5 with the designation for several HR diagram features. BHB, blue horizontal branch; RHB, red horizontal branch; RR Lyr, position of RR Lyr variables; AGB, asymptotic giant branch: RGB, red giant branch; SGB, subgiant branch; MS, main sequence; the transition between SGB and MS is the 'turn-off'.

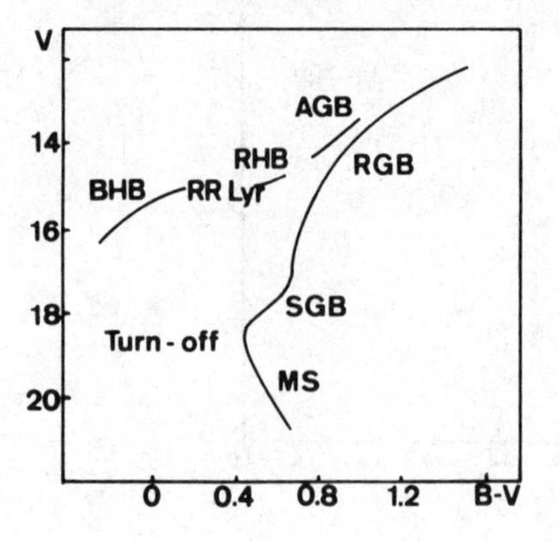

that λ Boo stars have large $V \sin i$ ($> 100\,\mathrm{km/s}$) and low radial velocity ($\rho < 20\,\mathrm{km/s}$), whereas for HB stars $V \sin i$ is low and the radial velocity high (as in all globular cluster stars).

A word of caution should be added regarding the B-type HB stars. Since these objects are defined through their He weakness to be HB stars, we can clearly not conclude afterwards that HB stars are He weak, without independent evidence. From a purely morphological point of view the HB stars, as well as the sdO and sdB stars, suffer from the lack of a classification based on homogeneous material. Stars are thus often studied in detail without previous knowledge of whether an object is really representative of the group: this is so because the faintness of most of these objects makes them accessible only to the largest telescopes. (For more details see Baschek 1975.)

In the UV ($\lambda < 3000$ Å) region, the spectra of the HB stars are characterized by the weakness of the metallic lines. In some stars flux depressions are also observed at λ1600 and λ3040, which appear in λ Boo stars (Jaschek, *et al.* 1985).

Photometry. As explained before, many of these objects were discovered because of their blueness on photographic plates. Because of the large possible errors inherent in photographic photometry, the stars must be reobserved photoelectrically. Figure 10.17 taken from Newell (1973) shows what happens when a sample of high latitude faint blue stars are observed in UBV.

It can be seen that except for a tendency to lie to the right of the main sequence by amounts which are larger than might be expected from reddening, most of the stars follow rather well the general pattern of the normal stars if we omit for the moment the white dwarfs and subdwarfs. The two gaps on the main sequence shown in the figure, which Newell believes to be real, are still under discussion. They are clearly seen in globular cluster HB stars.

The fact that in HB spectra the Balmer jump is large and that few metallic lines are visible can be used in ubvy photometry, and this was done by Philip (1978). In figures 10.18 and 10.19 are shown some of the results which may be compared with figure 9.30 of section 9.8. The same author has shown (Philip 1972) that for HB stars of type A, the Hβ index is smaller than for normal dwarfs. All this allows us to segregate with relative security HB stars later than A0, but this is not true for the B-type HB stars. We know little about the variability of these objects. McMillan *et al.* (1976) have investigated a small sample of seven stars; none was found to be variable at a $0^{m}.01$ level.

Figure 10.17. UBV photometry of high latitude faint blue stars. Continuous line, relation for dwarfs; top diagonal straight line, black body (BB); arrow, direction of reddening; shaded areas, region occupied by degenerates (wd) and subgiant subdwarfs of types F and G (sd); light lines mark two hypothetical gaps. From Newell (1973).

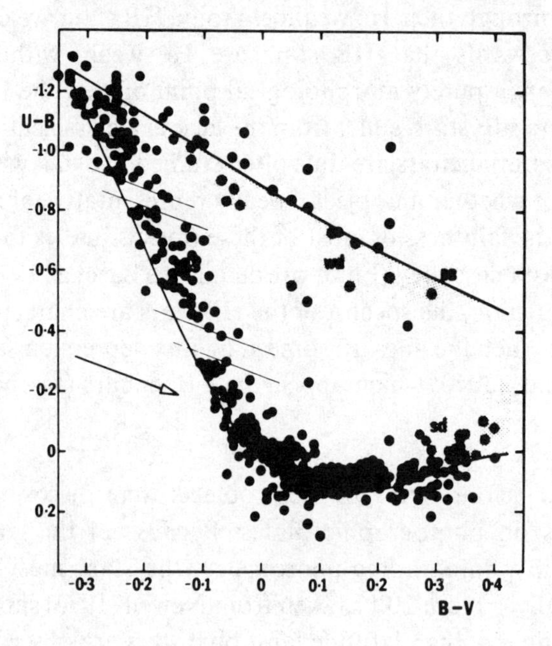

Figure 10.18. ubvy photometry of field horizontal branch stars. $(c_1)_0$ and $(b-y)_0$ denote de-reddened c_1 and b–y indices. From Philip (1972).

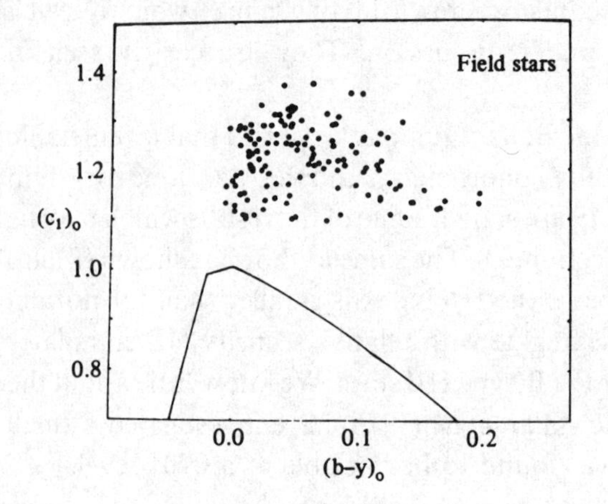

Spectrophotometry. Scans of a number of field HB A-type stars and globular cluster HB stars were measured by Philip and Hayes (1983) and Hayes and Philip (1983).

Rotation. This was measured by Peterson, Tarbell and Carney (1983) and Peterson (1983) and found to be small ($V \sin i < 30$ km/s).

Radial velocity. The radial velocities are large (Sommer-Larsen and Christensen 1985) in general and have a large scatter. From a sample of 30 stars, $|\bar{V}| \sim 80$ km/s.

Absolute magnitude. The absolute magnitudes come from HB stars in globular clusters (Hayes and Philip 1979). For B–V = 0.10, $M_v = +0.6$, and for U–B = -0.5, $M_v = +2.0$. This implies that early HB stars are two or three magnitudes fainter than RR Lyrae stars.

Number. About 120 HB stars are known from different sources, but only seven are brighter than $V = 9^m$.

Atlas. McConnell *et al.* (1971) reproduce the spectra of some field HB stars.

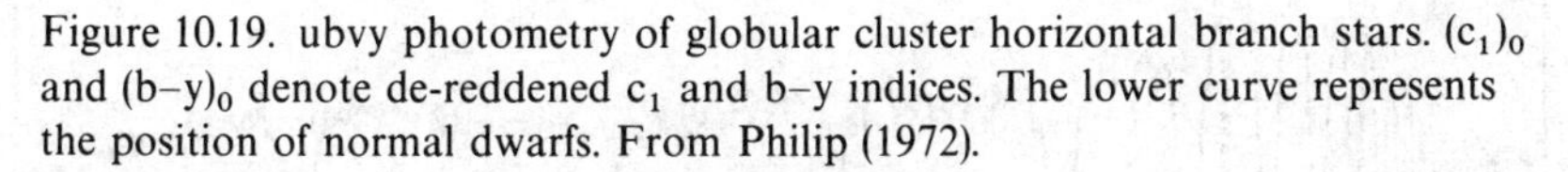
Figure 10.19. ubvy photometry of globular cluster horizontal branch stars. $(c_1)_0$ and $(b-y)_0$ denote de-reddened c_1 and b–y indices. The lower curve represents the position of normal dwarfs. From Philip (1972).

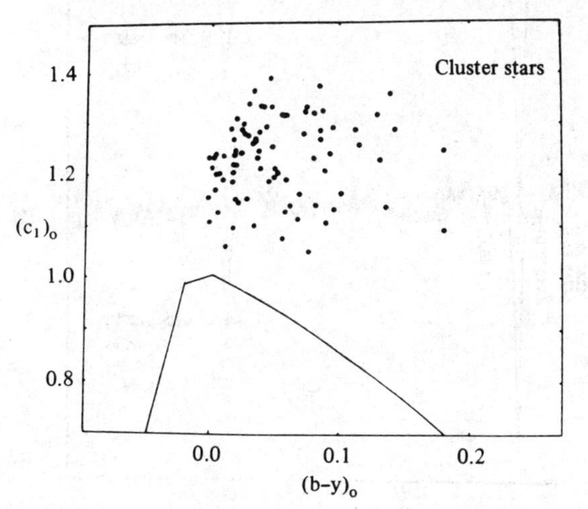

10.5 A-shell and Ae stars

In analogy with the terminology used for B-type stars, an Ae star is an A-type star with emission in the Balmer lines and an A-shell star an object in which two different types of line profiles coexist.

As a rule there are very few A-type stars with emission in the Balmer lines; the few cases correspond to Herbig Ae/Be stars (section 10.6) and to close (spectroscopic) binaries, where the emission is probably due to interactions in the system. This is in line with the rule followed by Be stars that emissions weaken with advancing type, so that in later B-type stars generally only Hα (and perhaps Hβ) are seen in moderate or strong emission (i.e. they rise above the level of adjacent continuum). In early A-type stars, the emissions very rarely rise above the adjacent continuum. For instance in a sample of 12 stars observed by Andrillat, Jaschek and Jaschek (1986) none showed this.

A-type shell stars were studied systematically for the first time by Abt and Moyd (1973) although isolated cases were known before. These authors found that shell lines – mostly of Ca II and Ti II – are present in a number of normal A-type stars of luminosity classes IV to V having $V \sin i > 100$ km/s (see figure 10.20).

Jaschek *et al.* (1986) studied a large sample – 28 stars – of stars which had

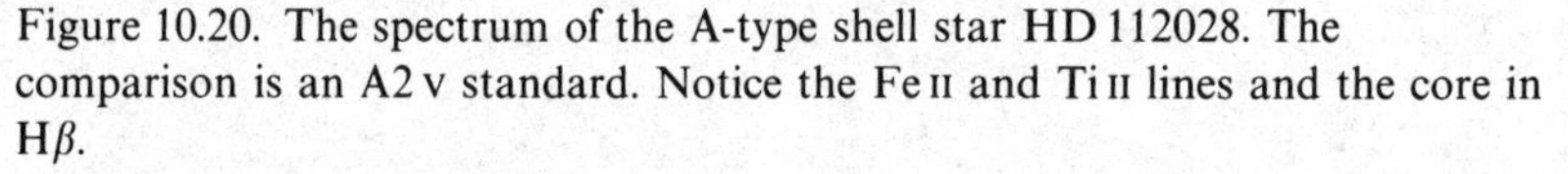
Figure 10.20. The spectrum of the A-type shell star HD 112028. The comparison is an A2 V standard. Notice the Fe II and Ti II lines and the core in Hβ.

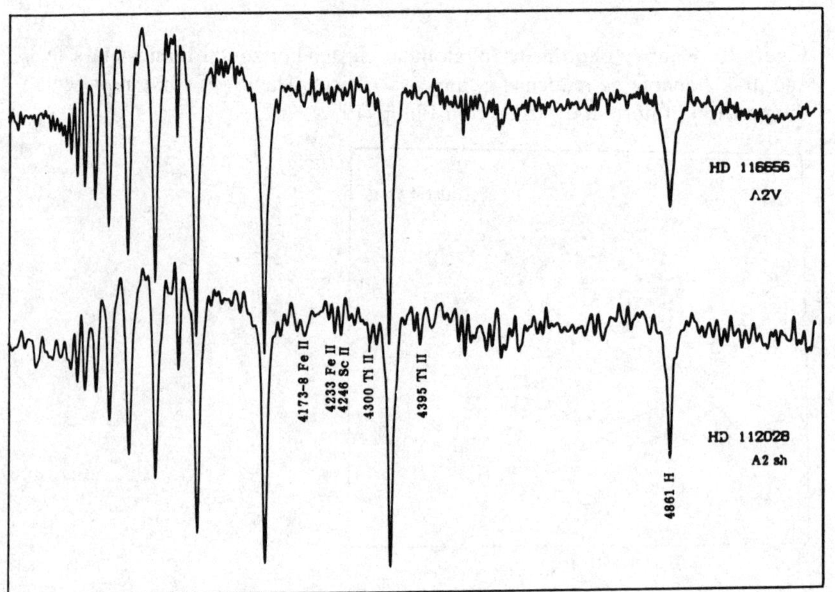

been announced at some time as belonging to this group. A first result is that some stars do not exhibit shell features at present; this implies that shell features are variable over time. It is not possible to specify the time scales; a general impression is that changes are slow, because at least in some cases the shell has not changed appreciably over thirty years.

When we examine those stars which show shell features, we find (as in B-type stars) a large variety of shells. Basically in the weak shells sharp absorption is seen in the Ca II (H and K) lines only, or in the Ca II and Ti II lines. In the stronger shells more elements are seen, especially Fe II and Sc II. The spectral types of the underlying stars go from A0 to late A, although 14 Com could also be called F0.

As far as the Balmer lines are concerned, only core structures are noticeable in the very early (A0–A2) objects, and here structures may appear up to H12 (HD 15253). Toward mid A-types the structures, if they exist, appear only in Hα, and after A5 they disappear completely.

Abt and Moyd (1973) suggested that all rapidly rotating A-type stars can develop shell structures at some time. This is an interesting suggestion; Jaschek *et al.* (1986) reobserved part of their sample and did not find new shell stars.

Photometrically not much is known about these stars. The location of the stars in a UBV diagram is shown in figure 10.21, from which it is evident that stars fall near (within $0^m.1$) the main sequence. No systematic study of the variability has been undertaken as yet.

Figure 10.2.1. UBV photometry of Ae and A-shell stars. The continuous curve represent normal dwarfs; points, weak shells; triangles, strong shells.

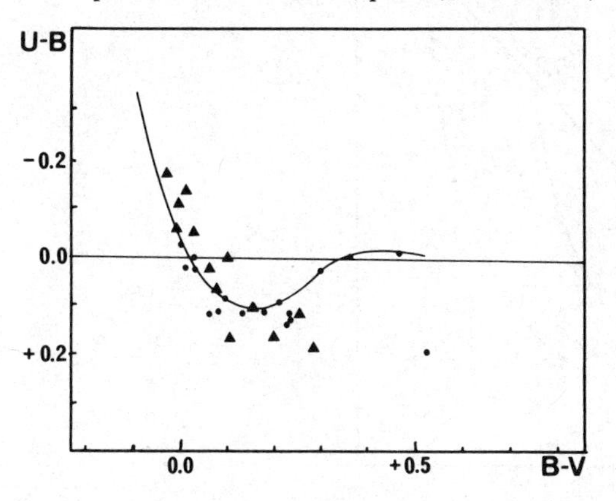

Figure 10.22. The spectrum of the Ae Herbig star HD 31293. The comparison is an A0 v standard. Notice the emission in Hβ and Hγ.

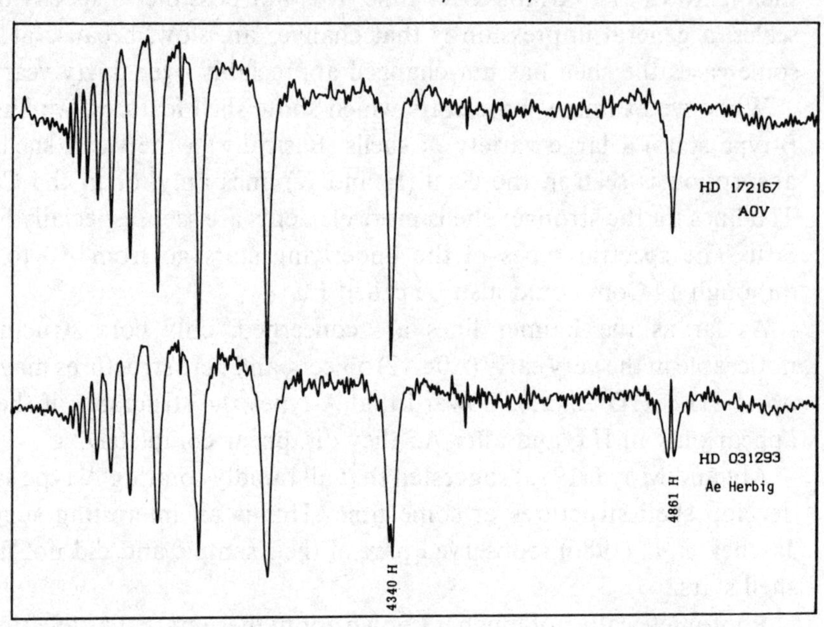

Figure 10.23. De-reddened broad band infrared colors for Herbig Ae/Be stars. Circles, classical Be stars; crosses, Herbig Ae/Be stars; big cross, typical errors; hatched bar, main sequence B0–F0; BB, black body curve; dashed curves, photosphere of an A0 star with thermal dust emission from bodies at 800, 1000 and 1500 K; arrow, reddening vector. From Finkenzeller and Mundt (1984).

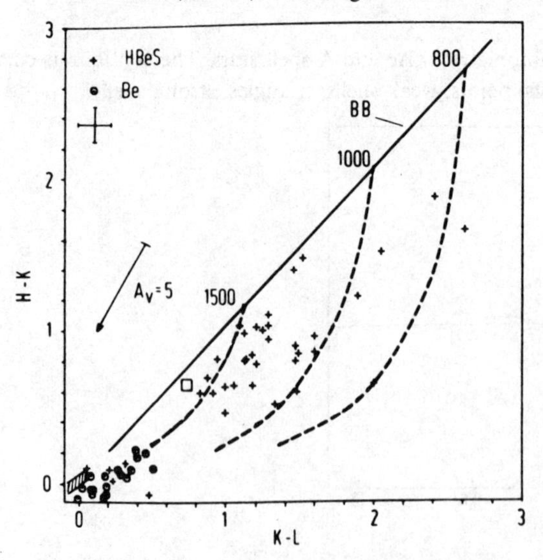

10.6 Herbig Ae/Be stars

Herbig (1960) studied a group of stars fulfilling the following conditions: (a) spectral type A or earlier, with emission lines, (b) the object lies in an obscured region, (c) the star illuminates fairly bright nebulosity in its immediate vicinity. Condition (d) should exclude those nebulous objects with envelopes probably produced by ejection, like planetary nebulae, WR stars and novae.

The distinction between a normal Ae, Be star and a Herbig Ae/Be star relies crucially on the association with nebulosity, and many authors have remarked that the two classes cannot be separated by spectroscopic criteria alone (see figure 10.22).

Recent work by Finkenzeller and Mundt (1984) has shown that a difference can be established when infrared photometry is used, because the Herbig objects show a large infrared excess at 3.5 μm (color 'L') illustrated in figure 10.23. These excesses are defined as $E = (V-L) - (V-L)_0$ where V–L is the de-reddened measured color and $(V-L)_0$ is the intrinsic color of a main sequence star with the same spectral type. Typical values of E are 3 to 5^m. Such a procedure does not however separate Herbig Ae/Be stars from [Be] stars (see section 9.2) which also show infrared excess.

Moreover since [Be] stars are defined by spectroscopic criteria alone, a certain overlap exists between Herbig Ae/Be stars and [Be] stars.

We may adopt as a working definition of Herbig Ae/Be stars 'stellar objects earlier than F0; associated with a region of obscuration and reflection nebulae; in their spectrum they exhibit emission lines of the Balmer series' (Bastian *et al.* 1983). Some 60 Herbig Ae/Be stars are known (Finkenzeller and Mundt 1984).

References

Abt H.A. (1961) *Ap. J. Suppl.* **6**, 37

Abt H.A. (1965) *Ap. J. Suppl.* **11**, 429

Abt H.A. (1979) *Ap. J.* **230**, 485

Abt H.A. (1984) in *The MK process and stellar classification*, p. 340, R. Garrison (ed.) Toronto

Abt H.A. and Bidelman W.P. (1969) *Ap. J.* **158**, 1091

Abt H.A. and Morgan W.W. (1976) *Ap. J.* **205**, 446

Abt H.A. and Moyd K.I. (1973) *Ap. J.* **182**, 809

Andrillat Y., Jaschek M. and Jaschek C, (1986) *AA Suppl.* **65**, 1

Ardeberg A. and Virdefors B. (1975) *AA Suppl.* **40**, 307

Arp H. (1962) *Ap. J.* **135**, 311

Baschek B. (1975) in *Problems in stellar atmospheres and envelopes*, Baschek B., Kegel W.H. and Traving G. (ed.), Springer Verlag

Baschek B., Heck A., Jaschek C., Jaschek M., Koppen J., Scholz M. and Wehrse R. (1984) *AA* **131**, 378
Bastian U., Finkenzeller U., Jaschek C., and Jaschek M. (1983) *AA* **126**, 438
Bidelman W.P. (1951) *Ap. J.* **113**, 304
Böhm-Vitense E. and Johnson P. (1978) *Ap. J.* **225**, 514
Borra E.F. and Landstreet J.D. (1980) *Ap. J. Suppl.* **42**, 421
Breger M. (1970) *Ap. J.* **162**, 597
Breger M. (1975) *Ap. J.* **201**, 653
Breger M. (1979) *PASP* **91**, 5
Burkhart C. (1979) *AA* **74**, 38
Conti P.S. (1965) *Ap. J.* **142**, 1594
Conti P.S. (1970) *PASP* **82**, 781
Conti P.S. and Strom S.E. (1968 a) *Ap. J.* **152**, 483
Conti P.S. and Strom S.E. (1968 b) *Ap. J.* **154**, 957
Cowley C.R. and Henry R. (1979) *Ap. J.* **233**, 633
Cowley A., Cowley C., Jaschek M. and Jaschek C. (1969) *A. J.* **74**, 375
Cucchiaro A., Jaschek M. and Jaschek C. (1978) *An atlas of ultraviolet stellar spectra*, Liège, Strasbourg
Eggen O. (1976) *PASP* **88**, 402
Fath E. (1935) *Lick Obs. Bull.* **479**
Ferrer O., Jaschek C. and Jaschek M. (1970) *AA* **5**, 318
Finkenzeller U. and Mundt R. (1984) *AA Suppl.* **55**, 109
Ginestet N., Jaschek M., Carquillat J.M. and Pedoussaut A. (1982) *AA* **107**, 215
Gomez A., Grenier S., Jaschek M., Jaschek C., and Heck A. (1981) *AA* **93**, 155
Greenstein J.L. and Sargent A.I. (1974) *Ap. J. Suppl.* **28**, 157
Grenier S., Gomez, A.E., Jaschek C., Jaschek M. and Heck A. (1985) *AA* **145**, 331
Hack M. (1959) *Mem. Soc. Astron. Ital.* **30**, 111
Hartoog M.R. (1976) *Ap. J.* **205**, 807
Hauck B. (1986) *AA Suppl.* **64**, 21
Hauck B. and Curchod A. (1980) *AA* **92**, 289
Hauck B. and Slettebak A. (1983) *AA* **127**, 231
Hayes D.S. and Philip A.G.D. (1979) *PASP* **91**, 71
Hayes D.S. and Philip A.G.D (1983) *Ap. J. Suppl.* **53**, 759
Heck A., Egret D., Jaschek M. and Jaschek C. (1984) *IUE low dispersion spectra reference atlas*, ESA SP-1052
Henry R.C. (1969) *Ap. J. Suppl.* **18**, 47
Henry R.C. (1978) in *Astrophysical papers dedicated to B. Strömgren*, Copenhagen Obs., p. 19
Henry R.C. and Hesser J.E. (1971) *Ap. J. Suppl.* **23**, 421
Herbig G. (1960) *Ap. J. Suppl.* **4**, 337
Honeycutt R.K. and McCuskey S.W. (1966) *PASP* **78**, 289
Houk N. (1978) *Michigan catalog of two-dimensional spectral types for the HD stars*, vol. 2, Univ. of Michigan
Humason M.L. and Zwicky F. (1947) *Ap. J.* **105**, 85

Jaschek C. and Gomez A. (1970) *PASP* **82**, 809
Jaschek C. and Mermilliod J.C. (1984) *AA* **137**, 358
Jaschek M. and Jaschek C. (1959) *A.J.* **62**, 343
Jaschek M. and Jaschek C. (1960) *Z. f. Astroph.* **50**, 155
Jaschek M. and Jaschek C. (1967) *The magnetic and related stars*, Mono Book Corp., p. 287
Jaschek M. and Jaschek C. (1974) *Vistas in Astronomy* **16**, 131
Jaschek M., Baschek B., Jaschek C. and Heck A. (1985) *AA* **152**, 439
Jaschek M., Jaschek C. and Andrillat Y. (1986) (to be published)
Keenan P.C. and Hynek J.A. (1950) *Ap. J.* **111**, 1
Landi J., Jaschek M. and Jaschek C. (1977) *Atlas de espectros estelares de red en mediana dispersion*, Cordoba Obs.
Lockwood W. (1968) *A.J.* **73**, 14
Luyten J. (1965) *First conference on faint blue stars*, Observatory, University of Minnesota
McConnell D.J., Frye R.I., Bidelman W.P. and Bond H.E. (1971) *PASP* **83**, 98
McMillan R.S., Breger M, Ferland G.J. and Loumos G.L. (1976) *PASP* **88**, 495
Mendoza E.E. (1971) *Bol. Tonantzintla y Tacubaya* **6**, 137
Mendoza E.E. (1978) in *IAU Symp. 80, The HR diagram*, Reidel D. Publ. Co., p. 289
Mermilliod J.C. (1980) *IAU Symp.* **85**, 129
Merrill P.W. (1934) *Ap. J.* **79**, 183
Morgan W.W. and Abt. H.A. (1972) *A.J.* **77**, 35
Morgan W.W., Abt H.A. and Tapscott J.W. (1978) *Revised MK spectral atlas for stars earlier than the sun*, Yerkes and Kitt Peak National Observatory
Morgan W.W., Keenan P.C. and Kellman E. (1943) *An atlas of stellar spectra*, Univ. of Chicago Press
Newell E.B. (1973) *Ap. J. Suppl.* **26**, 37
Osmer P.S. (1972) *Ap. J. Suppl.* **24**, 247
Parsons S. (1964) *Ap. J.* **140**, 853
Peterson R.C. (1983) *Ap. J.* **275**, 737
Peterson R.C., Tarbell T.D. and Carney B.W. (1983) *Ap. J.* **265**, 972
Philip A.G.D. (1972) *Dudley Obs. Rep.* **4**, p. 35
Philip A.G.D. (1978) *IAU Symp.* **80**, 209
Philip A.G.D. and Hayes D.S. (1983) *Ap. J. Suppl.* **53**, 751
Popper D.M. (1980) *Ann. Rev. AA.* **18**, 115
Roman N., Morgan W.W. and Eggen O. (1947) *Ap. J.* **107**, 107
Sargent W.L. and Searle L. (1947) *Ap. J.* **150**, L 33
Sargent W.L. and Searle L. (1968) *Ap. J.* **152**, 443
Schmidt-Kaler T. (1982) in Landolt–Börnstein, group VI, vol. 2b, p. 1
Slettebak A, Wright R.R. and Graham J.A. (1978) *A.J.* **73**, 152
Sommer-Larsen J. and Christensen P.R. (1985) *MNRAS* **212**, 851
Stickland D.J. (1973) *MNRAS* **161**, 193
Straizys (1977) *Multicolor stellar photometry*, Vilnius
Strömgren B. (1966) *Ann Rev. AA.* **4**, 433

Titus J. and Morgan W.W. (1940) *Ap. J.* **92**, 257
Uesugi and Fukuda (1979) Private communication
Winzer J.E. (1974) quoted in *PASP* **88**, 487
Yamashita Y, Nariai K. and Norimato Y. (1977) *An atlas of representative stellar spectra*, Univ. of Tokyo Press

11

F-type stars

11.0 Normal stars

According to the Harvard system an F-type star is characterized by strong Ca II (K and H) lines, which become much stronger than the hydrogen lines of the Balmer series. A multitude of fainter metallic lines accompanies both features. At F0, as already mentioned I(K) = I(H + Hε); at G0, I(K) ≫ I(H). Whereas at 100 Å/mm in A-type stars the Balmer lines are remarkable for their strength, in F-type stars they are no longer conspicuous. Another feature which appears at this dispersion is the G-band (near λ4300), which is due to the molecule CH; this feature appears around F3 and strengthens toward the later subtypes. The feature is constituted by the head of a molecular band and tends to dissolve when observed at lower plate factors.

In order to fix these ideas, table 11.1 provides the equivalent widths of some strong lines.

Besides the strong lines, there exists a host of weak lines, which produce, as we have seen, an increasing blocking. They also become so numerous that the undisturbed continuum is hard to see, except at low plate factors.

The spectral type is obtained from intensity ratios involving medium intensity features mostly from neutral elements. Attention has to be paid to two facts. The first is that because of the large number of lines present, any feature is a blend of several lines, except at very small plate factors. Therefore a feature used in classification cannot be identified with a single contributor; we must rather speak of a main contributor. The second is closely related to the first: all criteria depend upon the plate factor and cannot be applied at different plate factors without due precautions. Consider for instance the G-band of CH. This band is well seen as a band at 240 Å/mm at about F3 and at 100 Å/mm at about F5, whereas at 20 Å/mm it dissolves into isolated lines.

The Yerkes system uses line ratios given in table 11.2. The same criteria are usable at 73 Å/mm (Yamashita, Nariai and Norimoto 1977); at 40 Å/mm (Landi, Jaschek and Jaschek 1977) λ4226 Ca I/λ4481 Mg II may be added. The G-band as mentioned above becomes visible at about F5 in dwarfs, but only at G0 in supergiants.

Attention should be paid to the fact that not all decimal subclasses are used. For instance F1, F4 and F9 are seldomly used, and this has to be taken into account in statistical investigations.

Besides spectral type, luminosity effects are also observed. The line ratios used in the Yerkes system are summarized in table 11.3.

As can be seen, in three of the criteria the Sr II line is used, which has a pronounced positive luminosity effect (i.e. the feature gets stronger toward higher luminosity). Attention must be paid to the fact that strontium is an element which is definitely enhanced in some stars with peculiar spectra; if this is suspected, clearly we cannot use the line for luminosity classification. In such a case we should rely on the first criterion, or upon ratios involving Fe I and either Fe II, Cr II or Ti II lines. At 73 Å/m Yamashita *et al.* (1977) use for instance the ratios λ4179 Fe II/λ4144 Fe I, λ4216 Sr II/4226 Ca I and λ4554 Ba II/λ4481 Mg II, which all show a positive luminosity effect. (See figure 11.2.)

Table 11.1 *Equivalent widths (in Å) of some lines in F-typ stars.*

	λ3933 Ca II	λ4045 Fe I	λ4077 Sr II	λ4226 Ca I	λ4340 H I	λ4862 H I	λ5890 Na I	λ6562 H I
F0	6.5	0.1	0.2	0.25	8	7.0		5.5
F2	7.2	0.3	0.2	0.3	5.5	6.2		5.0
F5	9.3	0.5	0.3	0.5	4.8	5.0	0.6	4.3
G0	17.0	1.0	0.3	1.1	3.0	3.0	1.1	2.8

Table 11.2. *Spectral type criteria in F-type stars.*

λ4045 Fe I / λ4101 Hδ	λλ4030–34 Mn I / λλ4128–32 Si II
λ4226 Ca I / λ4340 Hγ	λ4300 CH / λ4385 Fe I

The Balmer lines vary little with luminosity, whereas Ca II lines show a positive luminosity effect. However both features are useless for classification, because they are very strong. Also the Balmer jump loses its usefulness with the progressive weakening of the Balmer lines and the progressive strengthening of the superimposed metallic lines.

Table 11.3. *Luminosity in F-type stars.*

$\frac{\lambda 4444\ \mathrm{Ti\ II}}{\lambda 4481\ \mathrm{Mg\ II}}$	$\frac{\lambda 4071\ \mathrm{Fe\ I}}{\lambda 4077\ \mathrm{Sr\ II}}$
$\frac{\lambda 4077\ \mathrm{Sr\ II}}{\lambda 4045\ \mathrm{Fe\ I}}$	$\frac{\lambda 4077\ \mathrm{Sr\ II}}{\lambda 4100\ \mathrm{H\ I}}$

Figure 11.1. Spectra of F-type stars: temperature effects. For explanation see text.

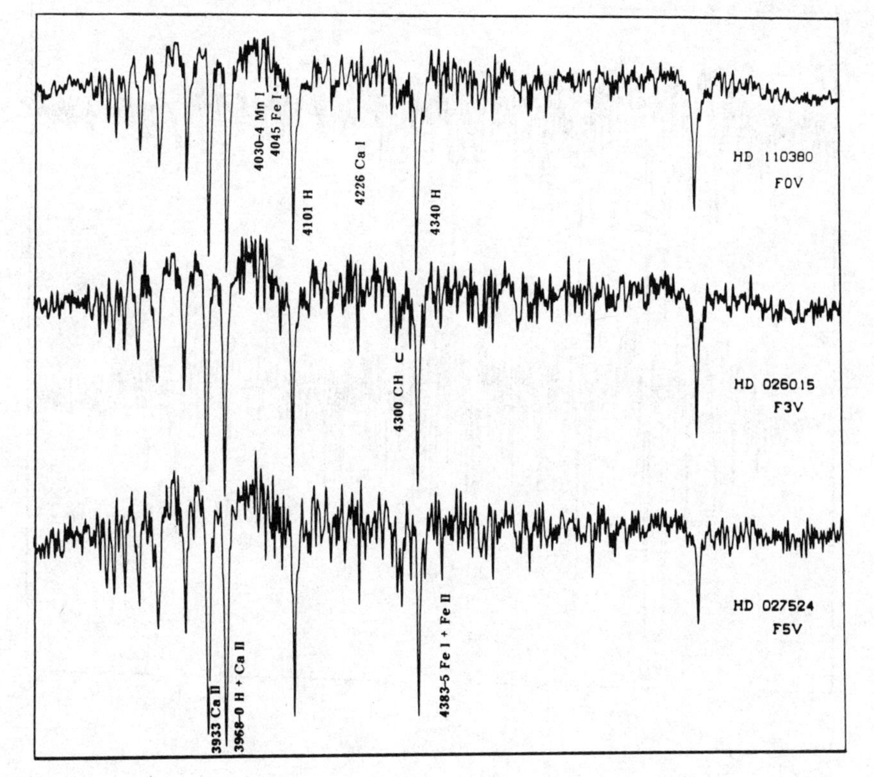

Figure 11.2. Luminosity effects in F-type stars. For explanation see text.

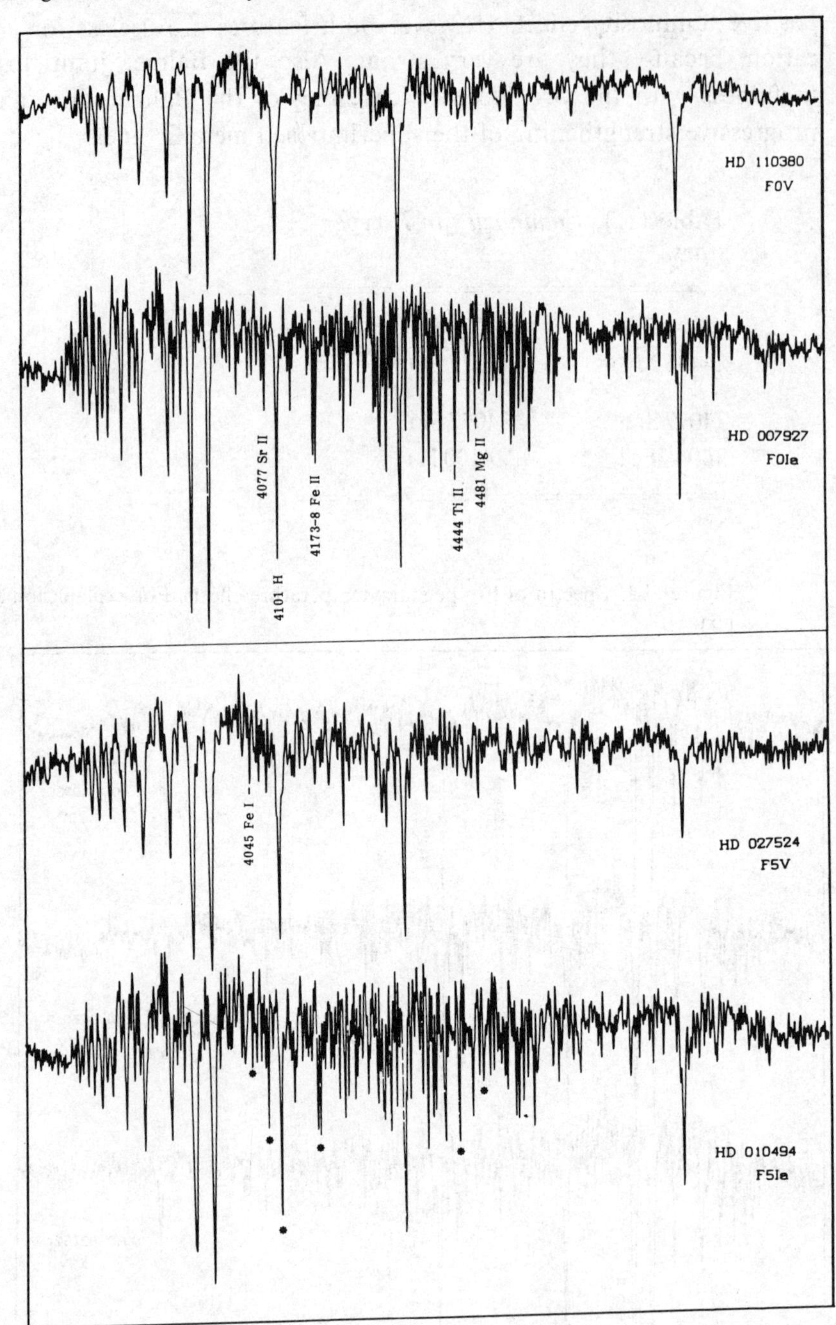

Ultraviolet. The region $\lambda < 3000$ Å is characterized by a rapidly declining continuum: at F0 the spectrum disappears at about λ1700 and at G0 at λ2000. This can be used as an easy temperature criterion, although it should be handled with caution when the object is weak lined.

Figure 11.3 illustrates the shape of some F-type spectra at a resolution of 36 Å (Cucchiaro, Jaschek and Jaschek 1977).

Infrared. The near infrared shows about the same features as in A-type stars. Paschen lines are however weaker. Luminosity effects are seen in O I (see figure 10.4, section 10.0) and in the Ca triplet (see figure 14.4, section 14.0).

Spectral peculiarities and emission lines. In a broad sense, F-type stars show few peculiarities. It is a matter of speculation if this is natural or due to a lack of appropriate studies. For instance a survey by Malaroda (1975) of 455 southern F-type stars produced only a few peculiar objects which are of the 'pec F', 'δ Scu' and 'δ Del' types described in sections 10.2 and 11.1. Both Am and Ap stars, so abundant among A-type stars, peter out in the early F-type stars, and emission line objects, abundant both in earlier and in later type objects, appear only in small numbers after F5 (see section 11.2).

Some weak line stars also exist (see section 11.3) although their number is rather small ($\lesssim 2\%$), if spectroscopic criteria are used. They are more numerous if photometric detection methods are used (section 11.4).

Rotation. Rotation continues to diminish with advancing spectral type, as happened in A-type stars. The average values are given in table 11.4.

As can be seen, the rotational velocities decrease to very small values at G0. The usual technique of estimating rotation from line widths has to be replaced by Fourier transform techniques (Smith 1980), in order to measure the very small values observed. The values in table 11.4 were compiled from Slettebak (1970), Altschuler (1975), Smith (1980) and Danziger and Faber (1972).

Magnetic fields. If magnetic fields exist, they fall below the precision of measurement of present-day techniques (Gray 1984).

Photometry. We have seen that the Balmer lines weaken progressively through type F, whereas the metallic lines strengthen in the opposite sense. This implies that the Balmer discontinuity on the whole should not change very much, because one effect offsets the other. Thus a broad band system

Table 11.4. *Average rotational velocities ($\overline{V \sin i}$ in km/s).*

	V	III	I
F0	70	130	30
F2	50		
F5	25	45	
F8	15	45	
G0	5	40	25

Figure 11.3. Low resolution dwarf F-type spectra in the $\lambda\lambda$1400–2600 region. The r_7 index measures the ratio between the fluxes at λ1850 and λ2400, and grows with advancing spectral type. From Cucchiaro, Jaschek and Jaschek (1977).

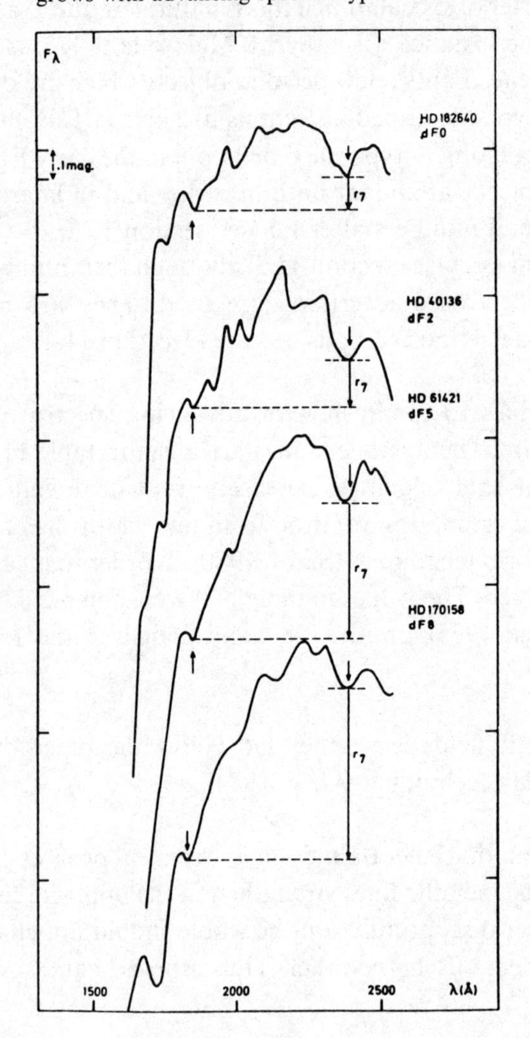

like UBV should show no large variation in U–B which measures the Balmer discontinuity, and this is what is observed globally. We have illustrated the behavior of UBV in figure 11.4 which contains also the values of the average colors for dwarfs in the Johnson RI...L system (Johnson 1966).

For early F-type stars the Balmer jump still differs for stars of different luminosity and UBV colors are thus sensitive to luminosity class. This is illustrated in figure 11.5. Observe however that in practice we must correct for reddening before using this fact. Similar effects are also observed in

Figure 11.4. Broad band photometry of F-type stars: UB...L colors. The mean wavelength of the different bands is indicated on the abscissa. V = 0.

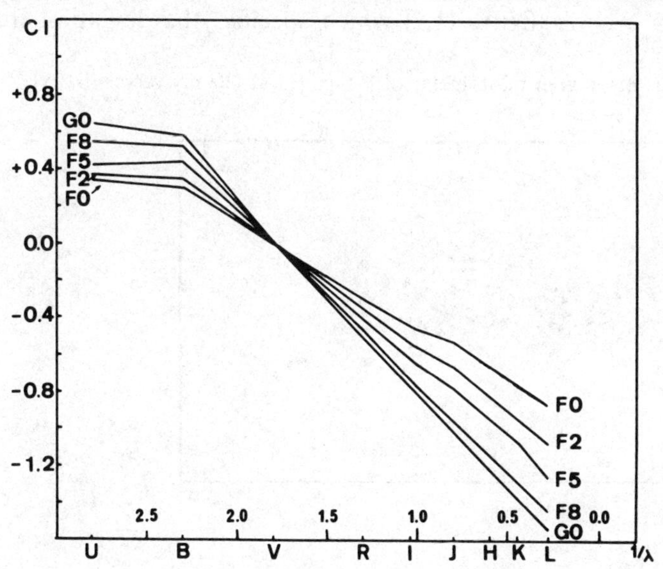

Figure 11.5. Luminosity effects in F-type stars in UBV photometry. M.S. = main sequence.

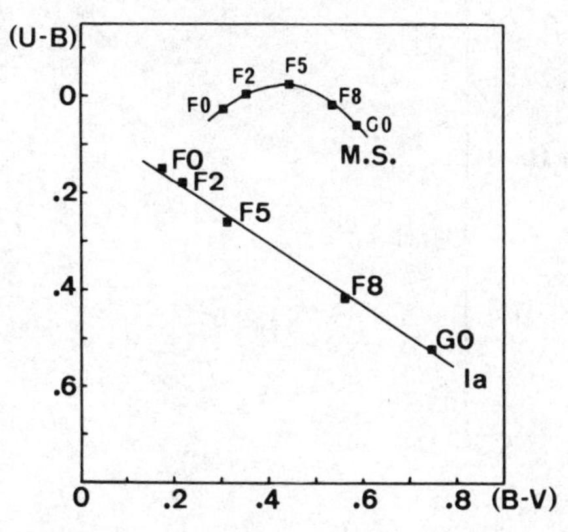

intermediate band photometries, as in the Strömgren system. Since in this system one more color index is available, we can define a 'strength of line' or 'metallicity' index which measures the accumulated effect of the lines. This is, as we have seen in section 5.2, the m_1 or $[m_1]$ index. We have shown in figure 5.17 of that section the general shape of the $(b-y, m_1)$ relation. Figure 11.6 reproduces the section of such a diagram corresponding to F-type stars (Crawford 1975). Such a diagram may be used to derive metal abundances. The first step is to introduce a convenient reference line; generally the Hyades sequence is used as such. This is equivalent to saying that we accept the stars of the Hyades cluster to be perfectly normal. The Hyades relation falls near the lower envelope, so we can then define a differential index δ or Δ (figure 11.7) which specifies the deviation from

Figure 11.6. Strömgren photometry of F-type stars. The m_1 versus $(b-y)$ relation.

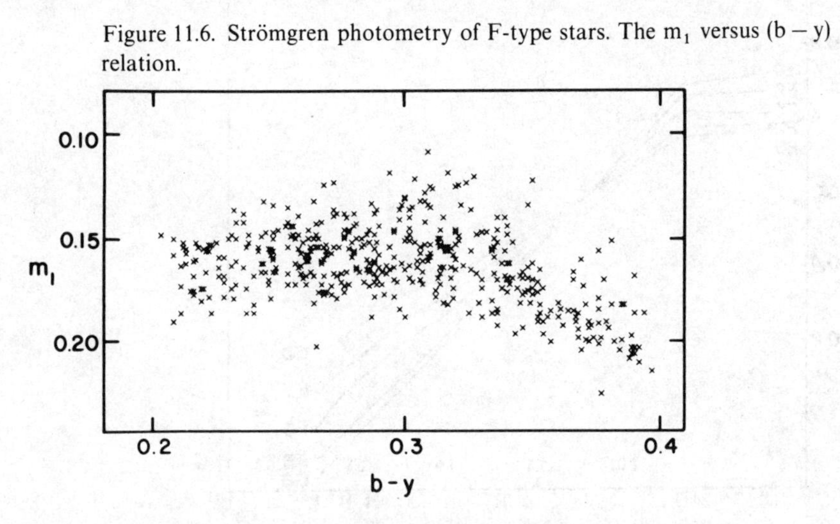

Figure 11.7. Definition of differential indices in Strömgren photometry. The m_1 versus $(b-y)$ diagram. For explanation see text.

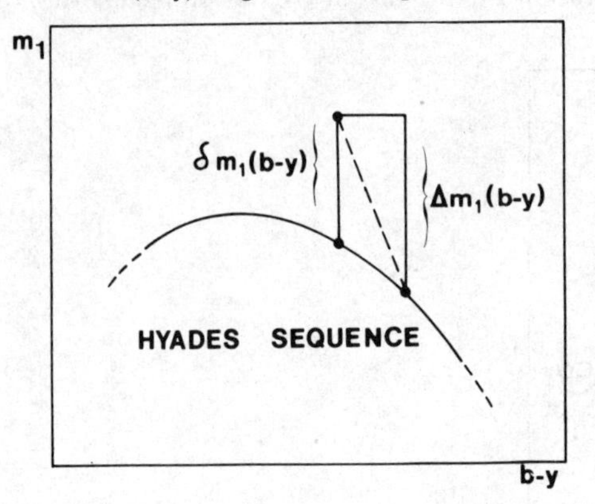

normality. Stars with fewer lines would lie above the sequence, so that a positive deviation means a strong line weakening, which can be expressed in terms of metal abundance obtained from stellar atmosphere analysis.

Existence of such quantitative methods to assess line weakening is very welcome, because spectroscopy is not well suited to deal with a general weakening of all lines; apart from extreme cases of strengthening or weakening the spectroscopist is unable to detect such anomalies.

Olsen (1979, 1980) has made extensive observations of F-type stars in both hemispheres, enabling the detection of a large number of anomalous objects. Only one reservation should be kept in mind, namely that a color anomaly is not related unequivocally to a specific spectroscopic peculiarity.

Another intermediate band photometric system conceived specifically to deal with later type stars, the so-called DDO system, becomes usable around mid F-type stars. (See chapter 12.)

Narrow band indices on the other hand become more difficult to interpret in F-type stars because of the many faint lines which accompany the strong lines we want to measure. So K line photometry (Henry 1979) (section 10.0) loses its usefulness around F5, and Hβ photometry (section 9.0) around F8. On the other hand, O I λ7774 photometry (Mendoza 1978) (section 10.0) is still usable, especially for luminosity effects (see figure 11.8).

Binaries. The proportion of binaries seems normal, both from the point of view of general statistical studies (Jaschek and Gomez 1970) and from detailed surveys (Abt and Levy 1976). The latter authors found, in a survey of 123 F3–G2 systems of luminosity classes IV and V, 88 companions, implying that 70% are binaries.

Absolute magnitude. F-type dwarfs are sufficiently frequent that their

Figure 11.8. Luminosity effects in F-type stars through photometry of the O I λ7774 line. Points, dwarfs: crosses, supergiants.

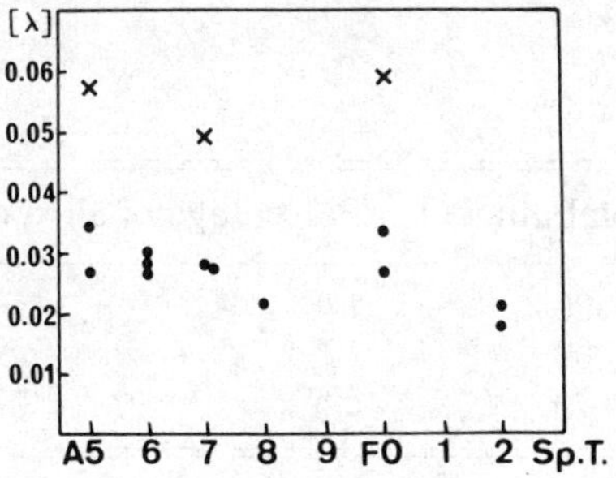

absolute magnitude can be derived from trigonometric parallaxes, whereas for other luminosity classes we are forced to use the usual variety of procedures. We summarize in table 11.5 the results of Schmidt-Kaler (1982). The absolute magnitudes for supergiants do have rather large uncertainties ($\pm 0^m.7$).

Number, distribution in the galaxy. F-type stars are moderately concentrated toward the galactic plane, with a halfwidth $\beta = 190$ pc. Within the galactic plane the distribution seems uniform, within a factor of three (McCuskey 1965). The number of F-type stars is given in table 11.6.

Despite their large number in magnitude limited surveys, they are rather rare in the close solar neighborhood, because of their relative brightness.

Of the 1300 F-type stars listed in the *Bright star catalog* (Hoffleit and Jaschek 1982), 2% are peculiar, 49% are dwarfs and the remainder are non-dwarfs.

F-type stars are found in moving clusters and open clusters. They may exist in associations, but most of them are beyond the reach of easy study.

11.1 Fp and Fm stars

As remarked in section 11.0, there are very few peculiar objects among F-type stars, and these few may be regarded as the tail of the Ap stars.

Table 11.5. *Average absolute visual magnitudes for F-type stars.*

	V	III	II	Ib	Ia	Ia0
F0	2.7	1.5	−5.1	−8.0	−2.5	−9.0
F2	3.6	1.7	−5.1	−8.0	−2.4	−9.0
F5	3.5	1.6	−5.1	−8.0	−2.3	−9.0
F8	4.0		−5.1	−8.0	−2.3	−9.0

Table 11.6. *Number of F-type stars.*

Survey limit		Total number	Percentage of all stars
Bright stars	$m \leqslant 6.5$	1300	14%
HD	$m < 9$	43 000	19%
Nearby stars	$r \leqslant 20$ pc	70	2.5%

We shall mention as an example the Fp star HR 4310 (Cowley 1976) with enhanced Sr II lines. Cowley comments: 'metallic line ratios are peculiar for a normal star. Many lines present; G band weak; λ4415 strong, but λ4173–78 not pronounced as in a giant; Sr enhanced, closest type is F5 III, but some individual lines wrong for this type.' This object could well be a late Ap star.

Sometimes very peculiar shell stars are also called Fp, for instance 14 Com. This is a misnomer; it is clearly 'F shell'.

The Fm stars are a prolongation of the Am stars toward later types. We have already remarked (section 10.1) that Am stars have hydrogen spectral types out to F2, but their number diminishes very rapidly after F0. Probably the latest Am stars – really Fm – can be described in the way Malaroda (1973) did for HR 3591: 'Hydrogen lines correspond to a F5 spectral type, the metallic lines to F8 and the K line to F0. The G-band is very strong and matches that of an F8-type star. λ4077 is too weak for a giant.' Houk (1978) describes the stars in similar terms: 'there are a few unusually late metallic line stars called Fm which have G-bands but also have line strengths (strong Sr λ4077 and λ4150, $\lambda 4416 \ll \lambda 4481$) similar to Am stars.'

Less than 1% of the F-type stars of a magnitude limited sample correspond to this type of object. Another group of stars that falls between the A- and F-types is the group of 'δ Del' stars, which Houk (1978) calls 'Fm δ Del'. We have already dealt with this in section 10.2.

11.2 H and K emission line stars and related objects

11.2.1 *H and K emission line stars = Ca II emission line stars*

A Ca II emission line star is any late type object (F4–M) which shows emission features in the Ca II H and K lines.

Schwarzschild and Eberhard (1913) discovered that in many stars of late spectral type emission is present at the bottom of the Ca II lines (λ3933, λ3968). Over the years a number of objects were added to the list of 'H and K emission line stars', as can be seen in the compilation by Bidelman (1954). The first to undertake a systematic study of the group was O. Wilson, whose papers on the subject cover practically three decades.

The profile of the K line (λ3933) is schematized in figure 11.9. If λ_0 is the laboratory wavelength of the line,

$$\lambda_v - \lambda_0 = \delta_v \quad \lambda_r - \lambda_0 = \delta_r \quad \lambda_a - \lambda_0 = \delta_a$$

and

$$\omega = \delta_r - \delta_v$$

Wilson obtained high dispersion material (plate factors 10 Å/mm or less) and measured with a micrometer the width (ω) of the emissions. He also provided line intensity estimates of the emission components on a mnemonic scale with 0 = no emission seen, 1 = weakest lines usable for measurements, ..., 5 = peak emission equal to or in excess of the neighboring continuum background outside the line.

An illustration of what the emission features look like is given in Wilson and Bappu (1957). They observed 185 stars later than G0 at 10 Å/mm; of these only 15 have intensity 0, and 9 have intensity 5, implying thus that H and K emission is a widespread phenomenon. Wilson (1963) gave statistics of emission intensity in field stars; his results are summarized in table 11.7.

Since the field stars are a mixture of all ages and compositions, surely we should find other results if we consider objects which are of similar age and/or composition. Table 11.8 (Wilson 1963) shows what happens in several clusters.

The result is clear; Ca II emissions are quite common and depend upon parameters other than spectral type. Of the various possible factors, one of the most obvious is age: younger stars show stronger emission than older stars.

Table 11.7. *Emission line intensity in field stars.*

	0–1	2–3	4–5	Total
G0–G4	27	0	3	30
G5–G9	27	7	0	34
K0–K2	26	8	6	40
Total	80	15	9	104

Figure 11.9. Schematic profile of the Ca II K line.

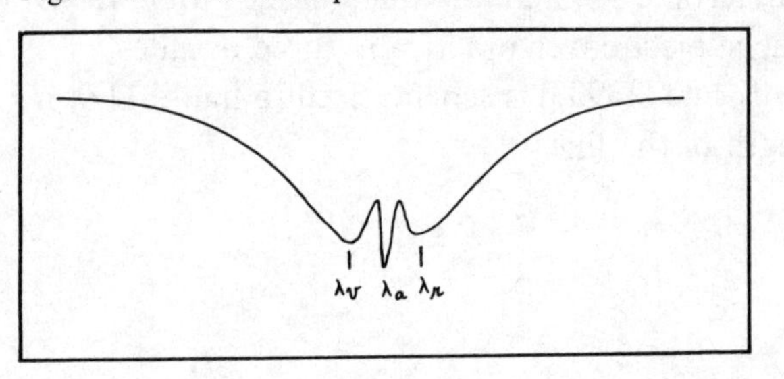

A second important result drawn from the emission widths was that the width depends upon the absolute magnitude M_v. This effect is the so-called Wilson–Bappu effect and it is best summarized in figure 11.10 from which the following relation

$$M_v = -15.8 \log \omega + 29.4$$

Table 11.8. *H and K emission line strength in various clusters.*

	Type	Number of stars	Line strength 0–1	2–3	4–5
Hyades	G0–K2	84	8	54	38
Praesepe	G0–K2	51	6	78	16
Coma	G2–G8	6	0	83	17
Pleiades	G0–K2	78	2	43	54
Field	G0–K2	104	77	14	9

Figure 11.10. Corrected Ca II emission line width (W_0) against absolute magnitude derived from trigonometric parallaxes ($M_v(T)$). From Wilson and Bappu (1957).

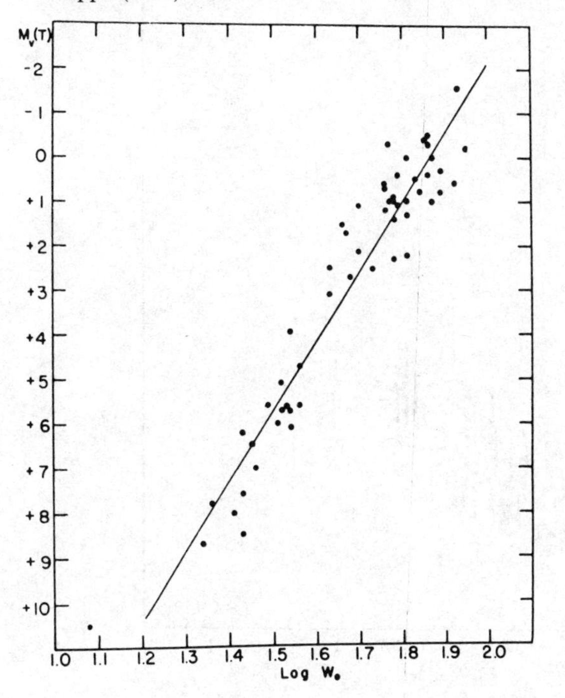

may be derived. Since according to Wilson ω is uncertain by about $\pm 10\%$, the possible error in the absolute magnitudes is about $\pm 0^m7$. Later studies produced small changes in the constants of the above-mentioned relation. Wilson (1976) quotes $M_v = -14.94 \log \omega + 27.6$ and the possible error is now given as $\pm 0^m5$.

The next step in Ca II emission line studies was the replacement of visual intensity estimates by photoelectric measurements. This was carried out by Wilson (1968); the precision of a single measurement is increased to about 2% (possible error). Subsequent changes in the apparatus (Vaughan, Preston and Wilson 1978) have left the possible error unchanged, but have helped to increase significantly the number of stars observed.

The new series of observations (Vaughan, Preston and Wilson 1978) extends down to M-type dwarfs, and confirms that H and K emission is a very widespread phenomenon in the whole range F4–M. When the intensity of emission is plotted against spectral type or color index, a considerable latitude in emission line strength is found, as can be seen in figure 11.11.

Figure 11.11. Intensity of the Ca II emission line as a function of B–V color. Bars indicate the variation range of the stars studied in detail. From Vaughan (1983).

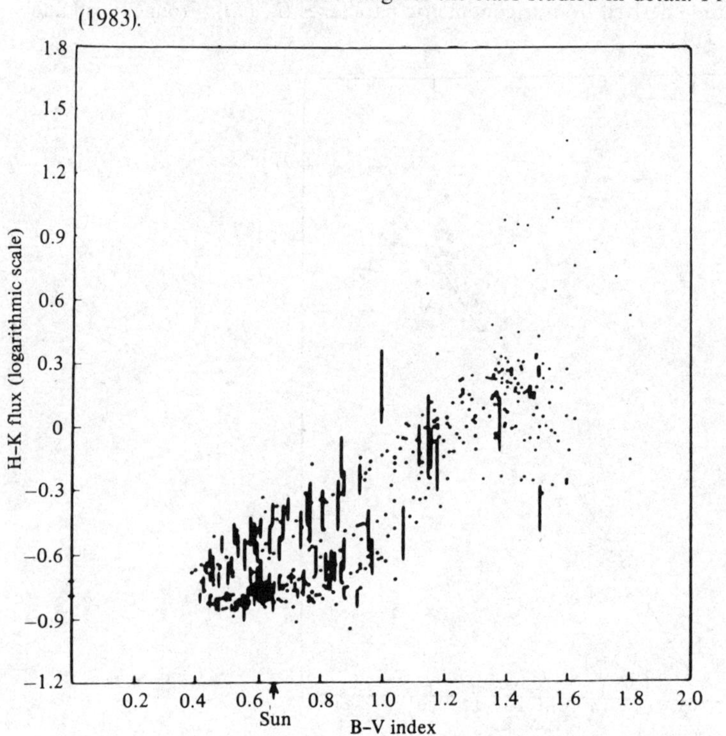

A second conclusion which emerges from the analysis of all observations is that if the emission strength is plotted as a function of time, there is very often a long term variation of the order of years, as illustrated in figure 11.12.

Obviously such long term variation suggests an analogy to the solar cycle, and Vaughan (1983) was able to list for 27 stars 'activity cycle lengths' whose average is about 10–12 years. The sun with its 11-year cycle is thus a very typical star.

Figure 11.12. The emission line flux as a function of time for three stars. Abscissa: years since 1900.

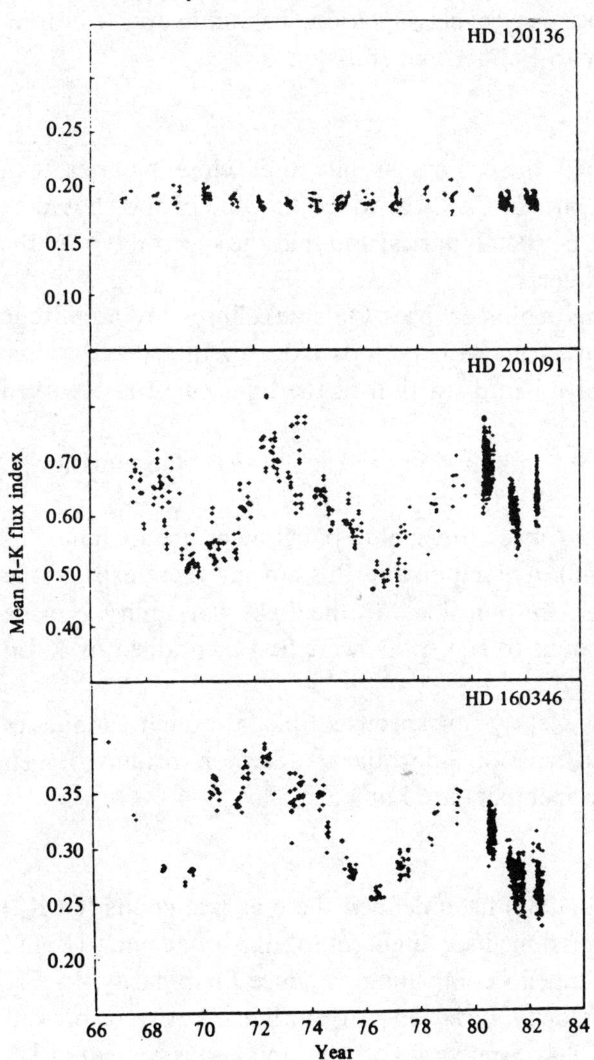

As can be seen from the figure, data obtained over short spans of time still show considerable scatter in the emission measures. This scatter can be interpreted as due to the rotation of the star which carries around the regions responsible for the emission. A detailed review is out of the question here; the reader may consult the various papers in *IAU Symp.* **102** for further information. The main result is that rotational periods of stars can be derived, and fall approximately in the range 7–50 days.

In conclusion, H and K emission lines are widespread spectral characteristics, which should *not* be used as classification criteria. This was stated by Bidelman (1954) but has often been ignored, with much ensuing confusion. If the letter e is appended to a spectral classification, it should never refer to H and K emission, but only to Balmer line emission.

11.2.2 *RS CVn stars*

Hall (1976) defined these stars as late type close binaries with emission in the H and K lines of Ca II. He added that the period should be longer than 2^d (to exclude contact binaries) and that the spectral type of the primary should be F or later.

Many but not all systems exhibit eclipses. Outside eclipses, low amplitude photometric variations are seen with periods close to the spectroscopic period, so that a slow wave-like distortion of the light curve is observed (Rodono 1981).

In some stars Hα is seen in emission, and in 20–30% the emission is sporadic (Bopp and Talcott 1978).

The stars are radio and X-ray emitters; both phenomena are attributed to an active envelope (corona). Accompanying this corona there exists large star spots which are held responsible for the light variability outside eclipses. Star spots, analogous to sun spots, were first mentioned by Kron (1947).

The group definition is clearly not spectroscopic, although candidates were found among Ca II emission line objects. (See for instance Eggen 1978.) About thirty group members are known.

11.2.3 *FK Com stars*

Bopp and Stencel (1981) have defined these as late giants (G–K2) with strong H and K emission lines, high rotational broadening ($V \sin i \sim 100$ km/s) of the lines and no compelling evidence for binarity. In FK Comae itself, no radial velocity variations larger than 5 km/s are present (McCarthy and Ramsey 1984). Sometimes erratic emission is present at Hα (Collier 1982). About ten members of this group are known.

11.3 Weak line stars

A 'weak line' star is one which, when compared with a suitable standard, shows a general weakness of most of the lines. Since the lines correspond mostly to metals, the term 'metal weak' stars is also used. One should however keep in mind that astronomers call metals everything with $Z > 2$ (Z = atomic number), a terminology at variance with that of chemists. 'Weak lined thus means weakness of all atomic lines except H and He. It is preferable to avoid the term 'metal weak' and use 'weak line' consistently, which has the merit of describing exactly what is seen.

The terminology was introduced by Roman (1950) when she discussed a group of F5–G5 stars at plate factors of 125 Å/mm. Since hydrogen weakens toward later types, whereas metallic lines strengthen, a weak line star shows up as a star where the hydrogen lines imply a later type than the metallic lines. The difference in behavior is then used to define 'weak line' (wl) stars, as opposed to 'strong line' stars i.e. normal line stars. The classification into 'weak' and 'strong' line stars was later extended to K-type stars (Roman 1954). Furthermore Roman showed that 'wl' characteristics may appear in luminosity classes V, IV and III.

The definition is very sensitive to misclassifications (Morgan 1958). If for instance an F7 star is classified as F9, it might later be called 'weak line', because for F9 the metallic lines are weak. Since misclassifications will happen even in careful work, it is certain that a number of spurious 'wl' stars will be created. On the other hand, a number of 'wl' stars will escape detection if the difference between H line intensities and metallic line intensities is small (see figure 11.13).

Keenan and Keller (1953) confirmed the existence of weak line stars, adding that the CH molecule (G-band) is strong – a fact which is in line with the strength of H. Thus H and CH show a parallel behavior, whereas all lines from other elements are weak. The molecular bands of CN should thus also be weak, and this is so. (See also section 12.2.)

Greenstein (1971) observed a number of possible weak line dwarfs of types K and M, and found weak TiO bands and very strong MgH bands – again as could be expected.

These procedures work well within certain limits, fixed essentially by the plate factors used. Whereas extreme cases can be detected even at 200 Å/mm (Graham and Slettebak 1973), less extreme cases can only be detected in the spectral interval in which the hydrogen lines are neither too strong nor too weak at a given plate factor. At about 80 Å/mm this restricts the interval from early F to early K. To reach late K and M stars the MgH band (λ6382 and λ6389) has to be used instead of the Balmer lines, as mentioned above. The number of weak line M-type stars appears to be very small, implying that either these stars do not exist or that we were unable to find them.

The next study of weak line stars was made by Bond (1970) who calls these objects 'metal weak' or 'metal deficient'. He used objective prism material (plate factor 108 Å/mm) obtained under good seeing conditions. He describes in some detail the procedure followed. He states that since metallic lines in F–K stars weaken with increasing temperature, the temperature

Figure 11.13. Spectra of two weak line stars. HD 140283 is a subdwarf, and HD 122563 an extreme weak line giant.

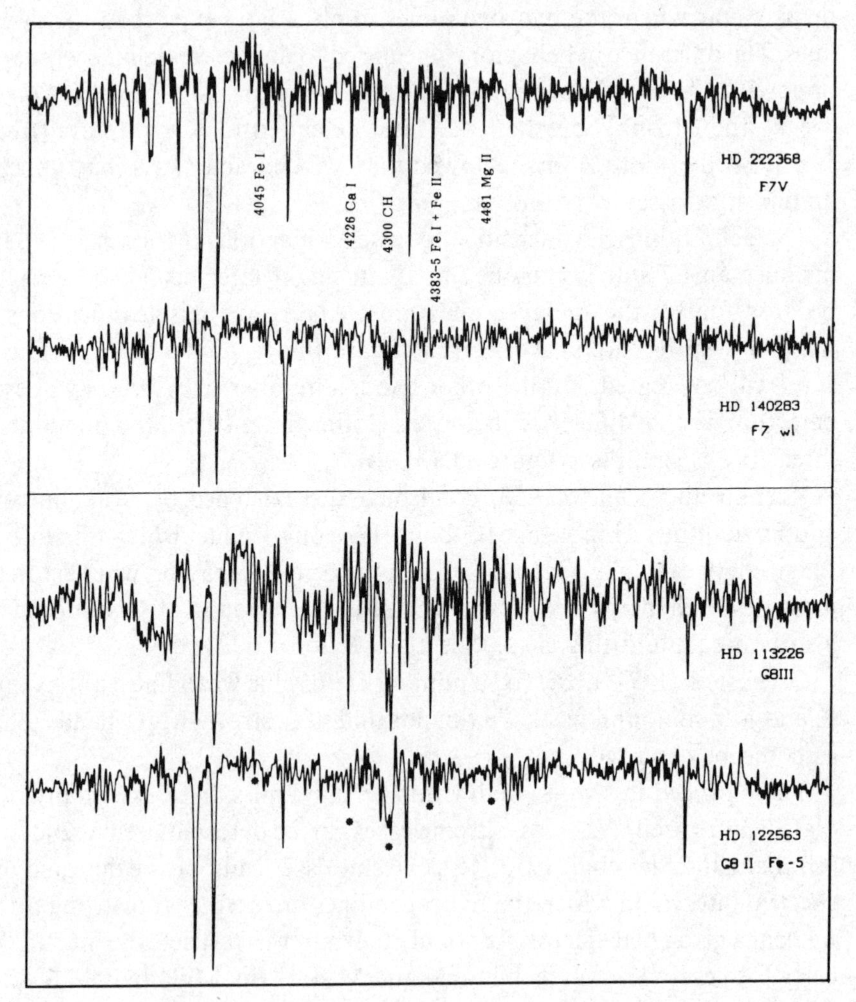

class of a star must be estimated first, before the weakness of metals can be decided. Bond used the strength of the Balmer lines of hydrogen, the Ca II lines and in cooler stars the strength and appearance of the G-band of the CH molecule. For luminosity classes the strength of Sr II λ4077 was used.

Bond found that poor seeing conditions can weaken the apparent strength of the metallic lines in a normal F star so that the spectrum becomes indistinguishable from that of a similarly affected weak lined star. In order to ensure that no bias was introduced this way, candidates were observed on slit spectrograms. Once the star's spectral type and luminosity had been classified, it was compared with a standard star to estimate the degree of metal weakening. Four degrees of metal weakening were used: none, slight (barely discernible), moderate (obvious weakening of even the strongest atomic lines) and extreme (all lines invisible except the very strongest).

Among the weak line stars a number of other peculiarities occur, for which the reader is referred to the relevant sections: λ Boo stars (section 10.3), (red) horizontal branch stars (section 10.4) and CH stars (section 12.3).

It is now time to interrupt our discussion and to introduce *photometry*. We will not duplicate here the discussion on photometry of section 11.5 on 'high velocity stars', but shall rather consider the detection limits of both methods. We quote for this the work of Bond (1970), who obtained Strömgren photometry for the stars he classified on slit spectrograms (plate factor around 100 Å/mm) according to his scale of metal weakening. Figure 11.14 summarizes his main results. He used Strömgren's m_1 index for the metallicity, and constructed a relation (b–y, m_1) for normal stars. From the observed m_1 index for a given star he subtracted the 'normal' m_1 index (at the same b–y) (for definition see section 11.7). These are the Δm_1 values used in figure 11.14.

The conclusion is that there is a considerable overlap between stars assigned to different groups of line weakening, although the stars with extreme line weakening all have large Δm_1 values – and no star with large Δm_1 has escaped spectroscopic detection. Thus it seems that photometric indices are capable of providing a more accurate assessment of line weakening than spectroscopic classification. We should however remember that this is true only when we know that the star is not abnormal in other respects. From ubvy photometry alone it is extremely difficult to distinguish between a weak line star and a Ba star, or a late Ap star. In other photometric systems similar difficulties exist. We can summarize by saying that probably the best technique is to use both spectral classification to single out peculiar objects and photometry to measure the amount of metal weakening.

With regard to the nomenclature of weak line objects, we start by observing that weak line dwarfs are often called subdwarfs (see section 10.4). Roman (1955) proposed calling them luminosity class VI, but this designation should definitely be abandoned. Extreme cases of weak line stars are also called 'population III' objects, a designation which is colorful but without a clear-cut definition. Houk (1978) proposes notation like G5 w F0. The spectral type based upon hydrogen line strength is given first, followed by 'w' for weak metallic lines, whose corresponding type is given next. (Sometimes a luminosity type is also included.) The difference between the two types indicates the degree of metal weakening. This notation breaks down for extreme metal weakness, or in early type stars where no metallic line type can be given; in such cases the notation is simply G0w, the spectral type corresponding to the hydrogen line appearance.

At lower plate factors (60–80 Å/mm, slit spectrographs) this notation can be considerably refined, to denote the strength of certain features. In general the notation uses a step scale from +3 (very enhanced) to −3 (very weakened); 0 represents normality and is not written. The scale itself is defined by the standard stars given by Keenan and Yorka (1985). These authors prefer to denote 'metal line weakness' by the Fe scale, rather than by

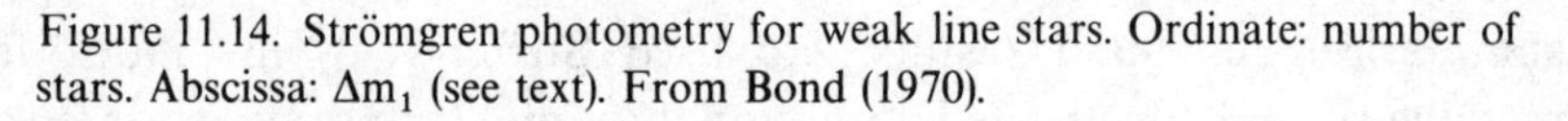
Figure 11.14. Strömgren photometry for weak line stars. Ordinate: number of stars. Abscissa: Δm_1 (see text). From Bond (1970).

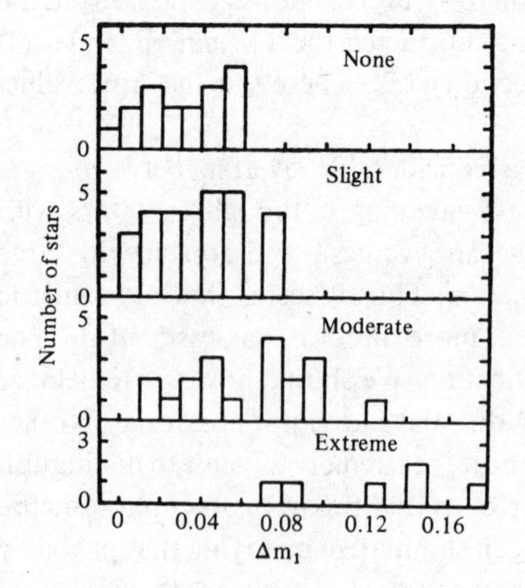

the CN scale as was done in the past. Thus weak line stars are denoted

K0 III Fe − 2

and not

K0 III CN − 2

Keenan and Yorka note that in general the CN weakness follows the Fe weakness, a fact which allows us to shorten the notation.

The question of the simultaneous variation can be analysed on the basis of the features listed in table 12.2. To do so, we use the 550 stars classified by Keenan and Yorka (1985). If all variations are simultaneous, we would not expect weak line stars to possess peculiarities other than a negative Fe index. A perusal of the list shows that this is not so. For instance, among stars with an Fe index − 1.5, we find 22 stars with no further peculiarity, but also 6 stars in which there are other peculiarities. These show up as CH indices between − 1 and + 1, which may or may not be further accompanied by Ca, CN and Hδ anomalies. This implies that even at classification dispersion the assumption of a simultaneous variation of all elements is invalidated by observation. Obviously the use of lower plate factors will further clarify the issue, but this is beyond the reach of classification methods.

Of the various anomalies mentioned above, we shall consider the case of the CH. It was Bond (1980) who called attention to the fact that in extreme weak line giants a wide range of CH strengths existed. He estimated visually the band intensity on a scale from 0 = no G-band visible, to 6 = extremely strong CH. 'Normality' for G and K giants is 3 or 4. Stars with estimate 6 should be, or are in transition to, CH stars (see section 12.3) – there is little doubt that they belong to the family of C-type stars. Stars of estimate 0 or 1 are reminiscent of the stars observed by Zinn (1973) in asymptotic giant branch stars of the globular cluster M 92 (figure 11.15). To forestall any hasty conclusions it should be added that *not* all globular cluster asymptotic branch stars have weak CH lines.

Frequency. Bond (1970) estimates that in the range F5–G5 the proportion of pronounced weak line stars is very low. He estimates that within the limits of his survey (one-fifth of the sky) there are about 4×10^4 stars of these types, whereas he found fewer than 40 weak line stars. Thus in magnitude limited samples, the proportion is about 1‰. In a distance limited sample, the incidence is even smaller because most of the normal stars are dwarfs, whereas many of the weak line stars found are giants. Since giants are much less frequent than dwarfs, the 1‰ is an overestimate. The most extreme weak line stars have been called population III stars and were expected to be frequent. At present all evidence points to the contrary.

A catalogue of metal weak stars detected by both spectroscopic and photometric criteria was compiled by Bartkevicius (1984).

11.4 Subdwarfs

Adams and Humason (1935) called attention to a group of (six) stars located between the main sequence and white dwarfs (degenerates), which they called 'intermediate white dwarfs'. They noted that at a plate factor of 120 Å/mm the hydrogen lines were narrow and sharp and the metallic lines faint. These stars were later called subdwarfs by Kuiper (1939) since he found them more similar to dwarfs than to degenerates. He defined them as stars not over 2–3 magnitudes below the main sequence and described the spectra of objects earlier than about G5 in terms similar to those of Adams *et al.* (1935).

These objects gained much more attention in the fifties, in the wake of Baade's discovery of stellar populations. All subdwarfs have high space velocities and were thus assigned to population II.

Roman (1955) classified a number of stars as subdwarfs based upon

Figure 11.15. Microphotometer tracings of the spectra of six stars on the asymptotic branch of M 92. A, C, D and F are normal G-band stars whereas B and E are weak G-band stars. From Zinn (1973).

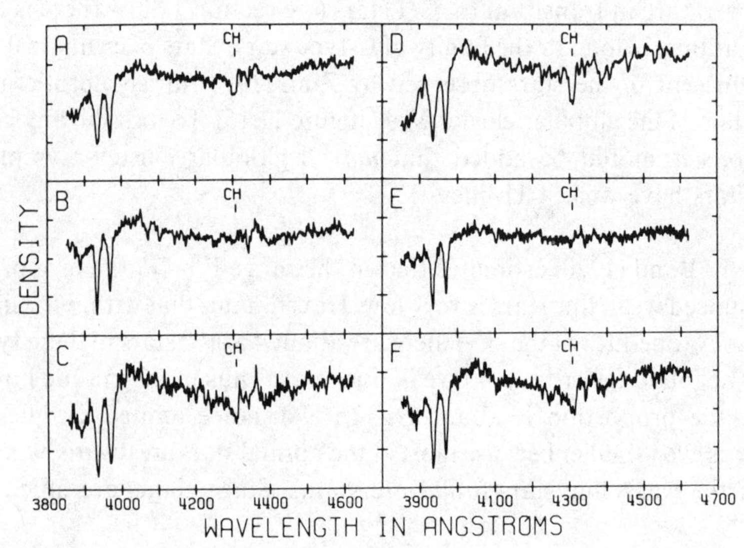

spectroscopic criteria alone (see section 11.3). It is fair to observe that most of the subsequent discussion on subdwarfs was done without (or almost without) spectroscopy, which was being replaced by photometry. It all started with the discovery by Roman (1954) that 'weak metal stars' exhibit an ultraviolet flux excess in the UBV system. This can be seen in figure 11.16, taken from Sandage and Eggen (1959). We have already discussed the situation in section 5.1 and we repeat here only that the effect of the 'weak lines' is to displace the object in the (U–B, B–V) diagram toward the upper left, and that the effect should be larger in U–B. This 'displacement' is taken with regard to the place which the object should occupy on the main sequence, if it had normal colors. Such an effect is obviously to be found in any photometric system using an ultraviolet color, and is called the 'ultraviolet excess'.

If the colors of a 'weak line star' and those of a 'normal' star are not identical, it is clear that we cannot plot both kinds of stars in an HR diagram without taking some precautions. If the weak line object is bluer than it would be if it had normal line strength, its color should be reddened and in this case the position in the HR diagram should be shifted (by means of a 'correction') to the right. Doing this, the 'underluminosity', which is of the

Figure 11.16. Photometric effects in subdwarfs: UBV photometry. The continuous line represents the relation for normal dwarfs and the broken line the giants and subgiants in M3. Points, circles and crosses represents field subdwarfs. From Sandage and Eggen (1959).

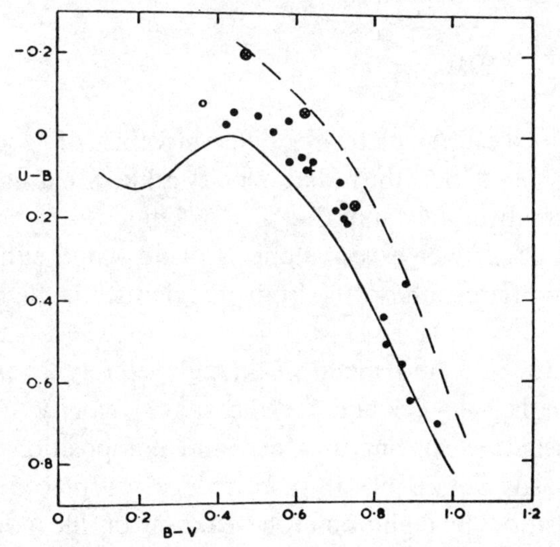

order of $1^m - 2^m$, diminishes or disappears entirely, as shown by Sandage and Eggen (1959) (see figure 11.17(a) and (b)).

To summarize subdwarfs are seen to share several interesting properties:

(a) a spectrum with weak metallic lines;
(b) an ultraviolet (photometric) excess;
(c) a position in the HR diagram below the main sequence, if uncorrected colors are used;
(d) a large space velocity.

In practice it is found that none of the four characteristics taken individually suffices to define 'subdwarfs'. The spectroscopic criterion is insufficient, because λ Boo stars and HB stars also satisfy it. If we consider the position in the HR diagram, we find that (usually) a trigonometric parallax is either unavailable or, if it exists, has such a large possible error as to make the 'underluminosity' rather indefinite. If the M is poorly determined, the space velocity suffers equally. Therefore many authors do not use space velocities but either (a) a very large proper motion or (b) a very large radial velocity. If we add the condition that the object must be a permanent member of our galaxy, its space velocity (v) should be smaller than 400 or 500 km/s.

Since the transverse motion T

$$T = 4.74 \frac{\mu''}{\pi''}$$

with $T \sim 474$ km/s,

$$\frac{\mu''}{\pi''} \leqslant 100 \qquad \text{or } \pi\mu^{-1} > 0.01$$

with μ known, the space velocity is calculated for different values of M and one chosen which fulfills $\pi\mu^{-1} > 0.01$. If the radial velocity is known, usually $\rho > 80$ km/s is used to make absolutely sure that $V > 62$ km/s.

We should add that an ultraviolet excess alone is of no value either, because it does not differentiate subdwarfs from subgiants, or highly reddened early type stars.

Eggen (1979) defines subdwarfs as old, metal weak, high velocity dwarfs. He suggests as a limit $[Fe/H] < -0.6$ and for the space velocity $V >$ 140 km/s. The definition seems clear but uses age and composition as parameters which are not easily obtainable. In practice 'age' is replaced by (space) velocity and 'composition' by a photometric weakness-of-line index,

Figure 11.17. The HR diagram for subdwarfs. From Sandage and Eggen (1959). (a) The B–V colors are the observed colors. Bars designate possible observational errors; continuous line, main sequence. (b) The stars are the same as in (a) but colors have been corrected (see text).

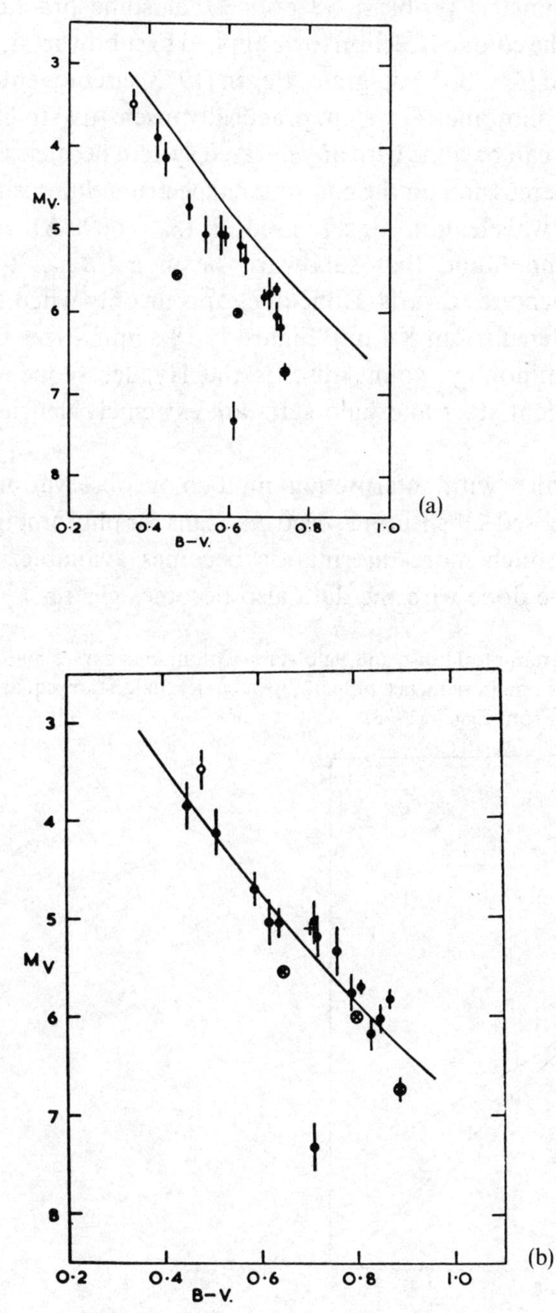

so that the definition itself is really not very helpful. Probably it would be clearer to use only the spectroscopic definition that a subdwarf is a weak line dwarf, a definition we proposed in the section on weak line stars.

Returning to the photometric problem we notice that some practical difficulties appeared with the color correction to be applied to subdwarf stars when they are plotted in an (M_v, B–V) diagram. Eggen (1973) circumvented these problems by using a photometric system practically insensitive to line weakness. In principle this can be done with any infrared system because the number of lines and its repercussion on the continuous spectrum diminishes steadily with increasing wavelength. Eggen used R ($\lambda_0 = 6000$ Å) and I ($\lambda_0 = 8000$ Å) indices and found that subdwarfs lie in a (M_{bol}, R–I) diagram on the same sequence as dwarfs. Difficulties appear only when K-type subdwarfs are considered (from K4 on). Figure 11.18 summarizes the situation. In Eggen's terminology 'young disc' is the Hyades sequence, subdwarfs are metal deficient stars and halo stars are extremely deficient stars.

In view of the difficulties with interpreting multicolor observations, Greenstein (1978) has proposed substituting spectrum scans for photometry. Figure 11.19 shows how much more information becomes available; he thinks that what should be done with the data also becomes clearer.

Figure 11.18. HR diagram of old disc and halo stars. Continuous curves, old disc (lower) and Hyades main sequence (upper); open circles, halo stars; squares, early type subdwarfs. From Eggen (1973).

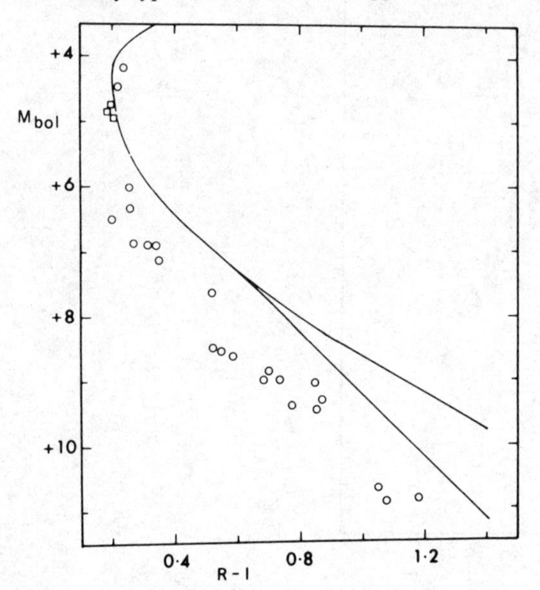

Binaries. A rather large number of binaries is known through photometry, mostly because the objects have very abnormal colors. Bessell and Wickramasinghe (1979) list some of these objects. Usually the combination consists of a white dwarf and a late subdwarf. In such a case the white dwarf provides the radiation in the blue, the subdwarf the radiation in the

Figure 11.19. Low resolution spectra of normal and subdwarf stars. Notice that the abscissa is λ^{-1}, the λ scale is given along the top. The position of certain spectral features is indicated. From Greenstein (1978).

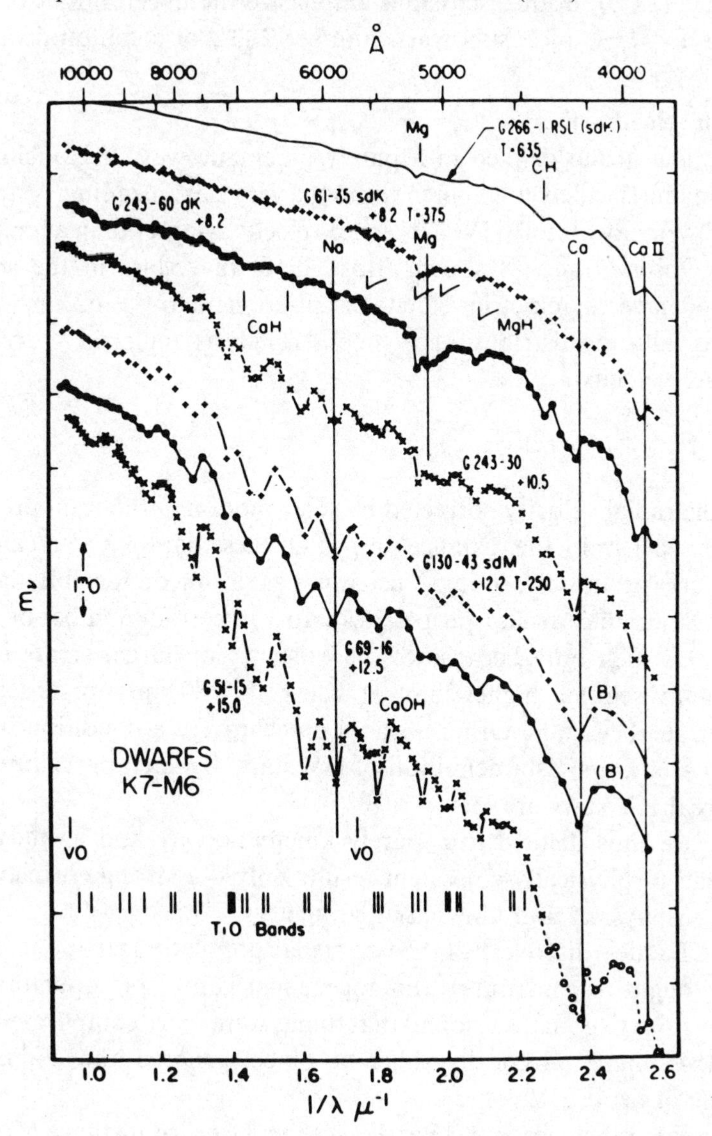

red. Such an object can only be detected with multicolor photometry. Due to the general faintness of these objects, very few spectroscopic binaries are known among them; Sandage (1969) mentions a few.

Number of subdwarfs. At present several hundred candidates for subdwarfs are known (Eggen 1979; Carney 1978) but fewer than fifty are brighter than $m = 9$.

The luminosity function of these objects was discussed by Schmidt (1975) and by Eggen (1979). Both discussions emphasize the uncertainties of the result; down to $M_v \sim +12$, subdwarfs number 2–3% of common dwarfs.

11.5 High velocity stars

This is a group defined in a purely kinematic way. It is included because the name is widely used, often without a very clear meaning. A star is called a 'high velocity object' (HV) if its space velocity is larger than a limiting velocity V. Observation has shown (Boss 1918) that stars in the solar neighborhood have a symmetric velocity distribution up to $V = 62$ km/s, but that after this value the distribution becomes strongly asymmetric. We recall that the space velocity

$$V = (\rho^2 + 4.74\,\mu^2 r^2)^{1/2}$$

where ρ is the radial velocity corrected by solar motion, μ the total proper motion (in ″/year) and r the distance (in pc). Of these data ρ and μ can be measured very accurately, but V usually has a uncomfortably large uncertainty. Since an error in r propagates into V, a certain number of HV stars appear as such only because of the existence of (large) errors in r. Authors usually select a higher limit of V (e.g. $V = 80$ km/s) in order to exclude spurious HV stars. Other authors prefer to use a condition on ρ alone (e.g. $\rho \geqslant 60$ km/s) as a definition of HV stars. By such procedures a number of real HV stars are lost.

HV stars are thus defined in a purely kinematic way and a study of them will lead to physically consistent results only if a strong correlation exists between physical and kinematic parameters.

Following Baade's discovery of the two stellar populations (I and II), HV stars were thought to constitute a (homogeneous) sample of population II stars. However very soon it was found that things were more complicated, in the sense that many physically different objects shared the common characteristic of being HV.

In recent years it has become generally accepted (see for instance Mould

1982) that high velocity objects may belong to two groups, namely

(a) old disc objects,
(b) halo objects.

Disc and halo refer to the form in which these objects are distributed in our galaxy. 'Disc' refers to objects confined to a layer around the galactic plane, 'halo' to objects contained in roughly a sphere around the center of the galaxy.

Such a terminology implies that type (a) stars and type (b) differ in location – type (a) should always be nearer to the galactic disc than type (b). Since however all stars are bound to cross the galactic plane at a given time, this means that some halo stars could be found near the sun, whereas no old disc stars should be found very far away from the galactic plane. Any segregation will thus be partial. On the other hand, since a star needs to have a large velocity component perpendicular to the galactic plane in order to reach a great distance from the plane, the $|Z|$ component could be used to disentangle the two groups: those with small $|Z|$ are bound to stay near the plane and are thus disc stars. (Z = velocity component perpendicular to the galactic plane.) The drawback of such separation is that a knowledge of the distance (or M_v) of the star is required to calculate $|Z|$, unless the star lies in the direction of the galactic pole, where Z coincides with the radial velocity component.

Eggen (1983) separates old disc and halo stars according to abundance, and defines a halo star – as mentioned above – as an object with $[Fe/H] \leqslant -0.6$. He further defines halo dwarfs as subdwarfs and calls objects intermediate between subdwarfs and Hyades main sequence 'old disc'.

The difficulty with such a definition lies in [Fe/H] which by definition is

$$\left[\frac{\text{Fe}}{\text{H}}\right] = \log\left(\frac{\text{Fe abundance}}{\text{H abundance}}\right)_{\text{star}} - \log\left(\frac{\text{Fe abundance}}{\text{H abundance}}\right)_{\text{sun}}$$

The only way of determining it is by the analysis of stellar atmospheres. As we remarked at the beginning of this book, at present we know the composition of only 2×10^3 stars, a fact that shows clearly that chemical composition is difficult to determine, and that such determinations cannot be carried out for a large number of objects. All authors therefore use a photometric approach to [Fe/H] which consists essentially of calibrating a relationship between a photometric index sensitive to metallicity (in a given system) and the quantity [Fe/H] obtained from direct stellar atmosphere analysis. Since calibrations can be done in many different systems, we quote only a few as examples.

We have already seen in the section on UBV photometry that weak line stars can be distinguished from normal stars by their position in the (U–B, B–V) diagram, provided that reddening has been taken care of (figure 11.16 in section 11.4). The δ(U–B) excess can then be correlated with known [Fe/H] values. Malyuto and Traat (1981) have summarized this for stars with [Fe/H] ~ -1.6; figure 11.20 reports the result.

As can be seen, the differences are not very large with respect to normal [Fe/H] ~ 0 stars.

Similar effects can be found in other photometries, for instance the Vilnius system (Bartkevicius and Sperauskas 1984) and can serve to single out metal deficient candidates or, if the metal deficiency is known spectroscopically, to derive the amount of metal deficiency.

The Strömgren uvby system can also be used for separating metal deficient stars from normal ones. Figures 11.21 and 11.22 illustrate this point rather well. c_1 characterizes the Balmer discontinuity, which is larger for stars with metal deficiency; m_1 characterizes the metal line strength and is smaller for metal deficient stars. The diagrams are taken from Bond (1980) who uses them to derive a photometric index of metal abundance.

A difficulty which appears in several of these calibrations is that they work

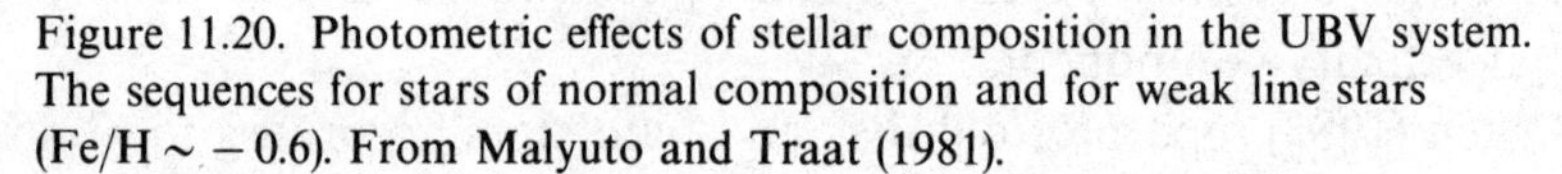
Figure 11.20. Photometric effects of stellar composition in the UBV system. The sequences for stars of normal composition and for weak line stars (Fe/H ~ -0.6). From Malyuto and Traat (1981).

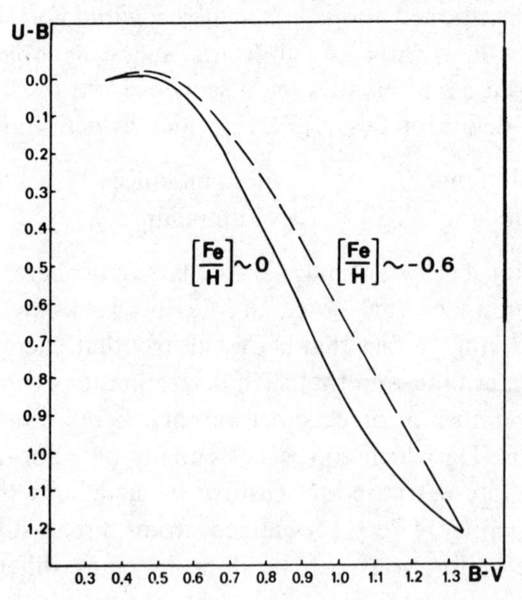

Figure 11.21. Strömgren photometry for metal deficient stars. The m_1 versus b–y diagram. Circles, metal weak stars; crosses, normal stars (× dwarfs, + giants). From Bond (1980).

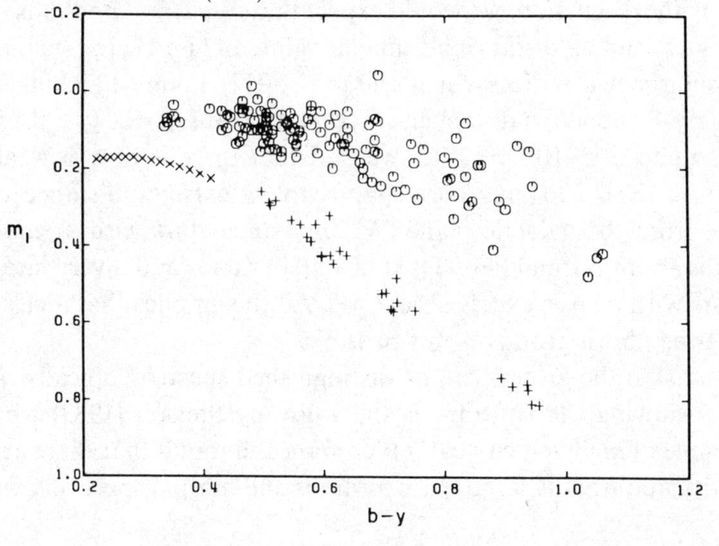

Figure 11.22. Strömgren photometry for metal deficient stars. The c_1 versus b–y diagram. Circles, metal weak giants: crosses, normal dwarfs. Stars with b–y ⩽ 0.6 and high c_1 are HB and AGB stars. From Bond (1980).

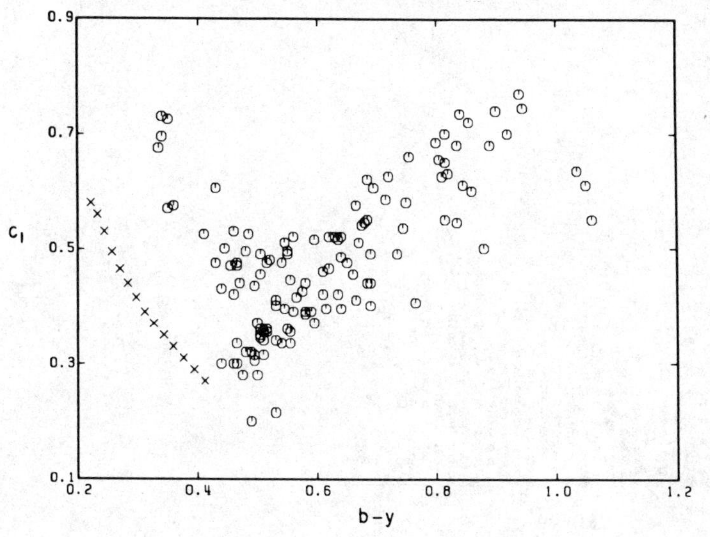

well between $0 > [Fe/H] > -2$, but not beyond. Stars with larger metal deficiencies practically all have the same photometric index.

Summarizing the discussion, we would expect that disc stars should have smaller $|Z|$ values than halo stars, and smaller values of [Fe/H] (or smaller values of the photometric indices which lead to [Fe/H]). Figure 11.23 taken from Carney (1984) shows what happens. Accepting that $[Fe/H] = -0.6$ is about equivalent to $\delta(U-B)_{0.6} = 0^{m}.12$, we find that there is only a weak correlation, since $|Z| = 200$ km/s corresponds to an extreme distance of about 6000 pc from the galactic plane. We find some stars with a small excess (and thus about normal [Fe/H]) at large $|W|$ values and a very large number of stars with a large excess at low $|W|$. All this implies that a clear separation between both groups is not possible.

We can then ask if the groups can be distinguished spectroscopically. A good example showing the contrary is the study by Stetson (1983) who analysed a sample of high velocity early type stars and found that there are A-type stars with abnormally large space motions and normal spectra. If we

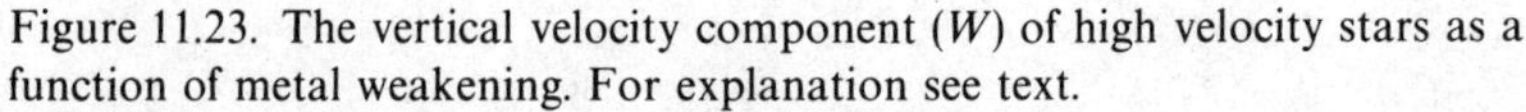
Figure 11.23. The vertical velocity component (W) of high velocity stars as a function of metal weakening. For explanation see text.

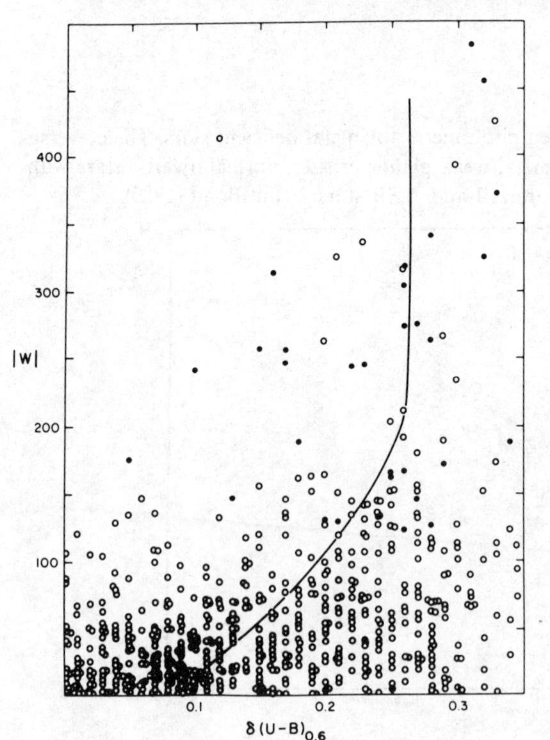

put this together with the discussion on runaway stars (see section 9.6) it is clear that no one-to-one relation exists.

What spectroscopy can provide are 'weak line spectra', and we refer readers to sections 11.4 and 11.5.

A good monograph on the subject is *Metal deficient stars* by Straizys (1982).

References

Abt H.A. and Levy S.G. (1976) *Ap. J. Suppl.* **30**, 273
Adams W.S. and Humason H.L. (1935) *PASP* **47**, 52
Altschuler W.R. (1975) *Ap. J.* **195**, 649
Bartkevicius A. (1984) *Bull. Vilnius* **66**, 79
Bartkevicius A. and Sperauskas J. (1984) *Bull. Vilnius* **63**, 66
Bessell M.S. and Wickramasinghe D.T. (1979) *Ap. J.* **227**, 232
Bidelman W.P. (1954) *Ap. J. Suppl.* **1**, 175
Bond H.E. (1970) *Ap. J. Suppl.* **22**, 117
Bond H.E. (1980) *Ap. J. Suppl.* **44**, 517
Bopp B.W. and Stencel R.E. (1981) *Ap. J.* **247**, 2131
Bopp B.W. and Talcott J.C. (1978) *A.J.* **83**, 1517
Boss L. (1918) *Dudley Obs. Annual Report*
Carney B.W. (1978) *A.J.* **83**, 1087
Carney B.W. (1984) *PASP* **96**, 841
Collier A.C. (1982) *MNRAS* **200**, 489
Cowley A. (1976) *PASP* **88**, 95
Crawford D.L. (1975) *A.J.* **80**, 955
Cucchiaro A, Jaschek M. and Jaschek C. (1977) *An atlas of ultraviolet stellar spectra*, Liège
Danziger I.J. and Faber S.M. (1972) *AA* **18**, 428
Eggen O. (1973) *Ap. J.* **182**, 821
Eggen O. (1978) *IAU Inf. Bull. Var. Stars* 1426
Eggen O. (1979) *Ap. J.* **229**, 158
Eggen O.J. (1983) *Ap. J. Suppl.* **51**, 183
Graham J.A. and Slettebak A. (1973) *A.J.* **78**, 295
Gray D.F. (1984) *Ap. J.* **277**, 640
Greenstein J. (1971) *IAU Symp.* **42**, 46
Greenstein J. (1978) in *IAU Symp.* **80**, 101
Hall D.S. (1976) *IAU Coll.* **29**, 287
Henry R.C. (1979) *Astron. pap. dedic. to B. Strömgren*, p. 19
Hoffleit D. and Jaschek C. (1982) *Bright star catalogue*, Yale Univ. Obs.
Houk N. (1978) *Michigan spectral catalog*, vol. II, Univ. of Michigan
Jaschek C. and Gomez A. (1970) *PASP* **82**, 809
Johnson H.L. (1966) *Ann. Rev. AA* **4**, 193
Keenan P. and Keller G. (1953) *Ap. J.* **117**, 24
Keenan P. and Yorka S. (1985) *BICDS* **29**, 25

Kron G. (1947) *PASP* **59**, 261
Kuiper G.P. (1939) *Ap. J.* **89**, 548
Landi J., Jaschek M. and Jaschek C. (1977) *An atlas of grating stellar spectra at intermediate dispersion*, Cordoba, Argentina
Malaroda S. (1973) *PASP* **85**, 328
Malaroda S. (1975) *A.J.* **80**, 637
Malyuto V. and Traat P. (1981) *Publ. Tartu* **48**, 199
McCarthy J.K. and Ramsey L.W. (1984) *Ap. J.* **283**, 200
McCuskey S. (1965) in *Galactic structure*, Univ. of Chicago Press, p. 1
Mendoza E. (1978) *IAU Symp.* **80**, 289
Morgan W.W. (1958) in *Stellar populations, Spec. Vaticana* **5**
Mould J.R. (1982) *Ann. Rev. AA* **20**, 91
Olsen E.H. (1979) *AA Suppl.* **37**, 367
Olsen E.H. (1980) *AA Suppl.* **39**, 205
Rodono M. (1981) in *Photometric and spectroscopic binary systems*, Carling E.B. and Kopal Z. (ed.), Reidel D. Publ. Co., p. 285
Roman N. (1950) *Ap. J.* **112**, 554
Roman N. (1954) *Ap. J.* **59**, 307
Roman N. (1955) *Ap. J. Suppl.* **2**, 198
Sandage A. (1969) *Ap. J.* **158**, 1115
Sandage A. and Eggen O. (1959) *MNRAS* **119**, 279
Schmidt M. (1975) *Ap. J.* **202**, 22
Schmidt-Kaler T. (1982) in Landolt–Börnstein, group VI, vol. 2b, p. 1
Schwarzschild K. and Eberhard G. (1913) *Ap. J.* **38**, 292
Slettebak A. (1970) *Stellar rotation*, Reidel D. Publ. Co., p. 3
Smith M.A. (1980) *PASP* **91**, 737
Stetson P.B. (1983) *A.J.* **88**, 1349
Straizys V. (1982) *Metal deficient stars*, Mokslas Publ., Vilnius
Vaughan A.H. (1983) *IAU Symp.* **102**, 113
Vaughan A.H., Preston G.W. and Wilson O.C. (1978) *PASP* **90**, 267
Wilson O.C. (1963) *Ap. J.* **138**, 832
Wilson O.C. (1968) *Ap. J.* **153**, 221
Wilson O.C. (1976) *Ap. J.* **205**, 823
Wilson O.C. and Bappu M.K.V. (1957) *Ap. J.* **125**, 661
Yamashita Y., Naria K. and Norimoto Y. (1977) *An atlas of representative stellar spectra*, Univ. of Tokyo Press
Zinn R. (1973) *Ap. J.* **182**, 183

12

G-type stars

12.0 Normal stars

G-type stars are characterized by weak hydrogen lines which become comparable in strength to the lines of some metals. Metallic lines increase both in number and in intensity toward later spectral subdivisions, and molecular bands of CH and CN become easily visible features.

In order to fix ideas, we quote in table 12.1 the equivalent widths of some strong lines.

We have not given the intensity of the G-band, which is easily observable at classification dispersion, but which breaks down on the low plate factor spectrograms needed to measure equivalent widths.

The spectral type is established by the comparison of hydrogen and metal lines, like Fe λ4143 and Hδ: they are about equally intense at G8 when seen at 80 Å/mm. Instead of this pair of lines, Fe λ4045/H λ4101 or Fe λ4384/H λ4340 and λ4921/H λ4861 may also be used. For types later than G5 the Ca I λ4226 line becomes sensitive to temperature and can be used for determination of spectral type as Ca I λ4226/H λ4101 (see figure 12.1).

If it is suspected that there are composition anomalies, the hydrogen-to-metal ratio should not be used but should be replaced by Cr λ4254/Fe λ4250 and Cr λ4274/Fe λ4271 (Keenan and McNeil 1976).

If for instance the star has weak metal lines, the ratio between hydrogen and metallic lines is earlier than it should really be, and only the ratio of two metal features can provide the right spectral type. For metal weak stars this can imply a shift of several tenths of spectral type.

It should be noticed that not all decimal subclasses are used. In the HD catalog only G0 and G5 existed, and the Yerkes system uses only G0, G2, G5, G8 and K0 as full types. The intermediate types are less frequently used, and this fact must be taken into account in statistics of spectral classes. Furthermore there are very few giants earlier than G5. This is the so-called ‘Hertzsprung gap’ in the HR diagram.

Table 12.1. *Equivalent widths (in Å) of some strong lines.*

	λ3933 Ca II	λ4045 Fe I	λ4101 H	λ4226 Ca I	λ4325 Fe I	λ4340 H
G0	17.0	1.0	4.2	1.1	0.7	3.0
G2	20.2	1.1	3.1	1.5	0.8	2.9
G5		1.8	3.2	1.8	1.2	3.1

Figure 12.1. Spectra of G-type dwarfs, temperature sequence. For explanation see text.

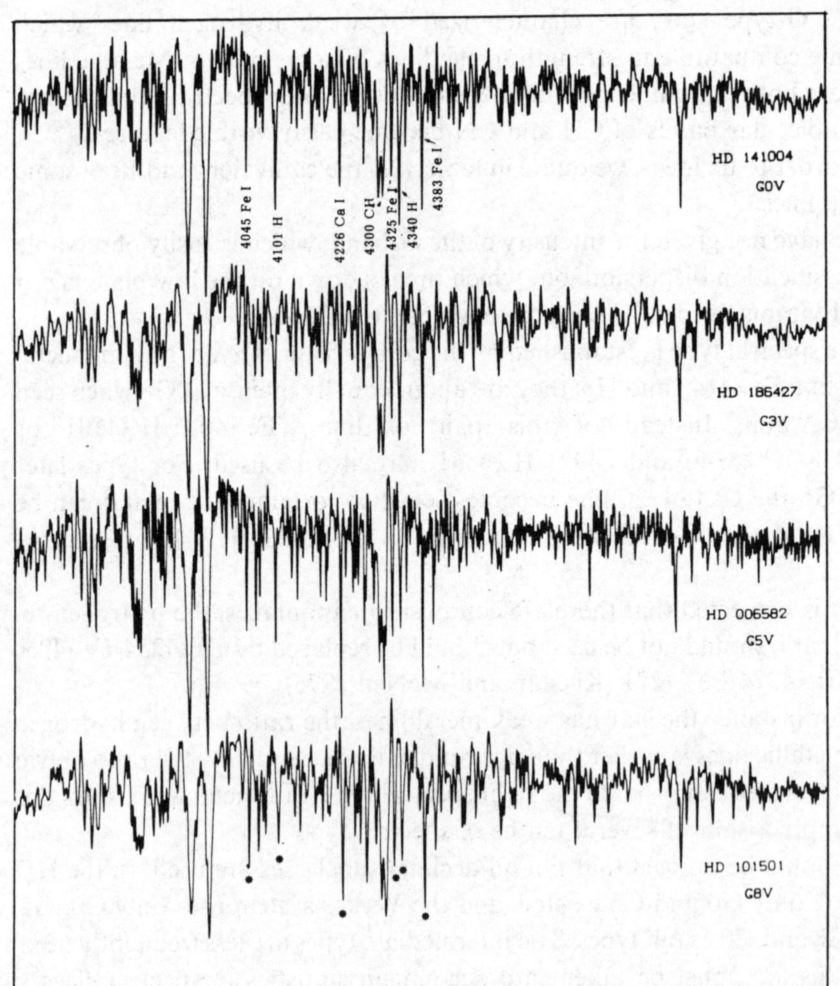

Luminosity effects in G-type stars are clearly visible, at low dispersion, in the CN bands. Lindblad (1922) used the band at $\lambda 4180$ and measured the depression produced by it on the pseudo-continuum; typically between giants and dwarfs the difference is of the order of 10–20%. At higher dispersions, the CN band is less useful because it tends to dissolve. Because of this, in the Yerkes system Keenan and McNeil (1976) use line ratios exclusively, although these should rather be called 'feature ratios'. For instance the Sr II line at $\lambda 4215$ is blended with Fe, so that we should speak of the Sr II, Fe $\lambda 4216$ feature. The feature is luminosity sensitive if the ratio $\lambda 4216$/Ca $\lambda 4226$ is considered. Similarly Sr II $\lambda 4077$/Fe $\lambda 4063$, $\lambda 4071$ can also be used. In the case of $\lambda 4216$ care must be taken if CN is anomalous, because of the band head of CN at $\lambda 4216$ (see figure 12.2). For supergiants Y II, Fe $\lambda 4376$/Fe $\lambda 4383$ or the blend $\lambda\lambda 4172+78$/Hδ may be used. Other criteria are Y II $\lambda 3983$/Fe $\lambda 4005$, Sr II $\lambda 4077$/H $\lambda 4101$ and Sr II $\lambda 4216$/Fe $\lambda 4144$. Care has to be taken when '*s* process elements' seem to be enhanced ($s =$ slow neutron capture). Since Sr is an '*s* type' element, it obviously cannot be used as a luminosity indicator in such stars. To repeat what we said in chapter 3, we have to use *all* features in the spectrum and definitely should not use a single criterion.

When lines of isolated elements deviate from normal behavior, this is indicated by special notation. Table 12.2 from Keenan and McNeil (1976) indicates the main features used and the notation followed.

Usually the index runs from 0 (normal) to -3 (pronounced weakness) and $+3$ (pronounced enhancement). Obviously this notation should be used only if its use is justified, i.e. to avoid lengthly descriptions. We shall consider in particular the abnormalities in the behavior of CN (see section 12.2) and CH (see section 12.3).

In the red region, the Mg I triplet ($\lambda\lambda 5167$–72–83), which is luminosity sensitive in the range G8–K5, may be used (Ohman 1936). In the interval B8–G8 the band strength increases slowly with temperature.

In the near infrared region $\lambda\lambda 7600$–9200, Barbieri *et al.* (1981) have illustrated the changes of line intensity with spectral type and luminosity class. The Ca II triplet ($\lambda\lambda 8498$–8542 and $\lambda 8662$) is still well visible. It weakens toward later spectral subdivisions and lower luminosity classes. Paschen lines weaken very rapidly and can only be seen in early subtypes and high luminosity classes; in such cases the O I $\lambda 7774$ line can also be used.

The sun. One of the reasons for the popularity of G-type stars is that the sun belongs to this type. The HD catalog classifies the sun as G0, and the Yerkes

Figure 12.2. Luminosity effects in G-type stars. For explanation see text.

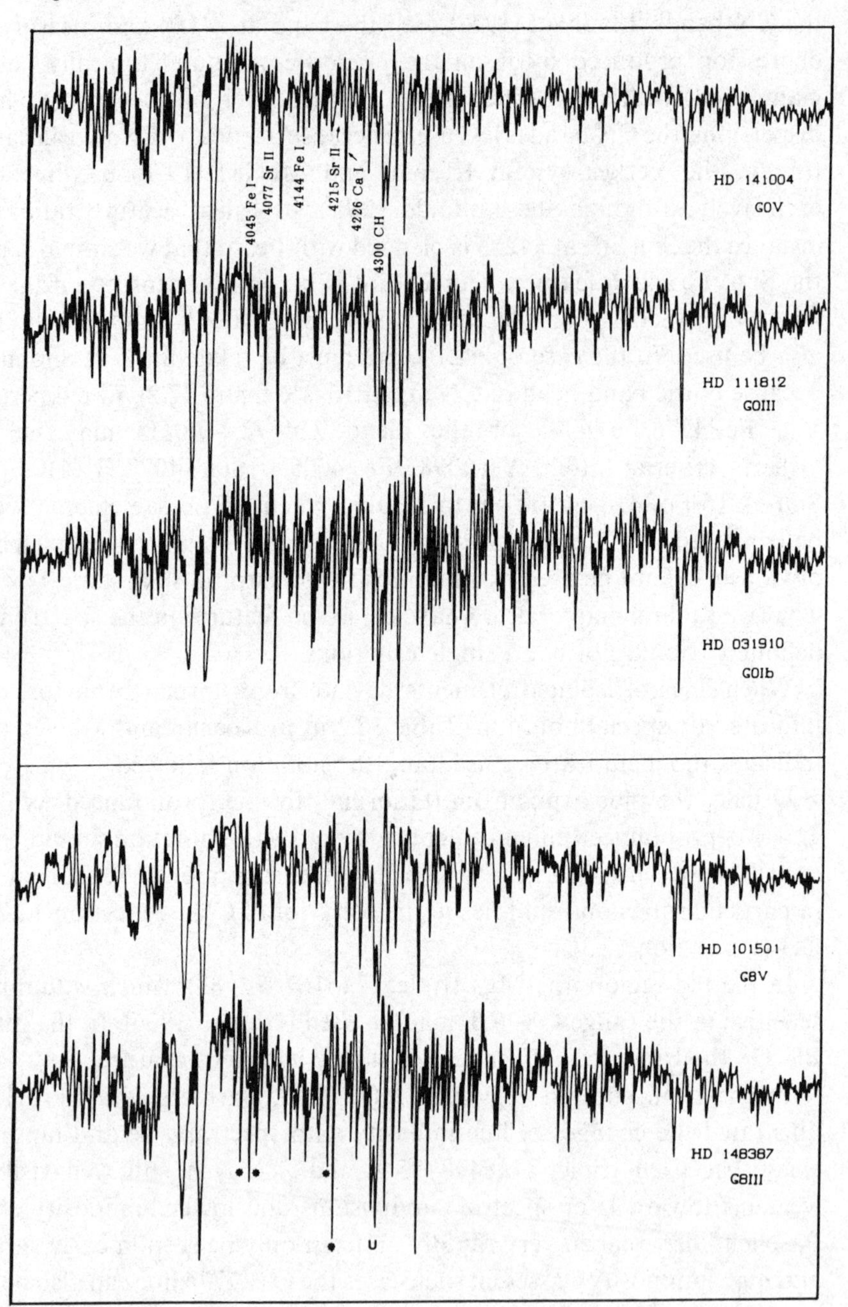

system classifies it as G2. The determination of the spectral type of the sun presents us with some difficulties, however, due first of all to observing conditions. In contrast to stars, the sun is very bright, it cannot be observed during the night and it is not a point source. The brightness of the sun forcibly produces a large quantity of stray light in the spectrograph, which means working under very different conditions than when observing stars at night. Some observers have used, instead of direct sun-light, reflected sun-light from the moon, planets or minor planets, or scattered sun-light from the 'blue sky'. Since the moon is also bright and not a point source, its use is objectionable. Planets on the other hand have atmospheres which modify the solar light they reflect. And the scattered sky-light is not a point source. What remains are the asteroids, whose spectrum should be identical to that of integrated sun-light. With hindsight it would have been much better if in the Yerkes system the type of object used for spectral classification of the sun had been specified. On the other hand, few classifiers used the sun as a classification standard – all astronomers used the standard stars given in the Yerkes system. The question of the spectral type of the sun came under discussion when Hardorp (1978) tried to find solar analogs, that is, stars as

Table 12.2. *Notation for abnormal line behavior.*

Observed feature	Classification symbol	Element or element group	Remark and examples
Balmer lines	Hδ	Hydrogen	Use only if Balmer much stronger or weaker, e.g. HD 49 500: CN-2 CH-1 Hδ I
Ca λ4226 or Na D at 5890	Ca or Na	Light alkaline	e.g. R Sct K0 Iab Ca-1, HR 918 G9 III Ca 1
Fe λ4263, λ4271, λ4325, λ4383 Cr λ4254	Fe	Iron peak elements	Used for 'strong line' or 'weak line' e.g. HD 221 170: K0 II: Fe-3
CN λ4216, λ3883	CN	C and/or N or total metals	E.g. 2 Ori K0 IIIb CN-2
CH G-band	CH	C or H	E.g. HR 6711 G8 III CH-3 CN-1, HR 8626 CH 2 Fe-1 G3 Ib-II

closely similar to the sun as possible. He looked for similarity of flux, measured either photometrically or spectrophotometrically. He started using spectral types of stars and plotted them against color indices and found that if he did so, the sun would not match the relations established. Part of his difficulty came from the fact that he used spectral classifications gathered from the literature, which are heterogeneous and thus insufficient for accurate comparisons. A second difficulty is that the sun's color index (B–V = 0.625) is difficult to measure because of the sun's brightness and may therefore be uncertain by $\pm 0^{m}.025$ (Chmielewski 1981). A third difficulty is that – as mentioned above – no canonical definition exists on how the spectrum of the sun (G2 by definition) should be observed. A long controversy followed, which can be retraced in Garrison (1979) and Garrison (1985), where the reader can find a complete bibliography.

Now that the dust is settling slowly, some issues have been clarified, or at least the problems have been brought into a sharp focus. Keenan and Yorka (1985) have considerably enlarged the set of standards in the critical region G0–G5. There are now 25 standards instead of 7, and they have reached a very high degree of consistency, as can be seen from table 12.3.

The color of the sun has still a large uncertainty, but one knows at least why the values obtained by different methods do not coincide, so that essentially the problem seems solved.

Spectral peculiarities. There are rather a large number of G-type stars with emission features in the cores of the H and K lines of Ca II (see section 11.2). Also a number of stars show abundance anomalies, for instance weak lines (see section 11.3) or strong lines (see section 12.2).

Table 12.3. *Standards of spectral type and colors around G2.*

HD	Sp. type	B–V value
115043	G1 va	0.60
10307	G1.5 v	0.62
186408	G1.5 vb	0.64
146233	G2 va	0.65
28099	G2 + v	0.66
186427	G2.5 v	0.66
140538	G2.5 v	0.68
Sun	G2 v	0.625 ± 0.025

Further problems of classification arise where heavy metals are examined; there are stars in which '*s* type elements' (Ba and Sr for instance) are enhanced (see section 12.1) and anomalies may also exist when the molecular bands (mostly CN or CH) are either strengthened or weakened (see sections 12.2 and 12.3). It should be added that most of these peculiarities are not limited to spectral type G, but are also seen in type K spectra.

We shall also discuss in this chapter the group of T Tau stars and their subdivisions (YY Ori, FU Ori) in section 12.4.

Rotation. If we consider the sun to be a typical G-type star, the sun's equatorial rotational velocity – 2 km/s – should indicate the order of magnitude of $V \sin i$ for G-type stars. Such a velocity is below the limit of detection of line broadening and is thus irrelevant for classification purposes. $V \sin i$ values for G-type stars have been derived from the study of star spots (see section 11.2). They are typically of the order of a few km/s.

The few G-type giants which exhibit high rotation ($V \sin i > 10$ km/s) are usually close binary systems in which the stars rotate in synchronization. Some of them belong to the FK Com group (see section 11.2).

As a rule any G-type star with broadened lines is an interesting or peculiar object.

Photometry. In UBV photometry, colors of G-type stars differ according to luminosity class. We have plotted in figure 12.3 the intrinsic colors for stars

Figure 12.3. Luminosity effects in UBV photometry for late type stars. The sequences correspond to de-reddened objects.

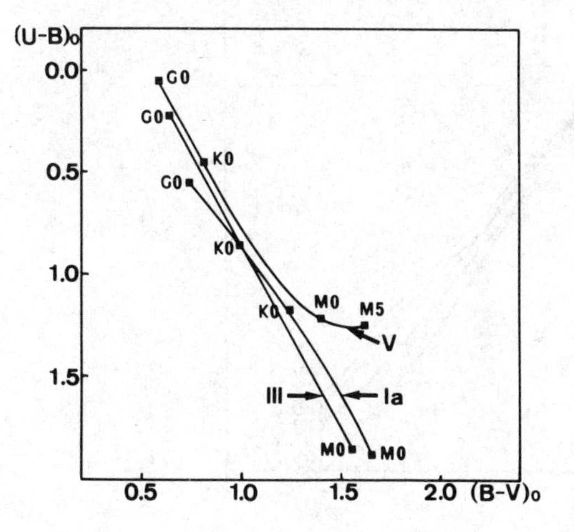

of luminosity classes I, III and V according to Straizys (1977). It can be seen that there are differences, although in practice the existence of corrections due to reddening and to metal content make their usefulness somewhat doubtful. Observe however that the same U–B (or B–V) corresponds to very different spectral types. For instance $(B-V)_0 = 0^m.8$ corresponds to K0 V, G7 III and G1 I.

We have already seen (section 5.1) that if the luminosity class is known, we may derive a metal strength index, which helps to single out stars in which the metallic lines are weak. This implies that an important new parameter is available, the metal strength, besides the other well-known ones (temperature, luminosity class, reddening) and that a system with more indices is needed to deal with these stars.

The easiest way to deal with extinction is through infrared photometry, because its influence is less on those bands. In figure 12.4 are illustrated the average colors for G-type dwarfs.

However infrared colors are insensitive to effects of line strengths and luminosity, so these colors are not used very much for studies of G-type stars. The next step is to turn to intermediate band photometry, for instance

Figure 12.4. Broad band photometry of G-type dwarfs: U, B...L colors.

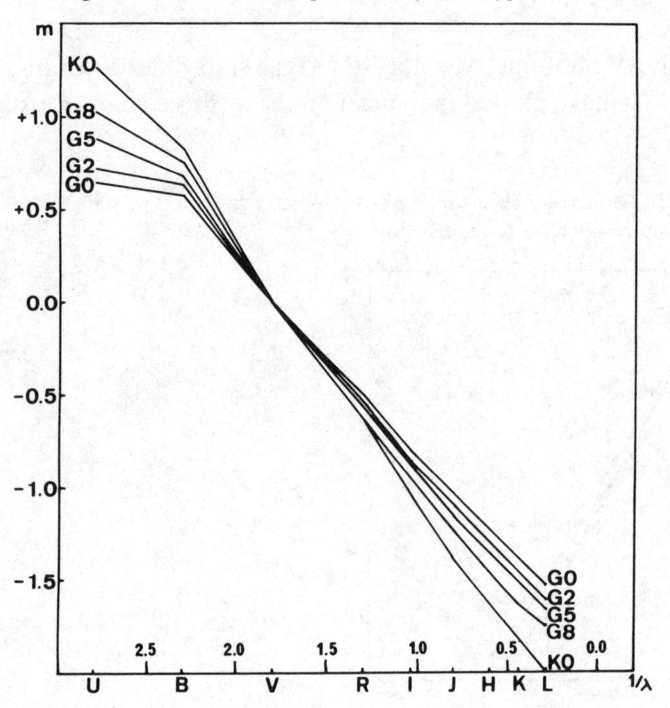

Strömgren. This system can still be used up to early K stars (Ardeberg and Lindegren 1981), although reddening poses problems. For unreddened stars we may use c_1 as a luminosity indicator and m_1 as a metallicity index. Figures 12.5 and 12.6 show some of the results.

Figure 12.5. Strömgren photometry for late type stars. The (c_1, b–y) diagram for MK standards. Points, class V; crosses, class IV; open circles, class III. The curves represent a tentative luminosity-class separation. From Ardeberg and Lindegren (1981).

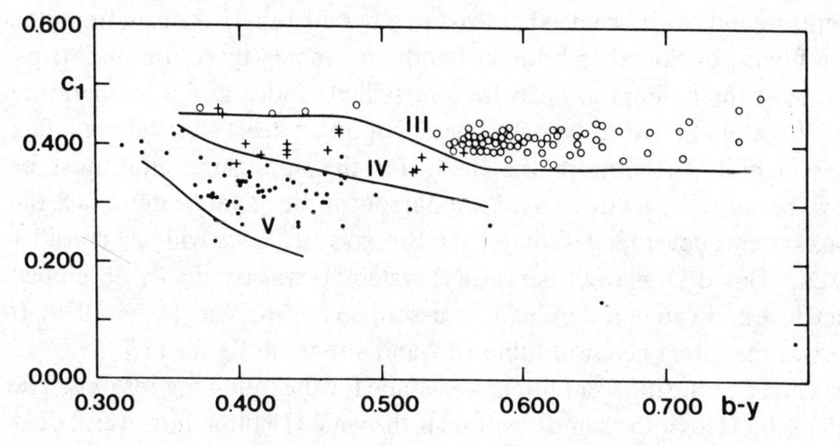

Figure 12.6. Strömgren photometry of late type stars. The (m_1, b–y) diagram for late type stars of different metal abundances. From Ardeberg and Lindegren (1981).

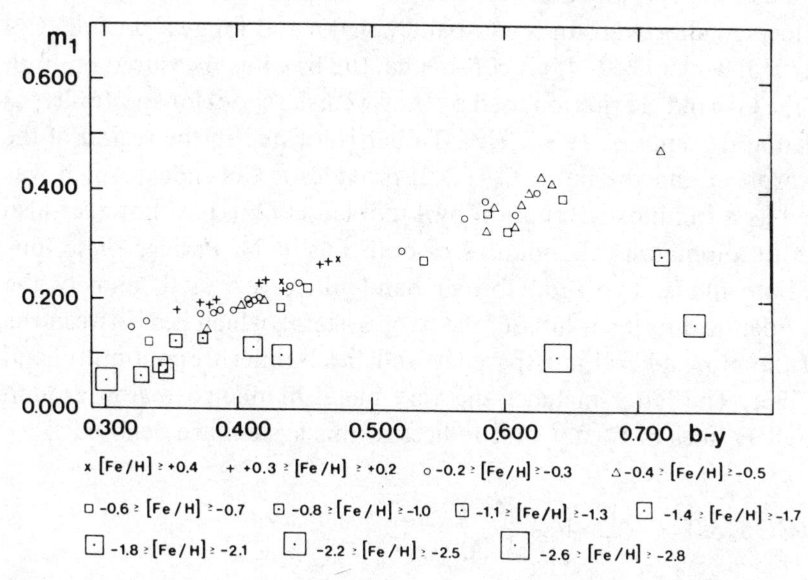

Interpretation of the results is difficult for figure 12.5. c_1 measures the Balmer discontinuity in earlier stars, which by now has practically disappeared so another explanation must be sought. Figure 12.6 on the other hand has the same interpretation as in earlier stars.

In view of the difficulties posed by the interpretation of figure 12.5, we can instead look for systems which measure those features which are known from spectroscopy to be sensitive indicators of temperature and luminosity. Among these we have the CH band for spectral type and the CN band for luminosity. Since we may also expect metal strength effects, and need a temperature indicator, we need at least two band strength indices, implying narrow filters positioned on both the bands and on nearby continuum strips, plus at least three filters to measure a metallicity index and a temperature index. We thus arrive at systems based upon at least six different flux measurements. Furthermore the filters for the band strengths must be narrow, because the features used are narrow and not very intense. Of the various systems developed (Golay 1974; Straizys 1977), we will just consider the DDO (David Dunlap Observatory) system. It was originally developed by McClure and van den Bergh and is described by McClure (1976, 1979). It comprises the filters given in table 12.4 and shown in figure 12.7.

The choice of the different filters was made for the following reasons. The 48-filter is located at the same position as the wide Hβ filter; however it does not measure Hβ (which is insignificant in these stars) but the MgH band. The 45-filter is located in a region relatively free of strong spectral features and is used as a comparison passband for both the 42- and the 48-filters. The 42-filter is located shortward of the G-band (λ4300) and longward of the CN band (λ4215); when C(42–45) is considered, the break is measured on both sides of the G-band, a criterion used by the Swedish school for spectral types (Lindblad and Stenquist 1934). The 41-filter is located in the region of the CN absorption and the index C(41–42) provides a CN index, which was proposed as a luminosity criterion by Lindblad (1922). It is however also sensitive to anomalous abundances of both C and N. Besides these four intermediate filters, two more broad band filters are used, namely the 35-filter (identical to the u-filter of the uvby system), which lies between the limit of the atmospheric transparency and the Balmer discontinuity, and the 38-filter. The latter includes the very metal blanketed region around the K and H lines of Ca II. Color indices in this system are denoted

$$C(35{-}38) = -2.5 \log \frac{I(35)}{I(38)}$$

Table 12.4. *Filters of the DDO system.*

Name	λ_0	$\Delta\lambda$	Name	λ_0	$\Delta\lambda$
48	4886	186	41	4166	83
45	4517	76	38	3815	330
42	4257	73	35	3460	383

Figure 12.7. DDO photometry. The bandpasses of the system. Abbreviations for bands are indicated inside each figure. The energy distribution of an K0 III star is given below to illustrate what is measured in the different filters. Filters are designated by the mean wavelength in hundreds of angstroms. Thus the filter centered at λ4200 is called 42.

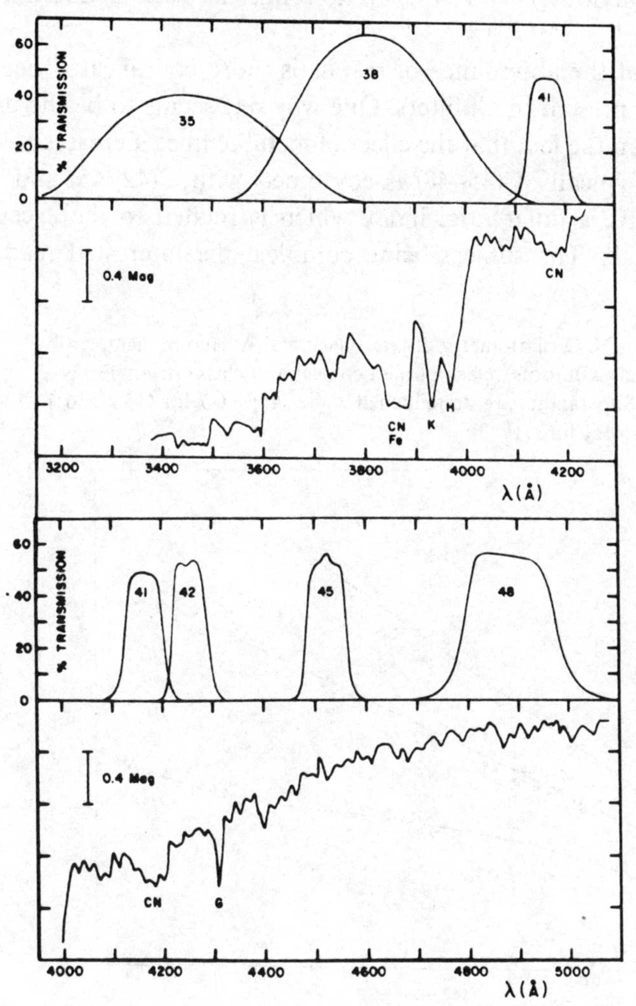

The choice of the passband brings out the meaning of some of the diagrams. From the explanations given, it is clear that C(42–45) should be related to spectral type, whereas C(41–42) should be related to luminosity. Figures 12.8 and 12.9, taken from McClure (1973), show this to be true for normal stars, provided supergiants are corrected for interstellar reddening. The latter are obtained in the standard way, i.e. using an average $F(\lambda)$ curve and taking the excess ratios from this curve.

Since there are five color indices available, we should be able to extract more information from the data. We can try for instance to use the C(41–42) index to select stars with strong and weak CN bands. This is shown in figure 12.10, also from McClure (1973), where we can see that the method works well, but obviously C(41–42) can no longer be used as a luminosity indicator.

The question of the abundance of metals is more complicated because metallic lines are present in all filters. One way out seems to be the use of indices based upon the fact that the effect of metallic lines increases toward the ultraviolet. Typically C(45–48) is combined with C(42–45) and with C(41–42), to derive a differential index which is related to the excess or weakness of metals. The subject being complex, the interested reader is

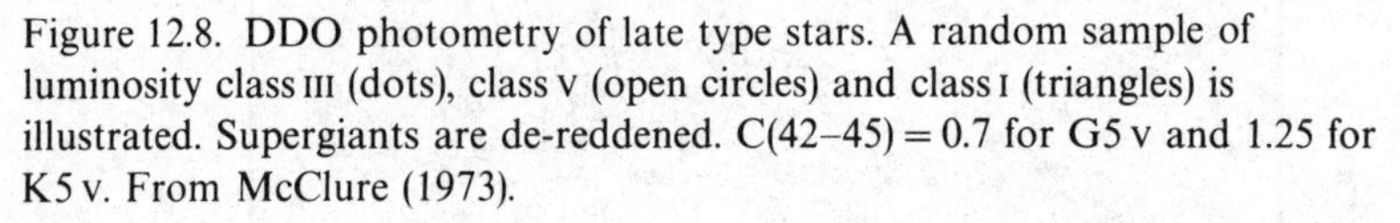
Figure 12.8. DDO photometry of late type stars. A random sample of luminosity class III (dots), class V (open circles) and class I (triangles) is illustrated. Supergiants are de-reddened. C(42–45) = 0.7 for G5 V and 1.25 for K5 V. From McClure (1973).

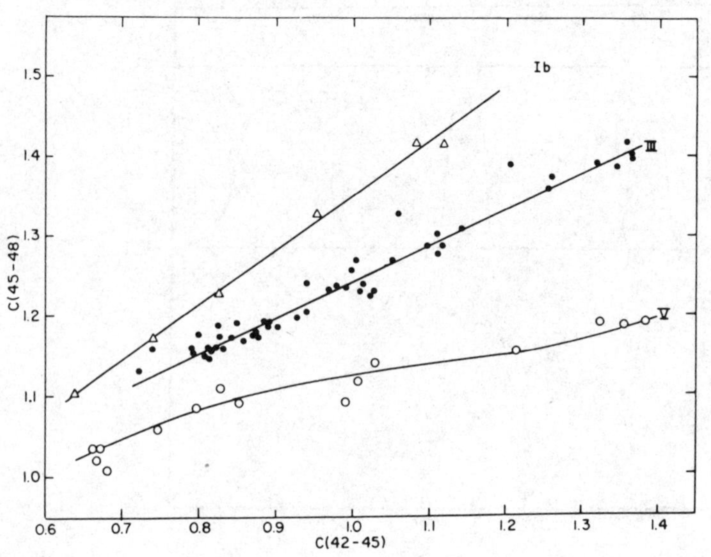

referred to McClure (1979), where a complete review of the system and a bibliography of its application is provided.

We mention in passing that much work has been done on the narrow band photometry of the Mg I red triplet ($\lambda\lambda$5167–72–83) which is luminosity sensitive in the interval G8–K5. See for instance Guinan and Smith (1984) and the references they quote.

Figure 12.9. DDO photometry of late type stars. (a) The observed color–magnitude diagram for the old open cluster NGC 2420. Large dots, giant branch stars, open circles, bright red giants. (b) The relative position of these objects in the C(45–48), C(42–45) diagram. From McClure (1973).

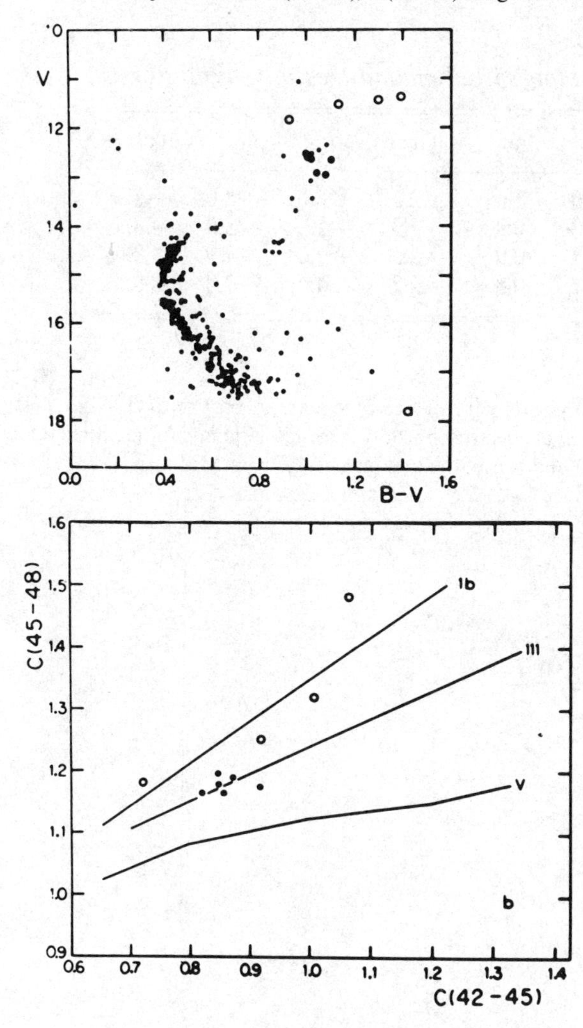

Absolute magnitude. The absolute magnitudes of G-type dwarfs can be calibrated from trigonometric parallaxes, whereas for giants statistical parallaxes are needed. For stars of higher luminosity the usual mixture of cluster parallaxes, galactic rotation parallaxes and so forth must be used, with subsequent uncertainties. Typical values are given in table 12.5, taken from Schmidt-Kaler (1982).

It should be noticed that this calibration is valid only for stars of normal type. Stars with spectral peculiarities (weak lines, for instance) have to be calibrated differently. See section 10.4 for such a case, namely the subdwarfs.

Presence in clusters. G-type dwarfs and giants can be observed in open and

Table 12.5. *Absolute visual magnitudes of G-type stars.*

	V	IV	III	II	Ib	Ia	Ia0
G0	+4.4	3.0	1.0	−2.3	−5.0	−8.0	−8.9
G2	4.7	3.0	0.9	−2.3	−5.0	−8.0	−8.8
G5	5.1	3.1	0.9	−2.3	−4.6	−7.9	−8.6
G8	5.5	3.1	0.8	−2.3	−4.4	−7.8	−8.5

Figure 12.10. DDO photometry of late type stars. A plot of C(41–42) versus C(42–48) for samples of spectroscopically strong CN giants (dots), high velocity giants (open circles) and normal giants (crosses). From McClure (1973).

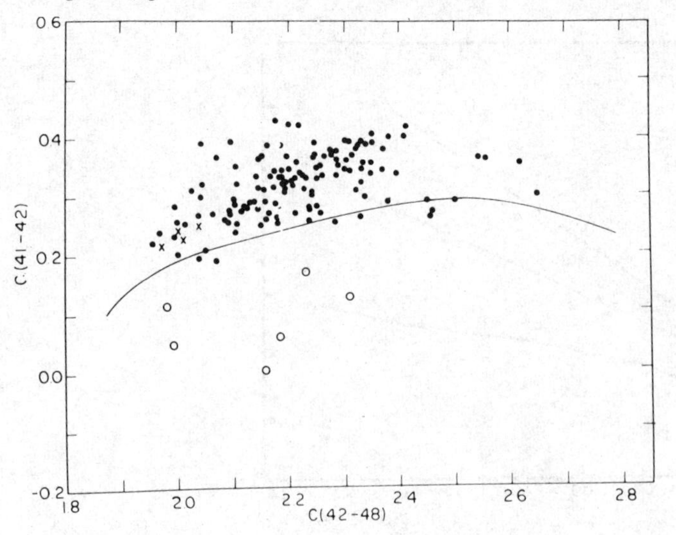

in globular clusters, but their relative faintness makes spectroscopic classification studies difficult.

According to theory, G-type dwarfs that formed at the time of formation of our galaxy can still be expected to be on the main sequence. If the spectra formed in the surface layers are unaffected by what happens in the deep interior, we should be able to see representatives of the different generations of stars formed since the formation of our galaxy. If the first generations of stars were metal poor, we would expect to find a (large) number of weak line stars. Observationally this has not been found (see section 11.3) and the divergence between the predicted and the observed number constitutes the so-called 'G–K dwarf problem'.

Number, distribution in the galaxy. G-type stars are moderately concentrated toward the galactic plane, with a halfwidth $\beta \cong 350$ pc. Their distribution is largely uniform, without specific concentrations. They are also rather common stars. The relevant figures are gathered in table 12.6.

Attention should also be paid to the distribution over luminosity class. Only 16% of the 'bright stars' are dwarfs – all the remainder are more luminous. This is essentially because the absolute magnitude of G-type dwarfs is faint so that at $m = 6^{\mathrm{m}}.5$ only a few nearby G-type dwarfs are visible. Moreover a large number of giants appear suddenly at G5. If both effects are taken together the first line of table 12.6 finds a satisfactory explanation. Notice however that among the 4% of nearby G-type stars we find *no* giant.

12.1 Barium stars

A 'barium star' (Ba star) is a giant (G2–K4) showing in its spectrum a very strong Ba II λ4554 line. The group was defined by Bidelman and Keenan (1951) based upon 80 Å/mm spectrograms. These authors also call attention to other features in the spectrum, namely (a) the enhancement of all Ba II lines, (b) the enhancement of the λ5165 band of C_2, and (c) the enhancement of the G-band due to CH. Point (a) shows that it is not just the resonance line

Table 12.6. *Frequency of G-type stars.*

Survey	Limit	Total number	Percentage of all stars
Bright stars	$m \leqslant 6^{\mathrm{m}}.5$	1200	13%
HD	$m \sim 9^{\mathrm{m}}$	42 000	19%
Nearby stars	$r \leqslant 20$ pc		4%

λ4554 which behaves abnormally, and the other two points link Ba stars to carbon stars. In addition to these bands the λ3883 and λ4216 of CN and possibly the one of C_2 at λ4737 are also enhanced. (See figure 12.11.)

The enhancement of CH is reminiscent of the CH stars; the difference is

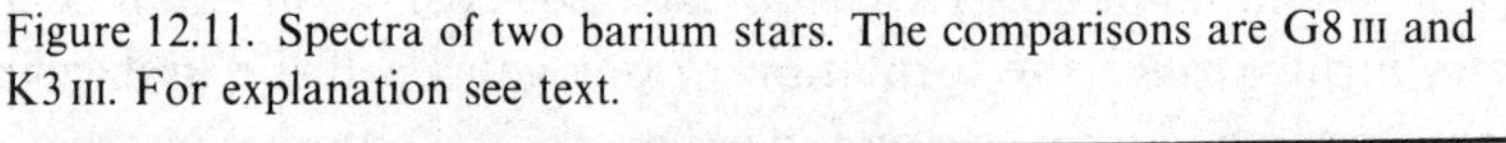

Figure 12.11. Spectra of two barium stars. The comparisons are G8 III and K3 III. For explanation see text.

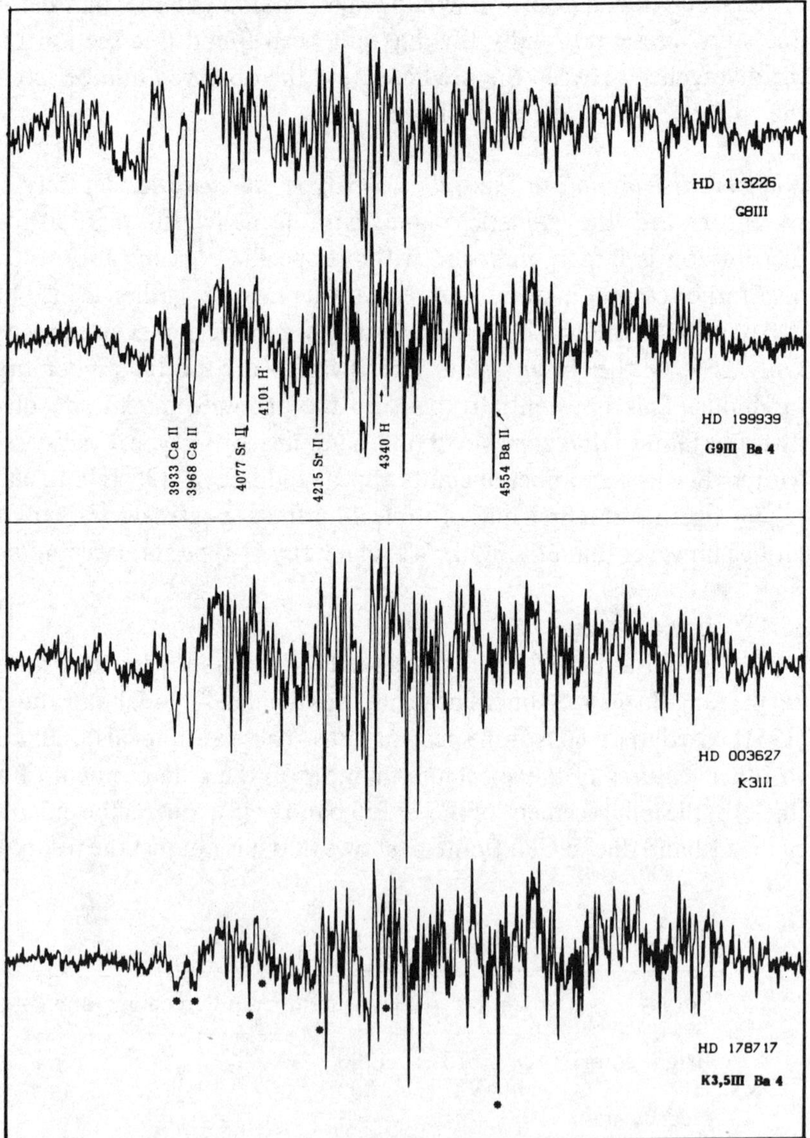

that in Ba stars CH is not as strong as in 'CH stars', whereas Ba is not enhanced in 'CH stars'. As mentioned above, the enhancement of C_2 relates Ba stars to C stars, but the strength of C_2 features in Ba stars is intermediate between that in normal giants and that in C stars.

Bidelman and Keenan noticed further a possible relation to S stars. In S-type stars, Ba II λ4554 is also enhanced, as are the Sr II lines. However, S-type stars are of later type (around M0) and, unlike the Ba stars, exhibit ZrO bands. From later work we retain an innovation introduced by Warner (1965), namely an index for the strength of the Ba II λ4554 line; the line strength is estimated on a scale from 1 to 5 (strong). A barium star is thus denoted K0–Ba 3, where K0 is the spectral type obtained from the strength of the lines of Ca and the iron peak elements. This practice has been followed by other authors, using different intensity scales however, so that we have to pay close attention to the scale utilized.

A word of warning should be inserted here with regard to spectroscopic luminosity classifications. As Keenan and Wilson (1977) have pointed out, line ratios involving elements enhanced in Ba stars (Sr, Ba, Y) should not be used in this way because the luminosity is falsified. A better set of luminosity criteria is given by Keenan and McNeil (1976).

Up to the seventies further studies of Ba stars were hampered by the small number of stars in the group. This changed radically with the paper by McConnell, Frye and Upgren (1972) who used the Michigan objective prism survey (110 Å/mm) to provide a coverage of a large part of the southern sky, resulting in the discovery of 200 new barium stars. They introduced two groups, the so-called 'certain' and 'marginal' barium stars 'based upon the prominence of the Ba II line at λ4554, the strength of λ4077 relative to Hδ, and the strengths of the CN band (λ4216) and the G band'. The fact that Ba stars were discovered from objective prism plates invited some criticism. Eggen (1975) and Ianna and Culver (1975) suggested on photometric grounds that only part of the stars discovered by McConnell, Frye and Upgren (1972) are true Ba stars. This was taken up by Catchpole, Robertson and Warren (1977) who reobserved a large sample of the McConnell stars, both spectroscopically and photometrically, and concluded that about two-thirds are indeed Ba stars. The remaining third is composed of classical CH stars or stars which have intermediate properties between classical giants and CH stars.

Morgan and Keenan (1973) introduced a group of stars with mildly enhanced Ba II λ4554 strength visible on spectrograms of lower plate factors. They call them 'Ba 0' stars, since the enhancement is barely noticeable on low or moderate dispersion, and later Keenan called them 'semibarium' stars. In the atlas of Keenan and McNeil (1976) further subdivisions like Ba 0.6 are used, but the remark is made that when the barium index is less than 0.3 the classification becomes uncertain.

Later high dispersion analysis of some semibarium stars by Pilachowski (1977) confirmed the enhancement of heavy elements with respect to iron peak elements and the moderate enhancement of CN, CH and C_2. These stars seem thus to be real intermediates between normal G and K giants and the classic Ba II stars.

The existence of such intermediate stars was confirmed by the spectrophotometric work of Williams (1975), who performed the photoelectric measurement of a Ba II index measuring the depression caused by the Ba II λ6142 line on the continuum defined by an average of two comparison regions each 10 Å wide centered at λ6072 and λ6207. The features at λ6142 (i.e. $\lambda\lambda$6140–44) are mostly Ba II, but also contain some Fe I and lines of Ni I, Si I and Zr I. Through an analysis of his measurements for 200 evolved G and K stars, Williams found that there is a continuous range in Ba II abundances, between Ba II stars and normal G and K giants.

A detailed analysis of Ba stars – see for instance the review by McClure (1984) – permits us to summarize the behavior of elements. Elements with $Z < 38$ behave as in normal giants, except carbon which is enhanced in Ba stars. Later elements are enhanced, up to Ba ($Z = 56$); still heavier elements are generally not enhanced, and in particular no technetium is found (Boesgaard and Fesen 1974).

Photometry. The first analysis of UBV photometry showed that Ba stars lie close to but not exactly on the giant sequence. Figures 12.12 and 12.13, from Catchpole *et al.* (1977), show the effect clearly. The UBV system was used, plus the I color from the Kron–Cousins system ($\lambda_0 \sim 8200$) (Cousins 1976). When plotted against V–I (which is a temperature index) the Ba stars have redder colors when compared to normal giants with the same V–I. From these figures we can derive a diagram giving Δ(B–V) and Δ(U–B) for each Ba star; the differences are taken with respect to a normal K giant having the same V–I color; i.e. a positive Δ implies that the Ba star is redder than the normal giant. If we apply a slight correction to make the Δ(B–V) free from interstellar reddening, we find the correlation given in figure 12.14. The correlation shows that if the Ba characteristics are very pronounced (large index), the Δ(B–V) is large. This permits the photometric recognition of pronounced Ba stars.

The interpretation of the diagram is not simple, since the Δ(B–V) cannot be due to the strength of the Ba II line itself, which has typically only an equivalent width of 0.4 Å. What is being measured must thus be something else. Bond and Neff (1969) showed that Ba stars possess a broad depression in the continuum centered around λ4100. This depression – called the Bond–Neff depression – may reach up to $0^m\!.2$ and thus provides the

Figure 12.12. Broad band photometry of Ba stars. (U–B) versus (V–I) diagram. Both colors are de-reddened. The continuous curve represents the relation for normal field giants. From Catchpole, Robertson and Warren (1977).

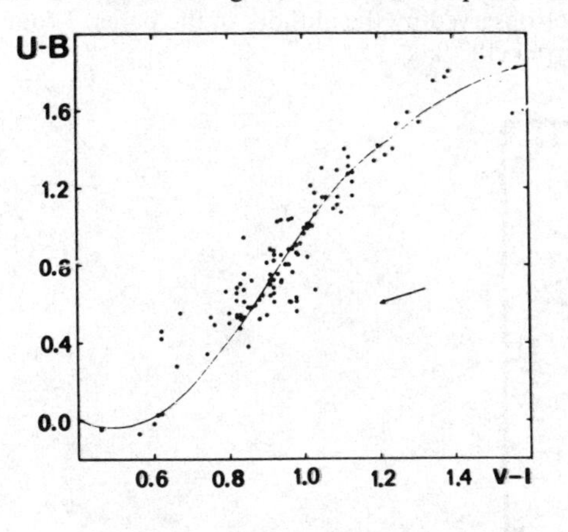

Figure 12.13. Broad band photometry of Ba stars. (U–B) versus (V–I) diagram. Both colors are de-reddened. The continuous curve represents the relation for normal field giants. Notice that Ba stars are redder than normal giants for the same (V–I) index, whereas from figure 12.12 they have 'normal' (U–B) indices. From Catchpole, Robertson and Warren (1977).

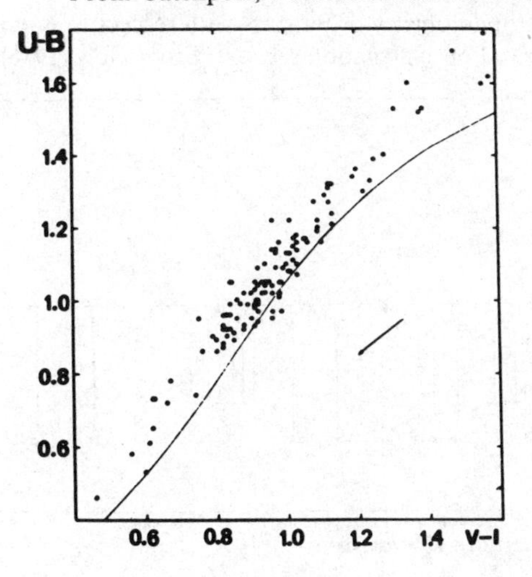

explanation of the (B–V) excess found. The general behavior of the continuous spectrum can be better seen on the spectrophotometry by Gow (1976) which covers the wavelength range λ3850 to λ5000. Figure 12.15 illustrates the mean relative spectral energy distribution of three 'Ba star–standard star' pairs at a resolution of 2.5 Å. Notice that heavy element lines are well marked. Since a comparison is made relative to normal giants, this

Figure 12.14. Color excess and barium line intensity. Abscissa, estimated Ba line intensity; ordinate, (B–V) excess from figure 12.13; circles, strong Ba stars; crosses, weak Ba stars; dots, not observed by the authors of the paper. From Catchpole, Robertson and Warren (1977).

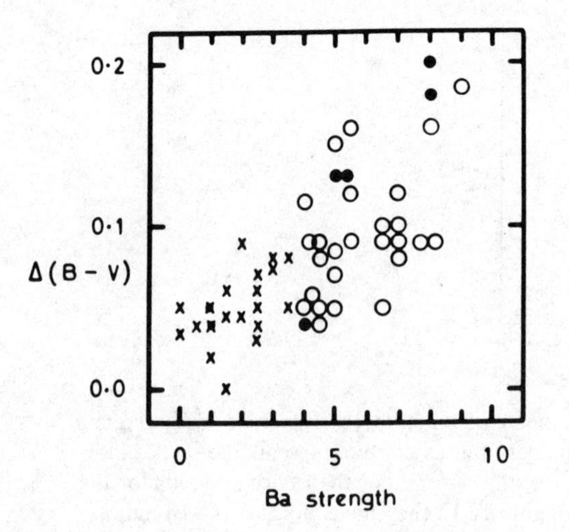

Figure 12.15. Differential spectrophotometry of Ba stars with respect to K-type giants, in the range $\lambda\lambda$3850–5000. For explanation see text. From Gow (1976).

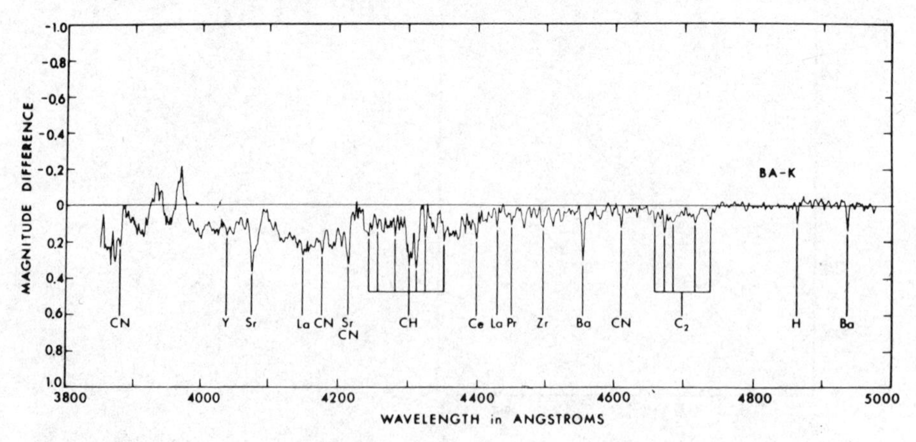

implies that these elements (i.e. Ba II) are much stronger in Ba II stars. See also the depression caused by the C_2 band at λ4737, CN at λ4216, CH at λ4315. However, two broad depressions – those at λλ3900–4100 and at λλ4100–4200 – remain unexplained (the two broad emission features in Ca II should be disregarded since they are spurious). Bond and Neff (1969) have suggested tentatively that the pseudo-continuous absorption of the molecule C_3 is responsible for the two depressions, but they doubt if sufficient C_3 exists in stars as hot as the Ba II stars. Another possibility is an absorption by the CH molecule, or enhanced metallic lines (Wing 1985).

The same depression found in spectrophotometry can also be seen in narrow or medium band photometries, like the Strömgren system or DDO photometry. Figure 12.16, taken from Lü and Sawyer (1979), illustrates the mean magnitude differences between Ba stars and normal stars at different wavelengths. It is clear that the depression at λ4100 exists, but its amount varies from star to star. Their main conclusion is that the depression is not closely related to either the Ba or the CN strength. Obviously the depression can be used to single out some Ba stars with any photometry which uses a filter at λ4100. We have seen this with the UBVI, but the same is true for UBVr (Mannery and Wallerstein 1970) or UBVRI (Eggen 1975). In view of the loose correlations pointed out in the preceding paragraph, the photometric procedures can only pick out the more extreme cases.

Feast and Catchpole (1977) have also observed a sample of Ba stars in the infrared, up to 2.2 μm(K). They found that when taken at equal V–I colors, Ba stars are brighter at 2.2 μm by amounts up to $0^m\!.2$, showing thus an

Figure 12.16. Magnitude differences between Ba stars and giants of the same spectral type. Each curve represents the average of several Ba stars, whose Ba strength is indicated. Data from Lü and Sawyer (1979).

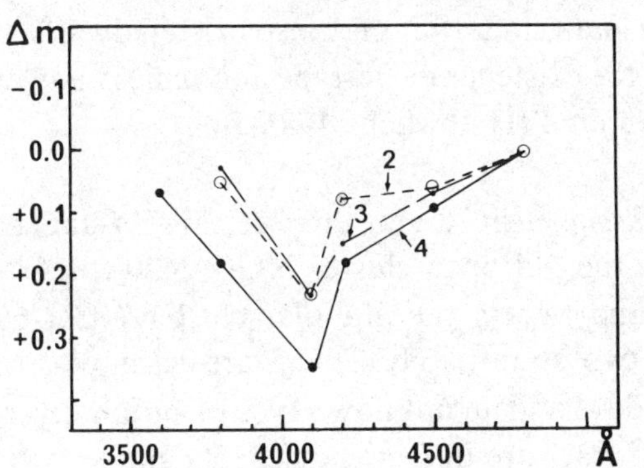

infrared excess. There exists a weak correlation between the amount of infrared excess and the flux depression at $\lambda 4100$.

Variability. No investigation seems to have been carried out on the variability of these objects.

Absolute magnitude. In the discovery paper, Bidelman and Keenan (1951) concluded from the existing evidence that the Ba stars should have the absolute magnitude of normal giants. This general conclusion has been shown to be true in subsequent work. The most recent work is by Jaschek *et al.* (1985), who obtained for certain Ba stars $0^{m}.0 \pm 1^{m}.2$, close to the value for normal giants ($-0^{m}.3$), whereas they found $-1^{m}.3$ for the marginal Ba stars. They also provide a summary of previous determinations.

What is striking is the very large dispersion ($\pm 1^{m}.2$) of the absolute magnitudes, a fact already known from previous work (for instance Kemper 1975; Culver, Ianna and Franz 1977). Ironically the brightest Ba star, ζ Cap, which has often been analysed as a 'typical' Ba star, has $M_v \sim -3$ and so in this respect is 'atypical'.

Binaries. McClure (1985) has shown that in a sample of 20 barium stars 17 show radial velocity variations. Although orbits have not been determined for all of them, it seems clear that for most of them the periods are very long ($P > 400^{d}$) and mass functions very small, a fact which suggests rather large separations. It is possible that at least some of these companions are white dwarfs (Böhm-Vitense, Nemec and Proffitt 1984).

Since for G- and K-type giants the percentage of spectroscopic binaries is ~20% (McClure 1985) it seems that all Ba stars are spectroscopic binaries. It has been suggested that Ba-type anomalies are due to mass exchange between the components at some evolutionary stage. Since mass exchange is not restricted to binaries in the giant stage, but may also exist in dwarfs, this suggestion leads necessarily to the existence of Ba-type anomalies in dwarfs. But up to now none has been found (Halbwachs 1985).

Ba stars in clusters, population assignment. So far three certain Ba stars have been found in clusters, one in the old open cluster NGC 2420 (McClure, Forrester and Gibson 1974) and two in the globular cluster NGC 6665 (Mallia 1976). However, the two stars of NGC6665 are variables, with periods of the order of 50–100 days and an unknown type of variability, and so it is not certain if these two stars are in fact classical Ba stars.

According to McConnell, Frye and Upgren (1972) barium stars belong to the same population as normal G and K giants. Eggen (1972) prefers to call them old disc population, but the evidence is not very conclusive; among 75 stars, seven have radial velocities larger than 60 km/s. It does not seem possible to decide the point, in view of the possible heterogeneity of the group.

Frequency. Bidelman (1985) and Holtzer (1985) conclude that Ba stars constitute about 1% of normal G- and K-type giants. This excludes the known semibarium stars, whose percentage is of about the same order. They confirm that more pronounced Ba stars are quite rare.

Down to $V = 6.5$ there are 65 Ba stars, and 300 Ba stars have an HD number.

12.2 CN anomalous stars

The CN bands at λ4216, λ4150 and λ3889 are strongly visible features in giants of types G7–K2, whereas they are never conspicuous in dwarfs. The history of the CN anomalies starts with the discovery of both weak and strong CN features when a suitable set of standards for classes V and III were defined in the Yerkes system.

12.2.1 *CN strong stars*

The CN strong stars were discovered by Morgan and Nassau on objective prism plates and were later called λ4150 stars by Roman (1952). She defined them as G and K giants with abnormally strong CN absorption for their spectral type and luminosity class.

Since usually the λ4216 CN band is more easily visible than both the λ4150 and the λ3889, Keenan and Keller (1953) used the ratio $\lambda 4216/\lambda 4172$ at a plate factor of 100 Å/mm. Both features are blends: λ4216 is a blend of λ4215 Sr II and λ4216 CN, whereas λ4172 is a blend of metallic lines. Thus a weak CN means a small index. Keenan (1958) recommended the introduction of the CN discrepancy $CN_* - \overline{CN}$, which is sometimes also called the 'CN anomaly' (CN_* = observed CN strength in a star; $\overline{CN}$ = average CN strength for a normal star of the same spectral type and luminosity class). He thinks that the index is a consistent indicator of metal abundance for spectral types later than G5. The observational status is defined by a list of stars (Keenan and Yorka 1985) with various degrees of CN discrepancy, between -3 and $+3$. Keenan defines the strength of λ4215 on the basis of the intensity ratio I (λ4211–13)/I (λ4219–21) (see figure 12.17).

Schmitt (1971) classified a large number of late giants and systematically estimated the CN strength. He uses the intensity ratio I (λ4210–16)/I (λ3215–27) for the definition of CN strength and defines a CN strong star as 'a star whose spectrum shows more absorption in the 0–1 sequence of the CN band with its head at λ4216 than the average MK standard shows at the same

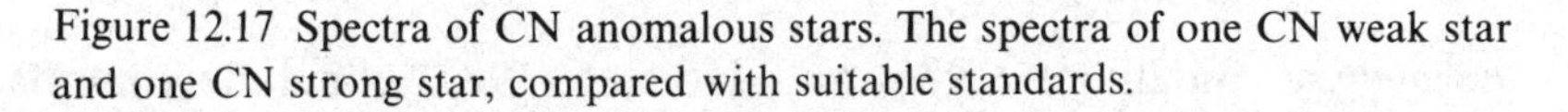

Figure 12.17 Spectra of CN anomalous stars. The spectra of one CN weak star and one CN strong star, compared with suitable standards.

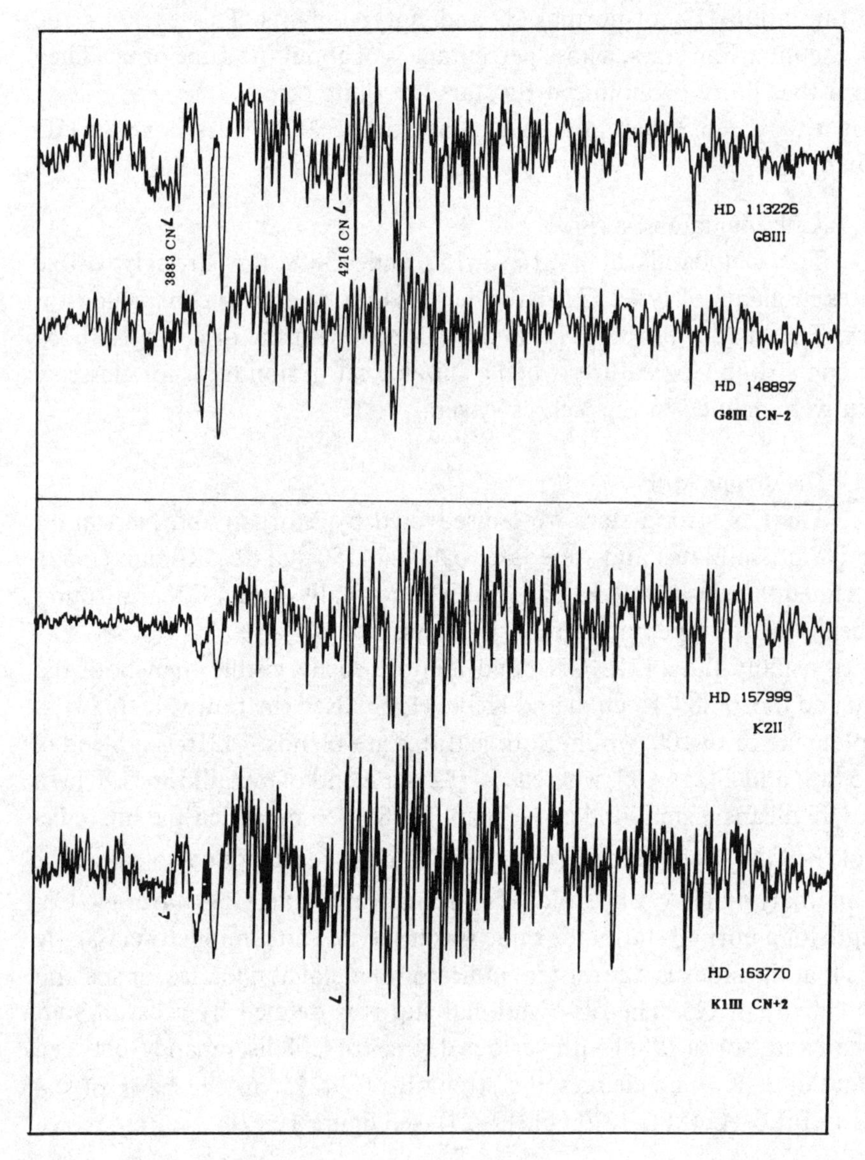

spectral type and luminosity class'. He uses three estimates of CN strength: 1 slightly stronger than normal; 2 a definite case of abnormal strength; 3 an excellent case of abnormal strength.

It is clear that the whole procedure rests on a good set of standards, which if possible should also illustrate different degrees of CN strengthening. As we have already remarked, CN is luminosity dependent – if a star with strong CN is classified as having higher luminosity, its CN strength becomes normal. On the other hand, a class II star could well be a CN rich class III object. Luminosity has thus to be assigned by atomic lines, independently of CN.

When improved standards became available, many of the stars formerly thought to be abnormal turned out to be normal stars. Keenan and Yorka (1985) have reclassified 12 out of the 25 stars from the Roman list and only 2 of the 12 may be considered CN strong. With regard to Schmitt's (1971) classification, out of the 13 stars reclassified, only 6 are considered CN strong by Keenan. This is of course no criticism of the older work, but demonstrates simply that progress comes from improved standards.

The fact that CN varies with luminosity is used by Houk and A. Cowley (1975) to denote CN anomalous stars – both weak and strong – by a 'pseudo'-luminosity as determined from the CN break and the CN band strength (for example K2 III, CN II and K1 III, CN IV).

CN strong stars became very popular for a while, because Spinrad and Taylor (1969) called some of them 'super metal rich' (SMR). The term 'super metal rich' means 'more metal rich than the sun'. If, on the other hand, the metal content of the sun is considered to be normal, the stars could also be called 'metal rich'.

As can be seen in Spinrad and Taylor (1969), SMR stars are the result of a rather complex procedure involving essentially spectrum scans and photoelectric photometry. The name SMR and its implications started a long controversy which is well reviewed by Taylor (1982). From a classification point of view, 'metal rich' stars are 'CN strong stars', or should be called at most 'strong line stars'.

Given the subtleties of the spectroscopic approach, many authors have sought photometric approaches, which are described in detail by Golay (1974). Earlier in this chapter we explained the DDO system, which uses one index specifically for the CN strength. But other medium or narrow band systems, such as the Vilnius, Geneva and Strömgren systems, are also capable of handling the problem. We could repeat almost verbatim what was said in section 11.3 about the power of photometric methods when dealing with minute changes in band intensity: they are definitively superior to spectroscopic methods. Nevertheless a spectroscopic approach is valid with one restriction: namely that it is known beforehand that no other peculiarity is involved.

CN stars have been found among field stars in open clusters (Smith 1982) and in globular clusters (Freeman and Norris 1981). The phenomenon is thus independent of population characteristics. In globular clusters CN and CH both show a large range of variation and some authors (for instance, Norris, Freeman and DaCosta 1984) suggest the existence of an anticorrelation. Radial velocities of field CN stars were observed by McClure (1985); he finds a normal percentage of spectroscopic binaries.

Estimates of the frequency of 'CN strong' stars are rather uncertain. From Boyle and McClure's (1975) photometric measures of a random sample of giants, a frequency $\sim 1\%$ can be derived, although Grenon (1985), who also uses a photometric approach but with a larger sample, arrives at 4%. His percentage refers to dwarfs and subgiants, and the author thinks that for giants the percentage should be less.

Finally Houk and Cowley (1975) obtained a value of 3% for CN anomalous objects (i.e. both strong and weak) detectable on objective prism plates.

12.2.2 *CN weak objects*

Morgan, Keenan and Kellman (1943) pointed out that in some high velocity stars there is a simultaneous weakening of metallic lines and of CN bands. This fact was confirmed by work on high resolution spectrograms, and it was found that both features weaken simultaneously: if metallic lines are weak, so is also the CN band, especially $\lambda 4216$ (see figure 12.17).

Keenan (1958) then introduced a specific notation for the weakness of the CN features, which has been discussed in section 12.2.1. Because of the general association of 'metal weakness' with 'CN weakness', later authors tended to speak of 'population II objects' or 'metal poor stars' rather than of 'CN weak' or 'weak stars'. Observe however that the group of 'weak line stars' is more general than that of the 'CN weak stars' because the former includes both dwarfs and giants and the latter only giants.

More recently Keenan and Yorka (1985) discarded the notation of CN weakness in many weak line stars, in favor of the (negative) Fe index. The group of 'weak CN' stars is thus restricted to stars (like 33 Piscium) where the metals are of normal strength and the only abnormal feature is weak CN.

12.3 **CH anomalous stars**

We have mentioned that the G-band of the CH molecule is a conspicuous feature in G (and K) stars, and that it does not vary greatly with luminosity. Variations in the strength of the G-band are thus likely to attract immediate attention. We have already mentioned (see section 11.3) that in weak line stars, CH is strong, but there is a group where this molecule becomes very strong – the so-called CH stars. We find also other groups related to this one, namely the 'CH like' and the 'subgiant CH' stars.

On the other hand we find stars in which the G-band is very weak or absent – the so-called 'weak G-band' stars. We shall describe each of these groups in turn.

12.3.1 *CH stars*

A CH star is a G-type giant (in the range G5–K5) in which the molecular bands of CH are very strong. The absorption due to the CH molecule is so great that it blots out (at plate factors 60–100 Å/mm) most of the features in the region below λ4300; even the Ca I λ4226 line becomes rather inconspicuous. Most of the metallic lines are weakened, but heavy elements are enhanced, as shown by the strength of the lines of Sr I λ4607 and Ba II λ4554. Other carbonaceous molecules, like C_2 and CN, are also intensified. The group was first described by Keenan (1942) (see figure 12.18).

Bidelman (1956) pointed out that the group was probably not homogeneous since CN band strength varies greatly throughout the group, as does the C^{13}/C^{12} ratio – in half of the stars C^{13} is absent and in the other half it is strong. The carbon isotope ratio is obtained by comparing the bands λ4737 $C^{12}C^{12}$, λ4744 $C^{12}C^{13}$ and λ4752 $C^{13}C^{13}$, which are visible at plate factors of 40–60 Å/mm or smaller.

The study of these stars is difficult because of their small number and their faintness; the brightest members are of magnitude eight. The number of members in the group increased slowly: Keenan (1958) gave a list of 11, a number which Yamashita (1975) raised to 18.

High resolution studies of the brighter members of the group confirmed the general picture given above: general metal weakness, strengthening of some heavier elements like barium, but absence of rare earths. Such a pattern is reminiscent of that of the Ba stars (see section 12.1).

The depression of the continuum for $\lambda < 4300$ Å due to the CH molecule can be used for the photometric detection of these stars. Since a similar depression is found in C stars, we shall refer the reader to section 13.1 on C stars for more details.

Figure 12.18. Spectra of CH anomalous stars. The spectra of one CH poor star and one CH star, compared with suitable standards.

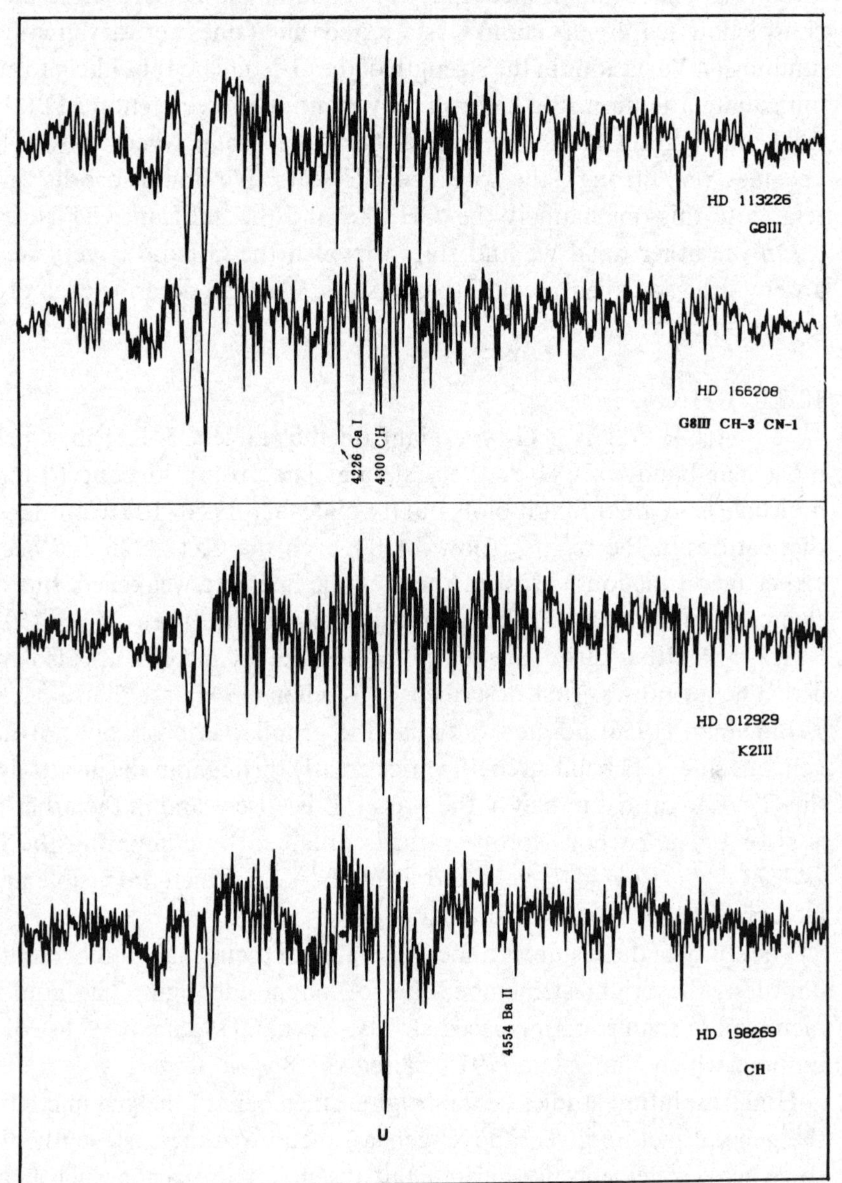

Because so few CH stars are known, the determination of the average absolute magnitude is difficult. Probably the safest result is that of Mikami (1975) who obtains $M_v = 1.6 \pm 0.3$ from statistical parallaxes. This puts the stars in the giant region.

The space velocity $|\bar{V}|$ is of the order of 250 km/s, a fact that categorizes CH stars as population II. Out of ten stars, seven show variable radial velocity and are thus spectroscopic binaries (McClure 1984).

CH stars are evenly distributed over the sky. They have also been found in globular clusters and in external dwarf galaxies (For references, see McClure 1984.)

12.3.2 *CH-like stars*

This group was defined by Yamashita (1972, 1975) as being a group of low velocity C0–C3 stars with a definite enhancement of Ba II λ4554, and often an enhancement of either Hβ or the G-band of CH. However, these two latter features are not considered to be a condition for being a CH-like star. These stars are like the two intermediate Ba–C stars found by McConnell, Frye and Upgren (1972). They list 16 stars of this type and illustrate some of them. They introduce the notation C 3.0 ch, where 'ch' stands for CH-like'.

The radial velocity of 'ch' stars is low, with $V < 60$ km/s, a fact which differentiates them from CH stars which are of high velocity. Yamashita adds that the C_2 bands are weak in most of the 'ch' stars, in the abundance range 0–3. In about half of them the C_2 bands are absent or quite weak, while in the rest C^{12}/C^{13} is as low as in most other carbon stars. He considers that the Ba II λ4554 enhancement is a real effect, since the line is blended with CN and can be seen only if really enhanced.

These stars can be regarded as low velocity old disc analogs of CH stars (Baschek 1979).

12.3.3 *Subgiant CH stars*

Bond (1974) discovered on objective prism plates (plate factor 110 Å/mm) a group of stars having late F- to early G-type spectra with somewhat weak metallic lines, very strong CH features and strong lines of Sr II at λ4077 and λ4215. He adds that several members of this class also showed a strong Ba II λ4554 line and were classified as 'weak lined Ba stars' by McConnell, Frye and Upgren (1972). In a few extreme cases, the metallic lines are so weak that the Sr II lines, CH features, the hydrogen Balmer lines and sometimes λ4554 were the *only* features strong enough to be seen longward of the Ca II H and K lines.

Luck and Bond (1982) confirmed the above description from a study of high resolution material: weak metallic lines, strengthening of heavier elements (Zr, Y, La, Nd, ...) and overabundance of carbon. It might be that such behavior can be expected in a star hotter than the (giant) CH stars.

From the kinematic data available, it seems that these stars have $\bar{M} \sim +2$, and are therefore intermediate between ordinary giants and main sequence stars, justifying the term 'subgiants'.

Bond thinks that the subgiant CH stars are more frequent than giant CH stars; he lists 11, but expects about 50 up to $m = 10$. More stars of this group have been found by Çatchpole, Robertson and Warren (1977) – they were listed there as 'metal weak' in the group of presumed Ba stars.

As can be seen, more investigation is needed to clarify questions concerning the group. It seems obvious that these stars are linked with both the giant CH stars and the Ba stars, although it would be premature to say anything more definite.

12.3.4 *Weak G-band stars*

A weak G-band star is a G- or early K-type giant with a very weak or absent G-band of CH.

The first star of this kind was mentioned by Cannon (1912) and discussed by Bidelman (1950), who adds that CN is also absent and that this distinguishes the stars from weak lined stars, in which CN is absent but CH normal or even strong. Greenstein and Keenan (1958) inferred that because of the simultaneous weakening of CH and CN, the stars must be C deficient.

For almost two decades the study of this group made little progress because of the small number of group members. Then Bidelman and McConnell (1973) found 34 new members in the southern hemisphere, which revived interest in the group.

Dean, Lee and O'Brien (1977) studied a sample of these stars both spectroscopically and photometrically. Spectroscopically they found that the stars fall in the interval G5–K3 and luminosity classes IV and III; the G5 limit is probably an artifact of the discovery technique, and more stars may be expected if lower plate factors are used. They confirmed the overall weakness of CN and CH, but CN is not always very weak.

Later high resolution studies (see for instance Cottrell and Norris 1978) confirmed this general picture and brought out more details. So Rao (1978) showed that the weakness of CH is accompanied by an enhancement of NH.

More work was done photometrically. Dean, Lee and O'Brien (1977) made UBVRI observations for a sample of stars. They reasoned that if both

CN and CH are weaker than in normal giants, the B band should be less affected, so that B–V should be bluer than in normal stars. Furthermore since both CH and CN intensify in normal stars toward later types, the difference between normal stars and 'no-G-band' stars should increase toward later types.

Figure 12.19 shows that both expectations are fulfilled, and shows a way of measuring the weakness of CH.

Hartoog, Persson and Aaronson (1977) tried to find out if the weakness of carbon in these stars is real. To do so, they used the CO band at 2.36 μm and the H_2O band at 2.2 μm compared to a continuum point at 2.2 μm. By measuring intensities at these three points with narrow band filters, they obtained a CO index. The results are displayed in figure 12.20. It is clear that CO is weak, and since CH and CN are also weak, the carbon weakness in the group is confirmed.

'No-G-band' stars were found in globular clusters (see figure 12.21 taken from Norris and Zinn 1977), where they constitute an important fraction of all giants. In some globular clusters, the lower part of the asymptotic giant branch is entirely populated by these stars. It seems that carbon weakness is not correlated with metal strength. For a summary of the observations see Cohen (1980).

The radial velocities of these stars are low, which could imply that these

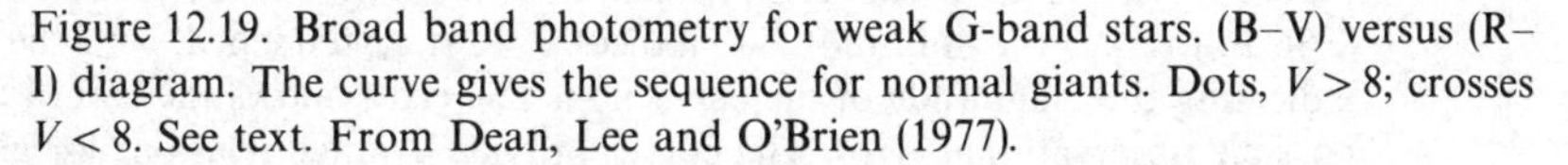

Figure 12.19. Broad band photometry for weak G-band stars. (B–V) versus (R–I) diagram. The curve gives the sequence for normal giants. Dots, $V > 8$; crosses $V < 8$. See text. From Dean, Lee and O'Brien (1977).

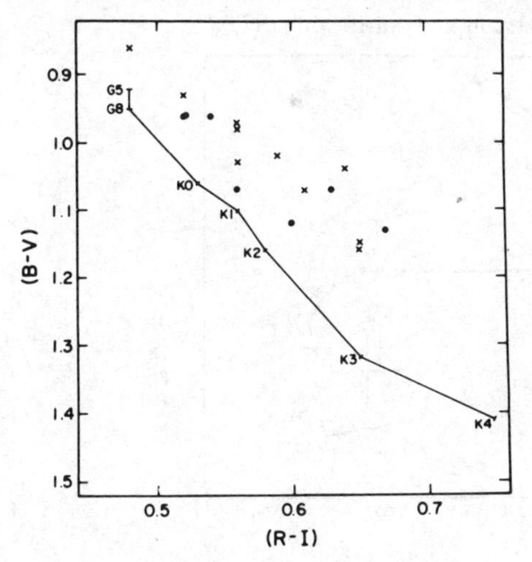

objects belong to population I. The percentage of spectroscopic binaries also seems normal (Tomkin, Sneden and Cottrell 1984).

Since we have seen that stars of this type appear in globular clusters – population II – and have low radial velocity – population I – the only safe conclusion is that the lack of G-band is independent of other characteristics.

The galactic distribution of the objects shows little concentration toward the galactic plane. About 50 field objects of this type are known.

12.4 T Tau and related stars

Although T Tau and related stars are variable and so lie outside the domain of this book, we have devoted this short section to them because spectroscopic elements in the definition of these groups need some clarification. We shall thus deal briefly with T Tau, YY Ori and FU Ori stars.

12.4.1 *T Tau stars*

A T Tau star is an irregular variable star associated with nebulosity (bright or dark) having certain spectral characteristics. The latter will be given in detail later, but it is important to remark from the start that this definition uses three different elements (photometric, spectroscopic and 'nebular'). This has produced some confusion, to say the least.

In addition the light curve characteristics are hard to systematize because

Figure 12.20. Carbon weakness in weak G-band stars. The CO band strength (2.36 μm) versus the (infrared) J–K indices. J–K is de-reddened, with the arrow indicating the magnitude of the correction. The cross shows the size of the possible observational error. The curves provide average relations for giants and dwarfs. From Hartoog, Persson and Aaronson (1977).

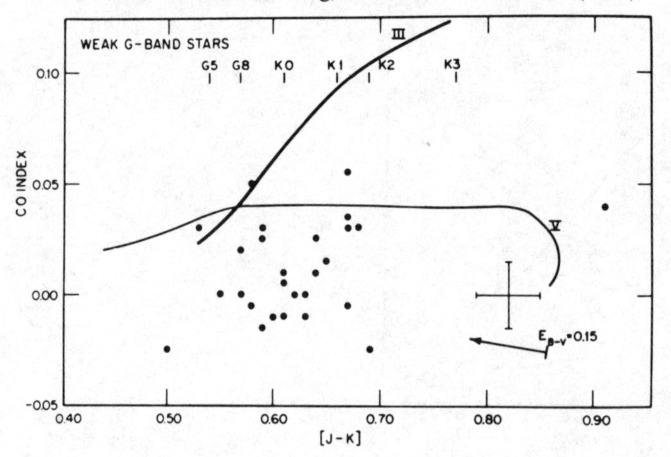

the stars present a bewildering variety of phenomena: from rapid fluctuation on an hourly time scale to long term erratic variations and sometimes to a practically constant light curve.

Association with nebulosity is a rule, but is insufficient by itself to qualify a star as T Tau, even if the object is also an irregular variable. This is the reason why spectroscopic specifications need to be added.

The spectroscopic characteristic is the existence of a late type absorption spectrum with no supergiant features, showing Ca II and Balmer lines in emission. The Hα emission must be strong ($W > 5$ Å) to differentiate these objects from UV Ceti (flare) stars.

The 'non-supergiant' specification segregates the Mira variables which can be described spectroscopically in a similar way. The 'late type' helps to discriminate these objects from Herbig Ae/Be stars (Bastian *et al.* 1983).

Figure 12.21. 'No-G-band' stars in globular cluster NGC 6397. The weak G-band stars are shown on the right side, whereas the left side shows some strong G-band stars. From Norris and Zinn (1977).

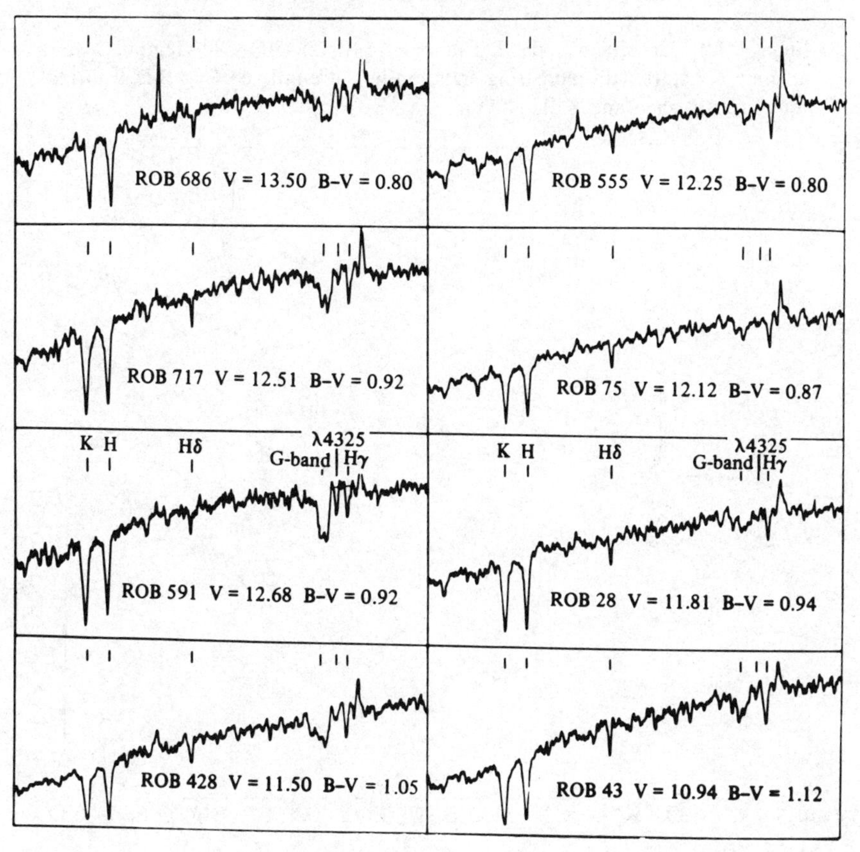

Let us hasten to add that these characteristics are variable to a certain degree as we shall see later, so that we should call T Tau objects all those which have shown these characteristics at least once.

We reproduce in figure 12.22 some spectra taken from Cohen and Kuhi (1979) which cover the $\lambda\lambda$4200–6800 range. These spectra show the Hα and Hβ emission lines clearly, as well as those of Na I (λ5889 + λ5896), [O I] λ6300, Fe II λ4924, He I λ5876 and a number of other emission features which, however, are not always present. The continuous spectrum behaves abnormally as can be seen by comparing RW Aur and 61 Cyg. B.

The strong UV continuum produces a 'veiling'. It is in general absent in stars with weak emission features, but can act as a blanket in other stars.

When large collections of spectra are available, authors have sought to determine emission classes, indicating the degree of emission strength. Herbig (1962) proposed five classes: from 1 (= weak Hα emission) to 5 (= well-developed metallic emission line spectrum).

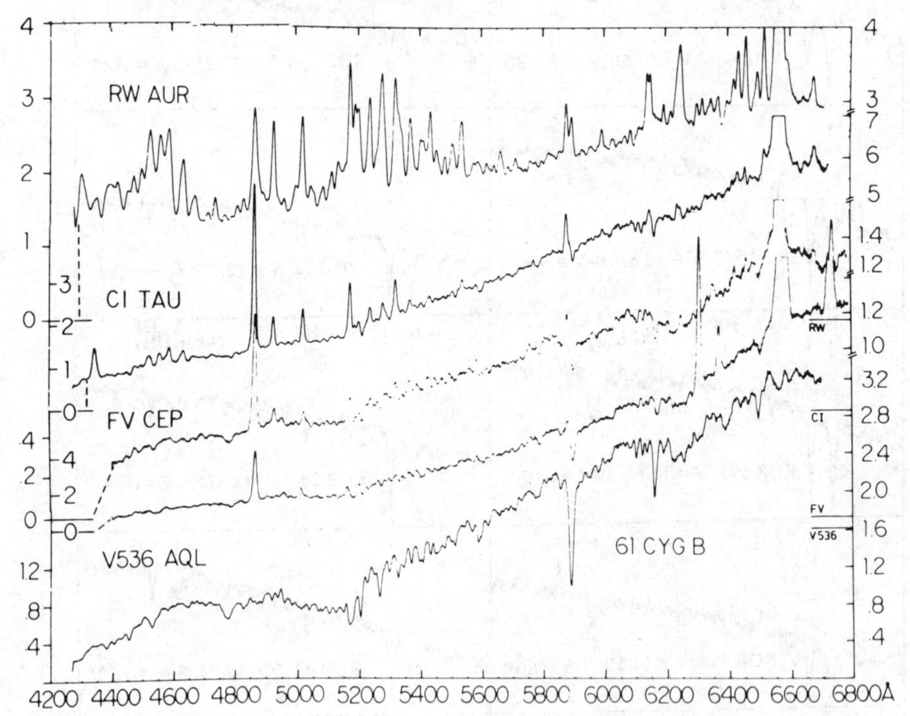

Figure 12.22. Spectra of some T Tau stars in the $\lambda\lambda$4200–6800 region. Spectra of some K7 stars with increasing emission line strength. 61 Cyg B is a normal star. From Cohen and Kuhi (1979).

Sun *et al.* (1985) provided the following descriptions, based upon a study of spectra covering the region $\lambda\lambda$3600–5200 (with a plate factor of 92 Å/mm).

In the weak emission stars only the Balmer series is seen in emission, with the rule that the emission is stronger in α than in β, and so on. In stars of stronger emission, the Ca II lines are seen in emission, increasing from a barely perceptible emission to strong ones. When Ca II is strongly in emission, usually Fe II also appears in emission – noticeably the M42 (λ4924, λ5018 and λ5169). When the emission is very strong, more Fe II lines (M27, 28, 37, 38 and 42) shortward of Hβ, and sometimes [Fe II] are additionally seen.

Of other elements, lines of [S II] (λ4068 and λ4076), [O II] ($\lambda3726 + \lambda3729$) and [O III] ($\lambda$5007, λ4959) are present, often without regard to the overall emission strength, and probably come from the associated nebulosity. In case of strong emission, He I lines and the fluorescent Fe I lines λ4063 and λ4132 also appear.

When equivalent widths are measured the meaning of the emission classes can be shown numerically. Table 12.7 is taken from Cohen and Kuhi (1979) and provides the relation with Herbig's classes, which are similar though not identical to the ones described.

For He I, λ5876 was taken as the representative line.

Cohen and Kuhi show further that Hα, He I λ5876 and Fe II λ4924 emission line strengths are closely correlated: all strengthen or weaken together.

There is, however, one general drawback in all such schemes, due to the fact that emission characteristics may change considerably in quite short periods of time. A description scheme, however elaborate it may be, thus has a limited significance.

Table 12.7. *Emission line strength in various classes of T Tau stars.*

Emission class	⟨EW(Hα)⟩	⟨EW(Hβ)⟩	⟨EW(He I)⟩
1	16	3.6	0.4
2	36	7	1.0
3	75	20	2.0
4	90	27	3.0
5	90	28	3.5

Note: ⟨EW(Hα)⟩, average equivalent width of the Hα emission line etc.

The variations are illustrated in figures 12.23 and 12.24 taken from Kuhi (1978). In case of RW Aur, the Na I D line varies in strength between two exposures two hours apart, whereas the Fe II lines vary from one day to the next. A few months later when the emission decreased, a K-type absorption line spectrum became apparent.

Kuhi concludes that there are certain empirical rules for changes in emission line intensity. In order of time scale, the changes occur in Na I D lines (hours), Fe II and He I (days), Balmer lines and veiling (weeks or months) and forbidden lines (years). We can add that Ca II also changes over days and in some instances the veiling has changed over days (Sun *et al.* 1985).

Because of these spectral variations, T Tau stars are very difficult to classify; obviously the best occasion to classify the underlying spectrum is when the emissions are weak. If they are strong, the veiling is usually also strong and the features below $\lambda 4200$ are washed out. Kuhi recommends using features in the Hα–Hβ region, where the veiling is less severe.

A similar difficulty exists in determining the luminosity class. In general, luminosity classes are III to V; Cohen and Kuhi (1979) derive a more precise class IV by using MgH bands.

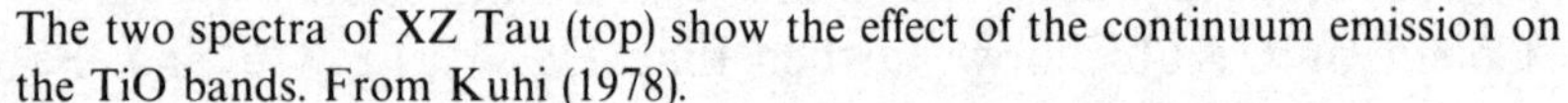

Figure 12.23. Spectra of some T Tau stars with emissions of moderate strength. The two spectra of XZ Tau (top) show the effect of the continuum emission on the TiO bands. From Kuhi (1978).

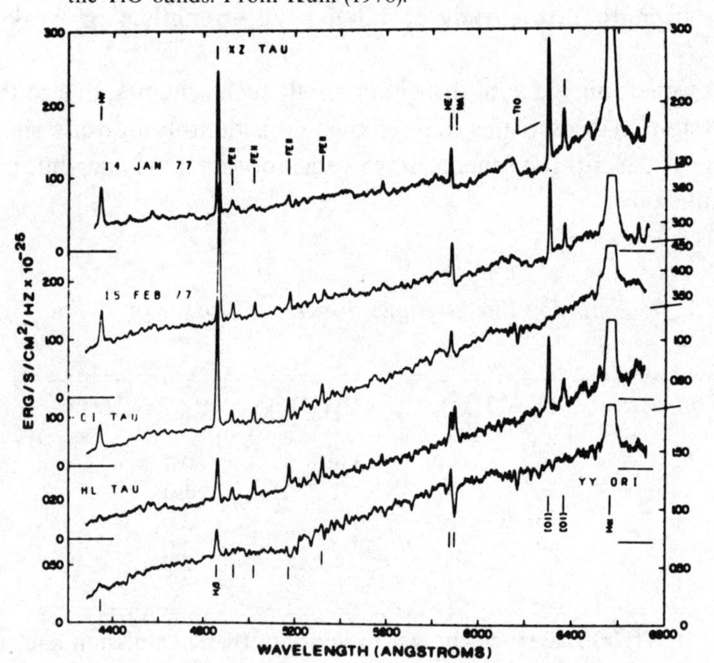

When examined at a higher resolution, the absorption lines are usually broad and a value of $V \sin i \sim 60$ km/s can be derived, which is very high for late type stars (G–M). When isolated lines are examined in detail, the variety of profiles illustrated in figure 12.25 is found, where the vertical line marks the rest position. Cases a, b and c – which are quite frequent – show a violet displaced absorption line, which has been interpreted as mass outflow. Case d is another frequent case, whereas e is less frequent. Since most of these line profiles are only seen on high dispersion material and all profiles are variable over time, we shall refrain here from further discussion.

The number of known T Tau objects is more than 10^3.

Figure 12.24. Spectra of two T Tau stars with strong emissions. The uppermost two spectra of RW Aur were taken two hours apart. Notice the variability of features on a short time scale. From Kuhi (1978).

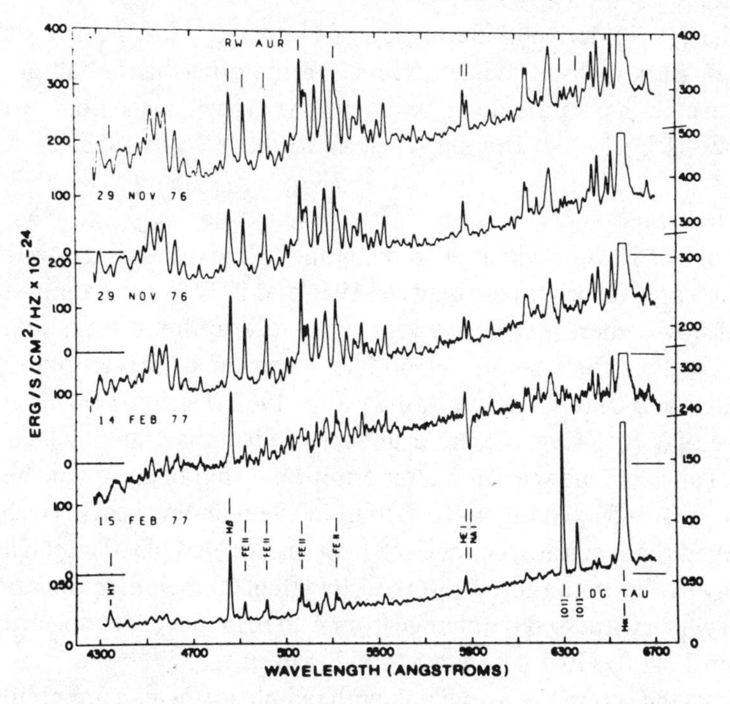

Figure 12.25. Variety of profiles of T Tau stars.

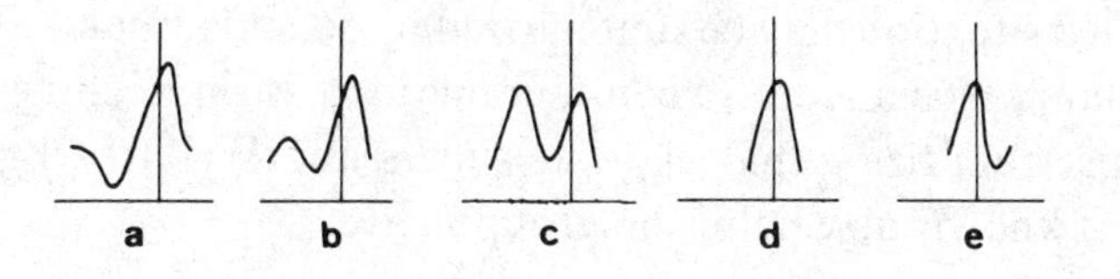

Different aspects of T Tau and related objects are considered in the paper 'Simposio sobre objetos Herbig–Haro, estrellas T Tau y fenómenos relacionados' (1983) in *Rev. Mex. AA*, **7**.

12.4.2 *YY Ori stars*

Walker (1972) introduced this designation for a group of T Tau stars exhibiting in their (Balmer and Ca II) emission lines an inverse P Cyg type structure. The latter implies that an absorption line occurs at the red side of the emission line feature. He added that most of the spectra show a strong veiling, which produces an ultraviolet excess shortward of $\lambda 3800$. In broad band photometry this translates into $(U-B) < 0$.

Later studies showed the rapid variability of the line profiles, in particular of the inverse P Cyg profiles, over periods of days. They even may disappear completely for several days.

Appenzeller (1977) derives a frequency of about 75% for YY Ori stars among T Tau stars with ultraviolet excess. On the other hand, half of all known T Tau stars have ultraviolet excess and he arrives at the final result that about 40–50% of all T Tau stars belong to the YY Ori subclass.

12.4.3 *FU Ori stars*

The object FU Ori which gave the name to the group is a variable star which increased its brightness between 1936 and 1937 by six magnitudes and declined slowly thereafter. Forty years after its eruption it was still five magnitudes brighter than before 1936. The spectrum before eruption is unknown but was F0 Iab in 1939 and F2 I–II in 1963 (Herbig 1977).

In the case of V 1057 Cyg, the pre-eruption spectrum is known which is that of an T Tau star. The spectrum after eruption is that of an A-type high luminosity star (A3–5 II) changing to F II–III in the next few years (Herbig 1977). However, the spectral types derived from the $\lambda\lambda 3900$–4300 region are systematically earlier than those from the $\lambda\lambda 6000$–6600 region: in the same time interval which witnessed the change from A to F, in the red the spectrum changed from F5 II to G0 Ib and later to G2–5 Ib or II.

All objects of the group are associated with nebulosity, which apparently was there before the outburst and is seen now because of the illumination from the star.

The complicated situation may be summarized by defining Fuors (FU Ori stars) as T Tau stars which undergo a considerable increase in brightness, the post-eruption spectrum being that of a late supergiant (Welin 1978).

The number of known objects of the group is five.

References

Appenzeller I. (1977) *IAU Coll.* **42**, Publ. Bamberg XI, N. 121, p. 80
Ardeberg A. and Lindegren H. (1981) *Rev. Mex. AA* **6**, 173
Barbieri C., Bonoli C., Bortoletto F., di Serego S. and Falomo R. (1981) *Mem. Soc. Astron. Ital.* **52**, 195
Baschek B. (1979) *22ème Coll. Liège*, p. 237
Bastian U., Finkenzeller U., Jaschek M. and Jaschek C. (1983) *AA*, **126**, 428
Bidelman W.P. (1950) *Ap. J.* **111**, 333
Bidelman W.P. (1956) *Vistas in Astronomy* **2**, 1428
Bidelman W.P. (1985) *Coll. Strasbourg, 'Cool stars'*, p. 43
Bidelman W.P. and Keenan P.C. (1951) *Ap. J.* **114**, 473
Bidelman W.P. and McConnell D.J. (1973) *A.J.* **78**, 687
Boehm-Vitense E., Nemec J.M. and Proffitt C.R. (1984) *Ap. J.* **278**, 726
Boesgaard A.M. and Fesen R.A. (1974) *PASP* **86**, 76
Bond H.E. (1974) *Ap. J.* **194**, 95
Bond H.E. and Neff J.S. (1969) *Ap. J.* **158**, 1235
Boyle R.J. and McClure R.D. (1975) *PASP* **87**, 17
Catchpole R.M., Robertson B.S.C. and Warren P.R. (1977) *MNRAS* **181**, 391
Cannon A.J. (1912) *Annals Harvard Obs.* **104**
Chmielewski Y. (1981) *AA* **93**, 334
Cohen J. (1980) in *IAU Symp.* **85**, 385
Cohen M. and Kuhi L.V. (1979) *Ap. J. Suppl.* **41**, 743
Cottrell P.L. and Norris J. (1978) *Ap. J.* **221**, 893
Cousins A.W.J. (1976) *Mem. RAS* **81**, 25
Culver R.B., Ianna P.A. and Franz O.G. (1977) *PASP* **89**, 397
Dean C.A., Lee P. and O'Brien A. (1977) *PASP* **89**, 222
Eggen O. (1972) *MNRAS* **159**, 403
Eggen O. (1975) *PASP* **87**, 111
Feast M.W. and Catchpole R.M. (1977) *MNRAS* **180**, 61P
Freeman K.C. and Norris J. (1981) *Ann. Rev. AA* **19**, 319
Garrison R. (1979) *Ric. Astronomiche. Spec. Vaticana* **9**, 23
Garrison R. (1985) *IAU Symp.* **111**, 17
Golay M. (1974) *Introduction to Astronomical Photometry*, Reidel D. Publ. Co.
Gow C.E. (1976) *A.J.* **81**, 993
Greenstein J.L. and Keenan P.C. (1958) *Ap. J.* **127**, 172
Grenon M. (1985) *Strasb. Coll., 'Cool stars'*, p. 147
Guinan E.F. and Smith G.H. (1984) *PASP* **96**, 354
Halbwachs J.L. (1985) *Strasb. Coll., 'Cool stars'*, p. 337
Hardorp J. (1978) *AA* **63**, 383
Hartoog M.R., Persson S.E. and Aaronson M. (1977) *PASP* **89**, 660
Herbig G.H. (1962) *Advances AA* **1**, 47
Herbig G.H. (1977) *Ap. J.* **214**, 714
Holtzer M. (1985) *Strasb. Coll., 'Cool stars'*, p. 153
Houk N. and Cowley A.P. (1975) *Michigan catalog of two-dimensional spectral types for the HD stars*, vol. I, Univ. of Michigan Press

Ianna P.A. and Culver R.B. (1975) *Bull. A.A.S.* **7**, 445
Jaschek C., Jaschek M., Grenier S., Gomez A. and Heck A. (1985) *Strasb. Coll., 'Cool stars'*, p. 185
Keenan P.C. (1942) *Ap. J.* **96**, 101
Keenan P.C. (1958) in *Handbuch der Physik*, Flugge S. (ed.), vol. L, p. 93, Springer-Verlag
Keenan P.C. and Keller G. (1953) *Ap. J.* **117**, 241
Keenan P.C. and McNeil R.C. (1976) *An atlas of spectra of the cooler stars*, Ohio State Univ. Press
Keenan P.C. and Wilson O.C. (1977) *Ap. J.* **214**, 399
Keenan P.C. and Yorka (1985) *BICDS* **29**, 25
Kemper E. (1975) *PASP* **87**, 537
Kuhi L.V. (1978) in *Protostars and planets*, Gehrels T. (ed.), Arizona Univ. Press
Lindblad B. (1922) *Ap. J.* **55**, 83, 85
Lindblad B. and Stenquist E. (1934) *Stockholm Ann.* **11**, N. 12
Lü P.K. and Sawyer D. (1979) *Ap. J.* **231**, 144
Luck R.E. and Bond H.E. (1982) *Ap. J.* **259**, 792
Mallia E.A. (1976) *MNRAS* **177**, 73
Mannery E.J. and Wallerstein G. (1970) *A.J.* **75**, 169
McClure R.D. (1973) *IAU Symp.* **50**, 162
McClure R.D. (1976) *AJ* **81**, 182
McClure R.D. (1979) *Dudley Obs. Rep.* **14**, 83
McClure R.D. (1984) *Ap. J.* **280**, L31
McClure R.D. (1985) *Strasb. Coll., 'Cool stars'*, p. 315
McClure R.D., Forrester W.T. and Gibson J. (1974) *Ap. J.* **189**, 409
McConnell D.J., Frye R.L. and Upgren A.R. (1972) *AJ* **77**, 384
Mikami T. (1975) *PAS Japan* **27**, 445
Morgan W.W. and Keenan P.C. (1973) *Ann. Rev. AA* **11**, 29
Morgan W.W., Keenan P.C. and Kellman E. (1943) *An atlas of stellar spectra*, Univ. Chicago Press
Norris J. and Zinn R. (1977) *Ap. J.* **215**, 74
Norris J., Freeman K.C. and Da Costa G.S. (1984) *Ap. J.* **277**, 625
Ohman Y. (1936) *Stockholm Obs. Ann.* **12**, n. 18
Pilachowski C.A. (1977) *AA* **54**, 465
Rao N.K. (1978) *MNRAS* **185**, 585
Roman N. (1952) *Ap. J.* **116**, 122
Schmidt-Kaler T. (1982) Landolt-Börnstein, group VI, vol. 2b, p. 1
Schmitt J.L. (1971) *Ap. J.* **163**, 75
Smith G.H. (1982) *A.J.* **87**, 360
Spinrad H. and Taylor B.J. (1969) *Ap. J.* **157**, 1279
Straizys V. (1977) *Multicolor stellar photometry*, Mokslas Publ. Vilnius
Sun Y.L., Jaschek M., Andrillat Y. and Jaschek C. (1985) *AA Suppl.* **62**, 309
Taylor B.J. (1982) *Vistas in Astronomy* **26**, 253
Tomkin J., Sneden C. and Cottrell P.L. (1984) *PASP* **96**, 609
Walker M.F. (1972) *Ap. J.* **175**, 89

Warner B. (1965) *MNRAS* **129**, 263
Welin G. (1978) *Protostars and planets*, Gehrels T. (ed.) Univ. Arizona Press. p.265
Williams P.M. (1975) *MNRAS* **170**, 343
Wing R. (1985) *Strasb. Coll., 'Cool stars'*, p. 61
Yamashita Y. (1972) *Ann. Tokyo Astron. Obs.* **13**, 169
Yamashita Y. (1975) *PAS Japan* **27**, 325

13

K-type stars

13.0 Normal stars

These stars are characterized by weak hydrogen lines, strong and numerous metallic lines and very strong Ca II lines. Molecular lines from the CH molecule (the G-band, for instance) are very strong.

The Harvard system used subdivisions K0, K2, K5 and, later, K7, and based the types upon the progressive weakening of the ultraviolet part of the spectrum and the appearance of the TiO bands at K5. In the Yerkes system line ratios are used, like

$$\frac{\text{Cr I }\lambda 4254}{\text{Fe I }\lambda 4250} \quad \text{or} \quad \frac{\text{Cr I }\lambda 4254}{\text{Fe I }\lambda 4260}$$

and

$$\frac{\text{Cr I }\lambda 4274}{\text{Fe I }\lambda 4271}$$

The line pairs used come from iron peak elements, a fact which avoids difficulties when the composition is anomalous. It is generally admitted that iron peak elements all vary simultaneously.

The Yerkes system uses subtypes K0, K1, K2, K3, K4, K5 and M0. Subtypes K7 and K8, although used occasionally, are not 'full' subtypes. The temperature sequence is illustrated in figure 13.1.

When working at lower plate factors ($\sim$ 70 Å/mm) Yamashita *et al.* (1976) proposed using

$$\frac{\text{Ti I }\lambda 3999}{\text{Fe I }\lambda 4005} \quad \frac{\text{Fe I }\lambda 4144}{\text{H }\lambda 4101} \quad \frac{\text{Ca I }\lambda 4226}{\text{Fe I }\lambda 4250}$$

Of the molecules, the G-band of CH dissolves at mid K-type into an array of separated lines. CN is no longer seen, TiO becomes visible at K7 and Mg H (λ4780) at about K5.

If on the other hand high plate factors (250–300 Å/mm) need to be used, very broad features like the Ca I λ4226/G-band are used instead (Seitter 1975).

Since all lines are blends, it is impossible to measure the equivalent width

Figure 13.1. Spectra of K-type giants; temperature sequence. For explanation see text.

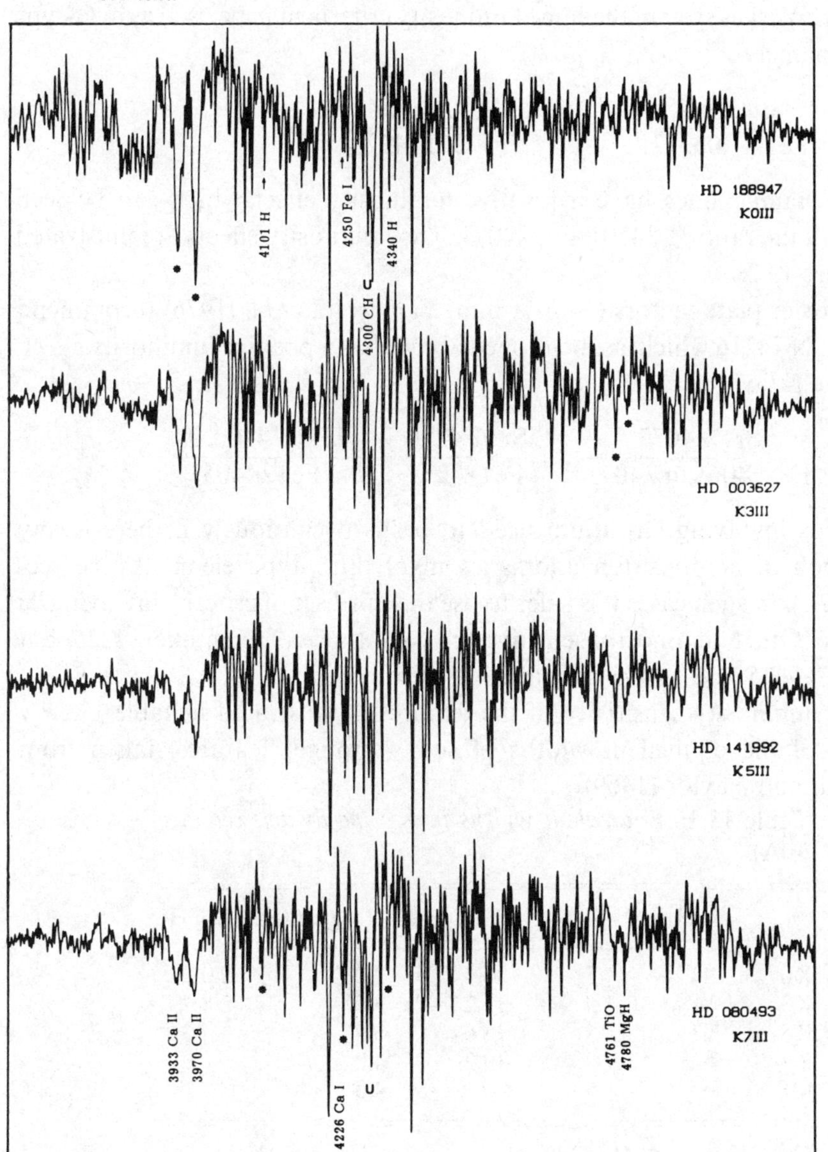

of 'isolated' lines. Rather, a mixture of many lines is always measured, except at very low plate factors. With this in mind let us quote nevertheless some equivalent widths from Taylor (1970) (see table 13.1).

Luminosity effects are clearly visible in K-type stars, even at low dispersion. Foremost among these is the positive luminosity effect of the CN molecule (Lindblad 1922) which we described in chapter 12; this is also valid in K-type stars.

In the Yerkes system the same luminosity criteria may be used as in G-type stars, namely

$$\frac{\text{Sr II } \lambda 4077}{\text{Fe I } \lambda\lambda 4063, 71} \quad \text{or} \quad \frac{\text{Sr II, Fe I } \lambda 4216}{\text{Ca I } \lambda 4226}$$

The Balmer lines have a positive luminosity effect which can be seen through the ratio H λ4101/Fe I λ4071. The luminosity effects are illustrated in figure 13.2.

At lower plate factors ($\sim$ 70 Å/mm) Yamashita *et al.* (1976) recommend using CN λ4216 which, as mentioned above, has a positive luminosity effect, and the ratios

$$\frac{\text{Sr II } \lambda 4077}{\text{Fe I } \lambda 4063 \text{ or } \lambda 4071} \qquad \frac{\text{Sr II } \lambda 4216}{\text{Fe I } \lambda 4271} \qquad \frac{\text{Ti II } \lambda\lambda 4400, 08}{\text{Fe I } \lambda 4405}$$

Ratios involving strontium need to be used cautiously if there is any suspicion of composition anomalies involving *s*-type elements (see section 12.1). In such cases it is safer to use the third set of criteria, involving Ti and Fe. Other luminosity sensitive lines are those of Ca I, like λ4226, but λ4425, λ4435 and λ4455 may also be used.

The luminosity sensitivity of the features is illustrated in table 13.2 by means of the equivalent widths of some stronger features, taken from Spinrad and Taylor (1969).

Table 13.1. *Equivalent widths in K-type dwarfs (values in Å).*

	Na I	G–CH	Ca I	CN λ4200	Mg H
K0	1.4	5	6.4	2.0	2.0
K3	2.1	5	11.2	3.0	2.8
K5	3.3	5	18.6	3.9	
K7	3.3	5	19.2	4.0	
M0	—	5	—	4.0	

Ca I λ4226 Mg H λ4780

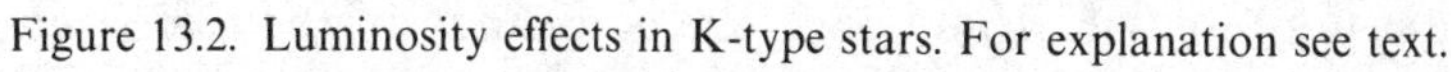

Figure 13.2. Luminosity effects in K-type stars. For explanation see text.

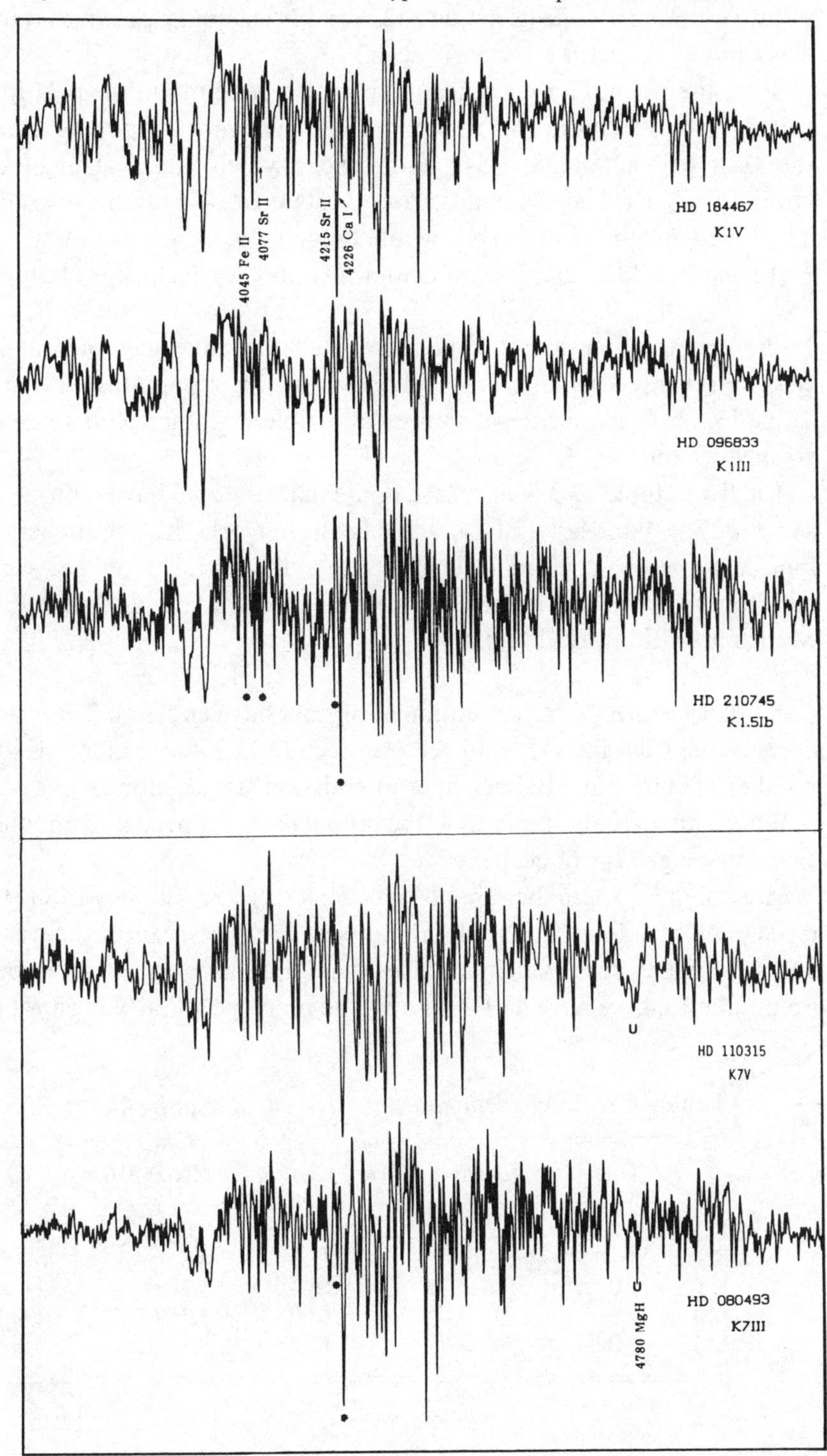

Comparing tables 13.1 and 13.2, the constancy of the CH strength, the negative luminosity effects in Ca I λ4226 and Na I and the positive luminosity effect in CN strength are clearly seen.

Some useful lines for classification can also be found in the near infrared ($\lambda\lambda$0.7–1 μm; see figure 14.4 in section 14.0). The most important feature is the Ca II triplet (λ8498, λ8542 and λ8662) which has a strong positive luminosity effect. The CN bands around λ8000 are also luminosity sensitive. The Ca II triplet is useful in the whole range between F- and M-type stars.

It must be added that stars of luminosity class IV disappear at about K1.

Molecules. Up to now we have mentioned the molecules present in late type stars only casually; we shall now try to systematize this a little. Table 13.3 lists the principal molecules visible, together with some of the strongest bands.

The list in table 13.3 is of course non-exhaustive; we have indicated only some of the bands useful for classification work. More discussion on molecular bands is given in chapter 15, and interested readers can find additional information in the review papers by Spinrad and Wing (1969) and Merrill and Ridgway (1979).

Spectral peculiarities. A large number of stars have emission features in the cores of the Ca II lines (H and K) – see section 11.2 – and in late K-type, a number of stars have Balmer lines in emission (see section 14.2).

We summarize in table 13.4 the anomalies connected with different elements or groups of elements.

Most of these anomalies are not specific to type K, but also affect stars of types G or M. Most of them do not appear in dwarfs, except the weak line and strong line stars, which are also found among dwarfs. As can be seen from table 13.4, we have had to discuss the many peculiarity groups present

Table 13.2. *Equivalent widths in K-type giants (values in Å).*

	Na I	G-band (CH)	Ca I	CN λ4216	TiO
K0	0.8	5.2	4.8	3.0	0.3
K3	1.2	5.2	9.0	4.5	0.7
K5	2.0	5.1	16.0	5.7	3.3
M0	2.0	5.0	17.0	5.7	4.5
M2	2.0	5.0	17.2	5.5	7.0

among giant stars in different chapters; here we shall give a brief unified view of these groups. The relative position of these groups is best visualized in a three-dimensional representation. We use the following parameters. On the *x* axis we represent the 'oxygen–carbon relation' in such a way that on the left we have stars with strong oxide bands whereas on the right we have stars with carbon bands. Normality (normal oxides) lies in the middle, at a point we call '*n*' (normal). On the *y* axis we represent the 'strength of metals', with strong metallic lines upwards, weak metallic lines downwards, and normality in the middle. On the *z* axis we represent the 'strength of the *s*-type elements', which can be 'strong' or 'normal'. Since

Table 13.3 *Molecules in late type stars.*

Molecule	λ	Comments
CH	4295, 4315	Visible from late F to late K, maximum around G8 (see 12.3)
CN	4216, 3883	Luminosity sensitive; visible from late F to late K (see 12.2)
TiO	4954, 4761, 4626, 4584, 4422	Visible from K5 to late M, with saturation around M3. In near infrared also $\lambda\lambda$7045, 7088, 7126, 7589. 8432 (see 14.0)
V0	7400, 7900	Visible in late M stars (see 14.0)
C_2	4395, 4697, 4737	In C stars (see 13.1)
SiC_2	4868, 4979	In C stars (see 13.1)
C_3	4053	In C stars (see 13.1)
ZrO	4641, 4620	In S stars. Also λ4493 (see 14.1)
LaO	7910, 7404	In S stars. Also λ5552 (see 14.1)
CO	1.56–2.35 μm	

Table 13.4 *Spectral peculiarities in K-type stars.*

Connected with	Name	Section
H	High luminosity H-poor stars	13.1.4
C	Carbon stars and related objects	13.1.1–3
	CH stars and related objects	12.3
N	CN anomalies	12.2.2
metals	Weak line stars	11.3
	Super metal rich stars	12.2.1
s-process elements	Ba stars and related objects	12.1

three-dimensional representations are difficult to accommodate on two-dimensional pages, we provide instead two figures which represent (x, y) plots at z 'strong', i.e. 'enhanced s-process elements' (figure 13.3(a)), and at z 'normal', i.e. 'no enhancement of s-process elements' (figure 13.3(b)).

The explanation of figure 13.3(a) is relatively simple. We find here Ba stars (normal oxides, normal metals and enhanced s-type elements), and C stars (strong carbon compounds, normal metals, enhanced s-type elements), CH stars and subgiant CH stars.

In figure 13.3(b) we find the 'weak G-band' stars, 'normal giants', 'early C stars' (which have no enhanced s-type elements), 'metal weak giants' and 'H deficient stars'. Between both planes (not plotted) on a line linking normal giants (b) and C stars (a) we find the S-type stars – therefore on the same line we have M, MS, S, SC and C stars.

The diagrams are from Jaschek (1985) where the reader can find a more detailed description.

Photometry. Broad band photometric systems like UBV continue to show the rapidly decreasing flux in the ultraviolet. Figure 13.4 shows the behavior of UBV ... L as a function of spectral type for dwarfs, and it is fairly obvious that spectral type or temperature indications can be supplied by any combination of indices. But the large variation of the color indices – (U–B) varies by $1^{m}\!.4$ between K0 and M0 – makes it difficult to detect the relatively small effects resulting from luminosity, reddening and metallicity. For luminosity the situation is illustrated in figure 12.3 of section 12.0, and it is clear that the effects in (U–B) are small ($< 0^{m}\!.1$).

To deal with both luminosity and metallicity, intermediate or narrow

Figure 13.3. Peculiarity groups among late type giants. (a) and (b) show two intersections of the three-dimensional (x, y, z) representation of peculiarity groups. (a) intersection at z 'strong'; (b) intersection at z 'normal'. For explanation see text.

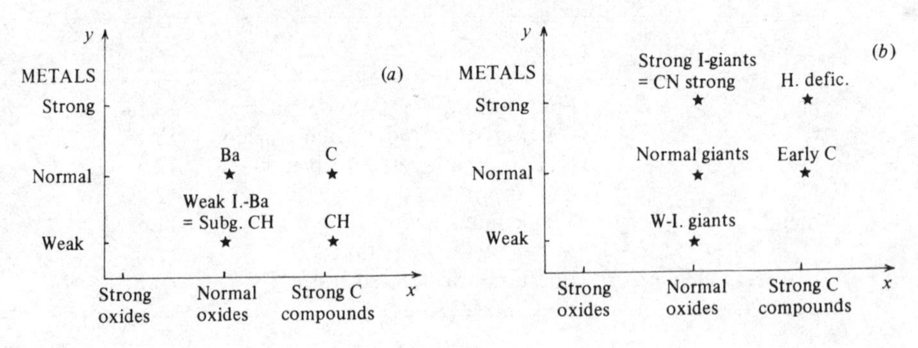

band systems, like the DDO, have to be used. Since this system was discussed in chapter 12, we refer the reader to section 12.0 and figures 12.7 to 12.10, which include K-type stars.

Another system which has been developed specifically for late type giants is the Copenhagen gnkmfu system. Since not very many stars (i.e. fewer than 10^4) have been measured in this system, we refer the reader to Hansen and Kjaergaard (1971) and Kjaergaard (1984). We shall say simply that the system measures the G-band, the CN band, a metallicity index and a temperature index, with a total of six narrow band filters.

Usually any of these systems works well for normal and for weak lined stars, but if other types of peculiar objects are observed, the interpretation of the measures becomes questionable. This can be illustrated for C stars (see section 13.1), which are characterized by the presence of carbon compounds like C_2, SiC_2 and C_3 (see table 13.3). These bands are strong and ubiquitous, so that all photometric bands are affected by them.

In constructing a photometric system capable of dealing with C stars, at least two more bands must be added: one centered on a carbon molecule

Figure 13.4 Broad band photometry of K-type dwarfs: U, B ... L colors. V = 0

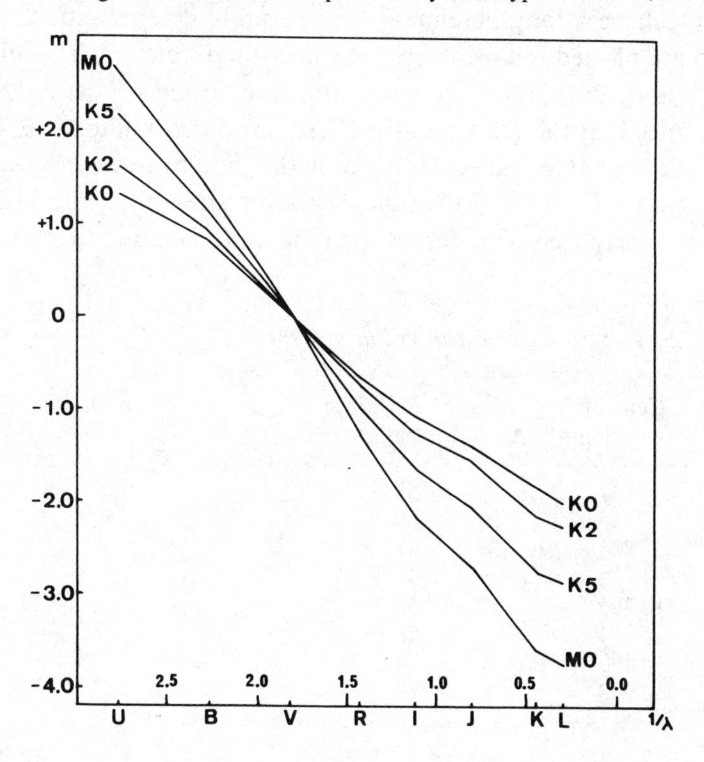

band and another on a band-free region, for comparison. The main problem resides in correctly placing the band-free comparison region. The blue region is full of atomic lines, and the yellow–red region of TiO bands. The best thing to do is to go to $\lambda > 0.7\,\mu m$.

This is also a reasonable choice because the maximum radiation of cool stars lies in the red or infrared region. As an example of a system created to deal specifically with late type stars, with both normal and peculiar spectra, we shall consider the Wing (1971) system. He uses an eight-filter system, whose main characteristics are given in table 13.5.

The choice of the filters is explained as follows. Filter 1 measures the TiO molecules, although there is some contamination by CN. Filter 2 is a continuum band suited for M0–M7 stars (although again there is CN present), whereas filter 3 is the continuum for G, K and C stars (in M stars it is contaminated by TiO). Filter 4 measures a CN band. Filter 5 is a continuum band (I(104)) apparently uncontaminated. Filter 6 measures the VO molecule. Filter 7 is a continuum band, specifically for C stars, and filter 8 measures a CN band. Band 7 may be contaminated to a small degree by the He I λ10 830 line which is erratic and appears strongly in emission in symbiotic stars. The filters fall into two groups, separated by a wide gap; this was done so as to have a long baseline for temperature determination.

In figure 13.5 are plotted the measures for some standard stars. The full curves are black body curves for the temperatures indicated on the right. Observe for instance that the K2 V and the K2 III star differ in filters 1, 2, 4 and 8. The much stronger CN molecular bands in the giant cause this. When passing from K2 III to K5 III and M2 III the increasing strength of the TiO bands in filter 1 is clearly seen. This depression (filters 1 to 2) grows to $2^{m}.5$ at M7.

Table 13.5. *Filters used in the Wing system.*

Filter	Central wavelength (Å)	Width at half power (Å)
1	7117	53
2	7545	50
3	7806	42
4	8122	43
5	10 392	55
6	10 544	58
7	10 800	74
8	10 968	73

From the definition of the passbands it is easy to understand some of the diagrams of the system. Spectral types should go with the strength of TiO; therefore a C(1–2) diagram should correlate well with spectral type. This can be seen in figure 13.6; in general C(1–2) works well from K4 to M8 – at M8 no continuum is left shortward of 1 μm. An important byproduct of this work is that TiO bands do not seem to be affected by metal abundance, since even well-known metal deficient stars exhibit normal bands of TiO. When considering very late stars (later than M6), the molecule VO can be used as a temperature indicator (filter 6).

The luminosities are more difficult to handle. Although the CN bands are known to be luminosity dependent, for giants a considerable spread exists in CN band strength, probably due to abundance effects. It seems, however, that the strength is never so great as to obliterate the difference between supergiants and dwarfs. White and Wing (1978) use a CN index defined as an average deviation of filters 4 and 8 from a black body curve defined by filters 2 and 6. They find that between K4 and M4, this index can

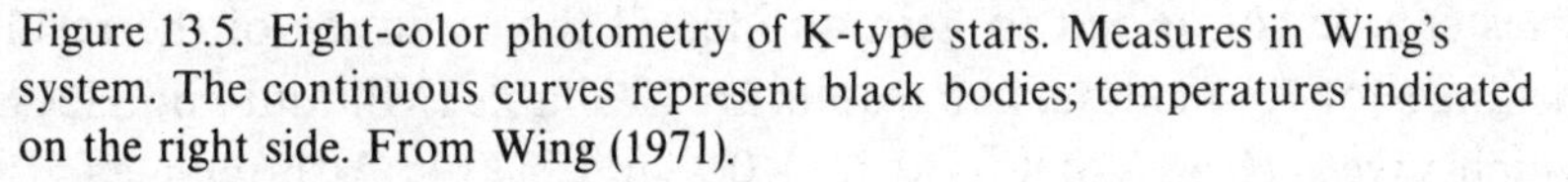

Figure 13.5. Eight-color photometry of K-type stars. Measures in Wing's system. The continuous curves represent black bodies; temperatures indicated on the right side. From Wing (1971).

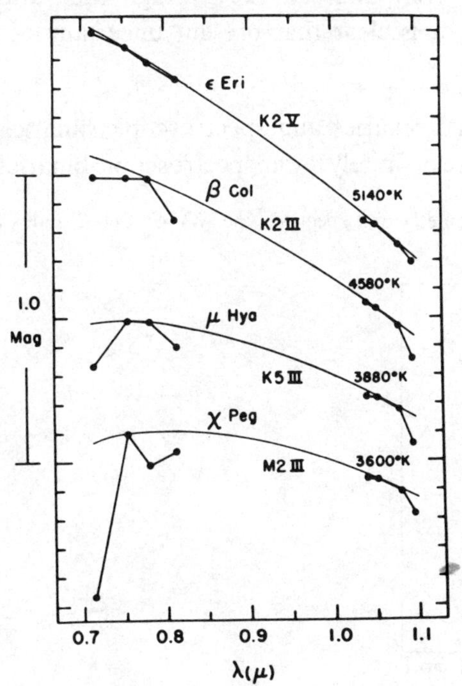

separate luminosity classes fairly well, and that the separation agrees well with the Yerkes luminosity classes.

Because of the fact that CN also influences filter 1 measures, it is necessary to introduce (small) corrections to the observed '1–2' color indices (Wing and White 1978).

Further applications of eight-color photometry concern stars with abnormal molecular bands. The best examples are the CN strong objects, such as the C stars, which show an enormous depression at filters 4 and 8 (see figure 13.7).

Absolute magnitude. The absolute magnitudes of dwarfs and giants is usually based on statistical parallaxes and trigonometric parallaxes, whereas for stars of higher luminosity the calibration rests on a mixture of cluster parallaxes, binary parallaxes and so forth, with subsequent uncertainties. Indicative values are given in table 13.6, taken from Schmidt-Kaler (1982).

The constancy of M for K-type stars of luminosity classes II, Ib and Ia0 illustrates the coarseness of our methods, rather than an observational fact. For giants the results of different authors using basically statistical parallaxes can be compared to gain an idea of the accuracy of the results. For instance, for K0 III stars, five results are available (Mikami and Heck 1982), which give $+0.4$, $+0.8$, $+0.8$, $+0.1$ and $+0.1$. Although each author claims a possible error of ± 0.3, it is clear that present uncertainties are greater than this value.

Binaries. The percentage of visual binaries and spectroscopic binaries in dwarfs is normal, whereas there are definitely fewer spectroscopic binaries in

Figure 13.6. Eight-color photometry and spectral type. Wing's C(1–2) index for dwarfs.

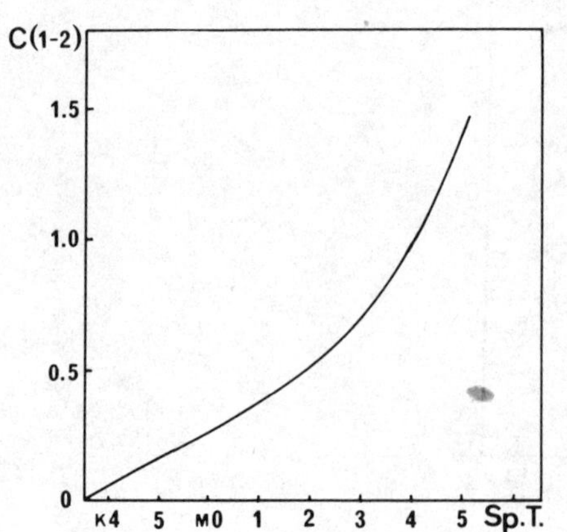

K-type giants. Harris and McClure (1983) quote 15–20% for spectroscopic binaries, which is roughly half of the value for dwarfs.

Number, distribution in the galaxy. K-type stars are moderately concentrated toward the galactic plane, with a halfwidth $\beta = 200$ pc. The distribution in the plane is largely uniform. K-type stars are a frequent type of star, as can be seen from table 13.7.

The distribution over luminosity classes is dominated by the fact that K dwarfs are intrinsically faint and have therefore to be very near to enter magnitude limited surveys: among the 'bright stars' only 2% are dwarfs. On

Table 13.6. *Absolute visual magnitudes of K-type stars.*

	V	III	II	Ib	Ia	Ia0
K0	5.9	0.7	−2.3	−4.3	−7.7	−8.5
K2	6.4	0.5	−2.3	−4.3	−7.6	
K5	7.3	−0.2	−2.3	−4.3	−7.5	
M0	8.8	−0.4	−2.5	−4.5	−7.0	−8.0

Figure 13.7. Eight-color photometry of two carbon stars. Measures in Wing's system. The continuous curves represent black bodies, temperature indicated in the middle. From Baumert (1974).

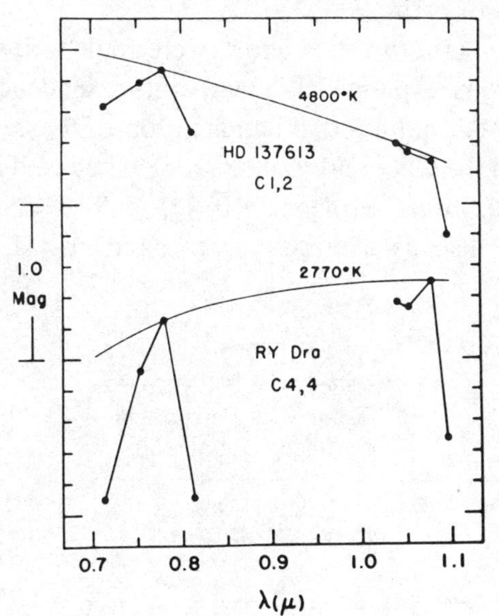

the other hand, in distance limited surveys practically no giants or supergiants are found because of their relative scarcity.

13.1 Carbon and related stars

13.1.1 *Carbon stars*

A carbon star is a late type giant with strong bands of carbon compounds and no metallic oxide bands. Thus very intense bands of C_2, CN and CH and no bands of TiO are seen.

Carbon stars were discovered by Secchi (1868). In the Harvard system (Cannon 1918), they correspond to two different spectral types, R and N. R stars are similar to G5–K0 stars but with the Swan bands of C_2 around λ4700 very strong, and the λ4395 as strong as the G-band. The continuous spectrum is visible up to the Ca II H and K lines. In N-type stars the Swan bands are still stronger, so that the spectrum is chopped up into sections of different intensity. The spectrum practically disappears for $\lambda < 4200$ Å. The R-type stars were divided into decimal subtypes, whereas N stars were divided into a, b and c.

The relation between types R and N and with normal stars had already been analysed by Rufus (1916), who studied the behavior of line intensities. He concluded that R stars were earlier than N (i.e. hotter) and that R and N stars constituted a sequence parallel to K and M stars. He judged correctly that the differences between the two sequences were caused by differences in the abundances of carbon and oxygen.

With the advent of Saha's theory in the twenties, efforts were made to see if the sequence of R and N stars could be explained as a temperature sequence.

Shane (1928) analysed the behavior of lines and bands in some fifty stars and concluded that the strength of the CN bands (λ4216, λ3883) behaved in such a way as to justify a continuous sequence R0, R1, ..., R10 = N0, N, ..., N7 (see figure 13.8). Shane also used the C_2 Swan band at λ4737

Table 13.7. *Frequency of K-type stars.*

Survey	Limit	Total number	Percentage of all stars
Bright stars	$m \leqslant 6\overset{m}{.}5$	2200	24%
HD	$m \sim 9^{m}$	61 000	27%
Nearby stars	$r \leqslant 20$ pc		11%

and found a more complex behavior (see figure 13.9) with a gradual increase, a maximum around R5, a minimum at N0 and further strengthening toward later types. Shane comments that such behavior is difficult to understand if it were a temperature effect. He concludes that 'temperature may not be of prime importance in determining the spectrum of carbon stars'.

In subsequent work it became clear that the branching of spectral types into the two sequences R and N was due to abundance differences of the elements C and O. If oxygen is more abundant than carbon, the spectrum

Figure 13.8. Intensity estimates of CN bands in carbon stars. From Shane (1928).

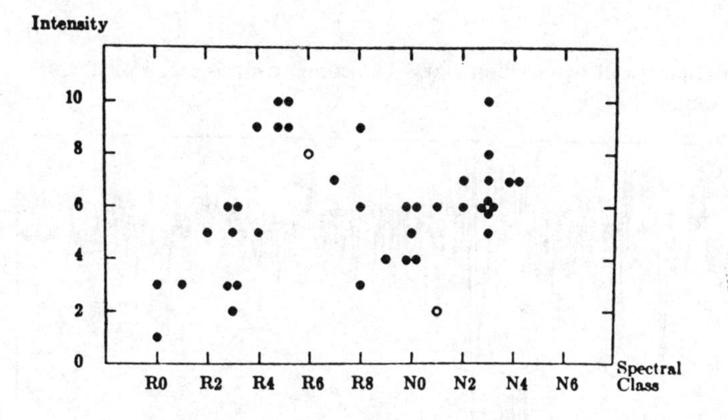

Figure 13.9. Intensity estimates of the C_2 $\lambda 4737$ band in carbon stars. From Shane (1928).

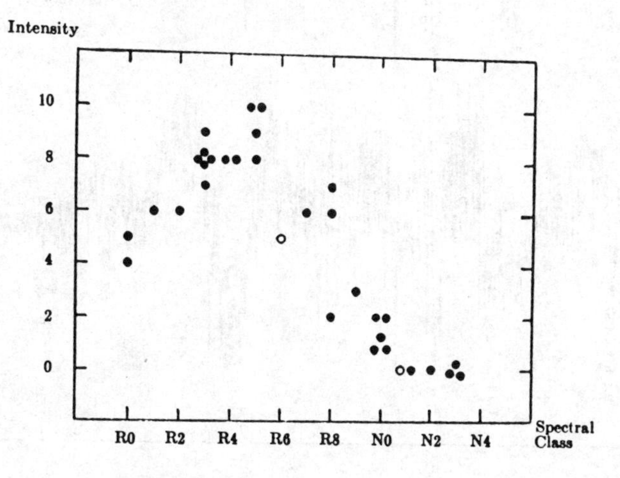

of the star is dominated by oxides like TiO; if the reverse is true, the spectrum is dominated by carbon compounds like C_2, CN and CH.

The next important contribution to the classification of carbon stars was by Keenan and Morgan (1941). These authors used the term 'carbon stars' rather than 'R and N stars' and considered that since carbon is overabundant in C stars, it would not be surprising if its abundance varied from one star to another. In this way they were led to introduce – besides a temperature index – another index to qualify the strength of the carbon compounds (see figure 13.10). They established the sequence of C stars by means of the temperature index of Cr I λ4254/Fe I λ4250 and the strength of the line Na I $\lambda\lambda$5890–96, as summarized in table 13.8.

The other parameter (C strength) is obtained from the strength of the C_2

Figure 13.10. Spectra of two carbon stars. The comparison is a K 1 III star, which corresponds to C3.

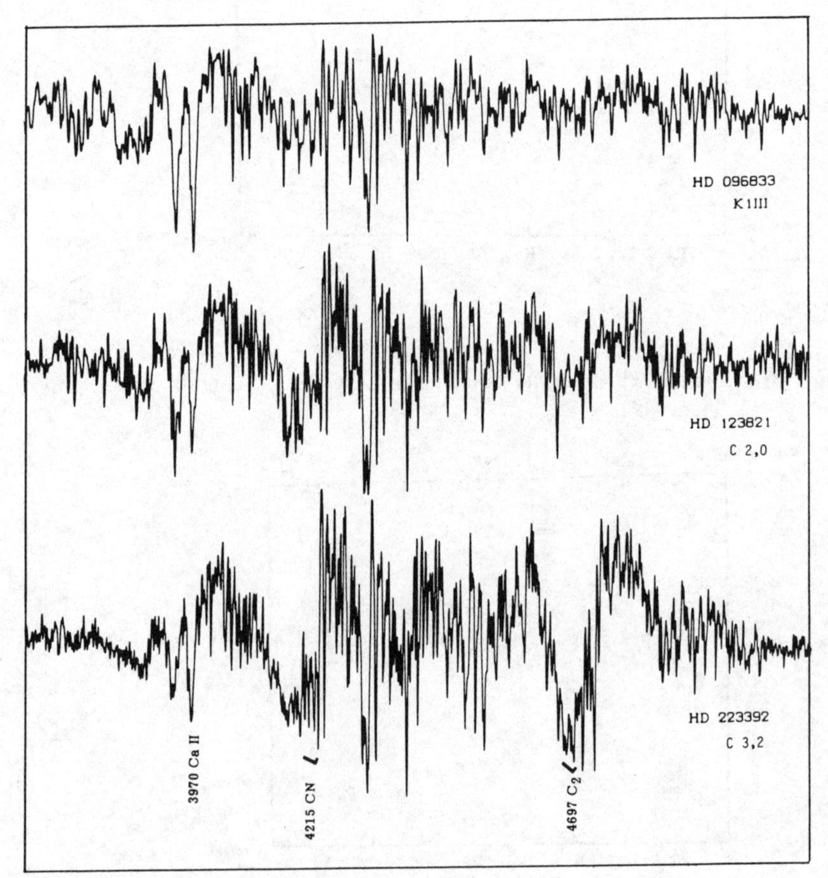

bands estimated on a scale from 1 to 5 (strongest). The C index is given as a subscript following the temperature type, for instance $C3_2$. Later on this was changed to C3, 2, which designates a K1–K2 giant with moderately strong C_2 bands. The equivalence with the Shane system is given in figure 13.11. There is good agreement in the R-types and very poor agreement for N-types. This can be explained by the intensification of the C_2 bands in some of the stars. Since Shane's system has no place for the variable strength of C_2, he classifies a hot star with strong C_2 bands as much cooler than it really is, which explains the scatter in N stars. Since, on the other hand, C_2 band intensities

Table 13.8. *Temperature sequence of C stars.*

C type	Equivalent type
C0	G4–G6
C1	G7–G8
C2	G9–K0
C3	K1–K2
C4	K3–K4
C5	K5–M0
C6	M1–M2
C7	M3–M4

Figure 13.11. Comparison of R-types and N-types with C-type classifications. From Keenan and Morgan (1941).

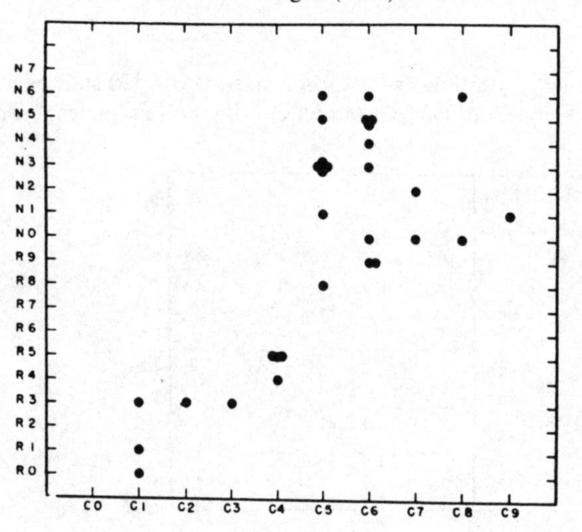

depend both on temperature and on overabundance, it is clear that the first C-types will not have very large C_2 intensities – and this is what figure 13.11 shows.

Support in favor of the Keenan and Morgan scheme was provided by Yamashita (1967). He estimated visual line strengths of eleven spectral features, using lines of H I, Na I, Ca I, Fe I, Ti I, Ba II and Sr I. Figures 13.12 and 13.13 show that the line intensities vary smoothly with the C classification, as expected if the C classification is a temperature ordering. Yamashita also studied the behavior of molecular bands. Figure 13.14 illustrates convincingly the large scatter in the intensity of the C_2 band, which was at the origin of the introduction of the second classification index. For CN the situation is

Figure 13.12. Line intensity estimates for carbon stars. Cross, Hd star; open circle, CH star; filled circle, ordinary C star. (a) Hydrogen lines. (b) Na I lines. From Yamashita (1967).

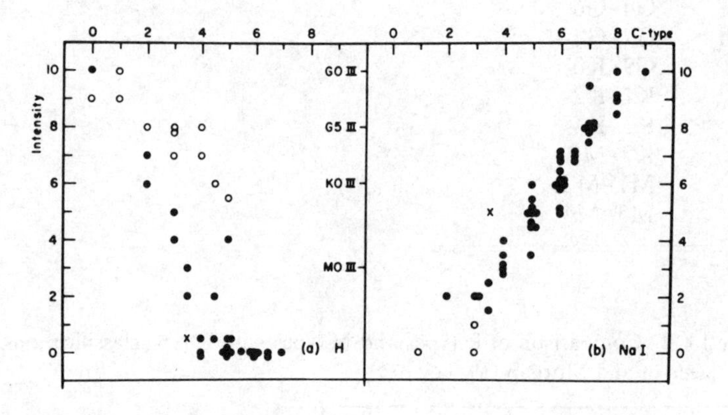

Figure 13.13. Line intensity estimates for carbon stars. Cross, Hd star; open circle, CH star; filled circle, ordinary C star. (c) Ca I lines (d) Fe I lines. From Yamashita (1967).

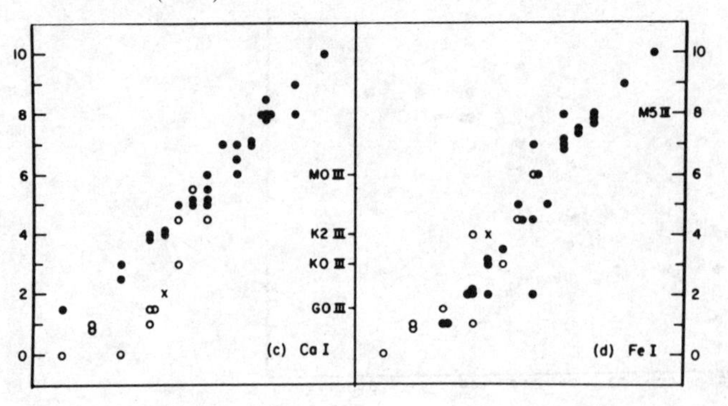

better, although the variation in carbon strength from one star to another causes some scatter in the strength of the carbonaceous molecular bands. The so-called Merril–Sanford bands at λ4640, λ4866, λ4905, λ4977 and λ5192, due to SiC_2, behave essentially in an erratic way.

The line intensity estimates were later extended by Yamashita (1972) to a total of 180 carbon stars. Afterwards Yamashita *et al.* (1977) replaced some of their estimated intensities by measures with narrow band filters. As expected, this reduces the scatter in the graphs, but can only be used for strong isolated features that have some undisturbed continuum nearby for comparison purposes. We have seen in section 13.0 that the latter is the main difficulty of the procedure.

Emission lines are seen sometimes, mostly in the Hβ and Hα lines. They are present in about 10% of the carbon stars later than C4, and they are denoted by an 'e' appended to the spectral classification. In some cases the emission strengths are variable.

As we have seen in section 13.0, the yellow and red regions of the spectrum are full of different molecular bands. To find less 'polluted' regions, attempts were made to develop a classification system in the near infrared. Richer (1971) did this in the $\lambda\lambda$7500–8900 region, at a plate factor of 125 Å/mm. In this region the important features are the Ca II infrared triplet (λ8498, λ8543 and λ8662), the CN bands and K I λ7699. Richer used CN as a luminosity indicator and Ca II as a temperature indicator, and established a temperature classification similar but not identical to that of Keenan and Morgan. It turns out, however, that the Ca II triplet lies in a region affected by CN bands. This implies that temperature effects and carbon abundances cannot be separated easily at the dispersion and in the wavelength region

Figure 13.14. Line intensity estimates for carbon stars. (a) C_2 bands. (c) CN bands. From Yamashita (1967).

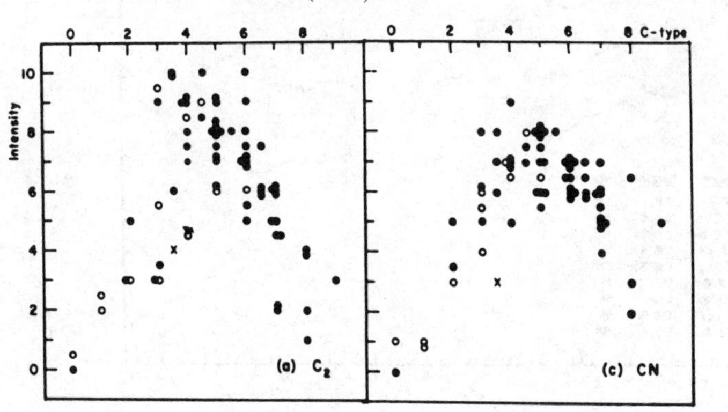

used by Richer. Nevertheless, it can be said that the bands permit the easy recognition of C stars. In fact, much higher plate factors can be used; some of the surveys for C stars were successfully made at 3400 Å/mm (Nassau 1956).

With the improvement of detectors, it has become possible to explore even longer wavelength regions, from 1 to 15 μm. We quote some results from Merrill and Stein (1976), who normally found in C stars a broad emission feature about 3 μm wide, at 11.5 μm, due to SiC. In most stars a strong absorption feature, about 0.2 μm wide, is present at 3.09 μm, due to $HCN + C_2H_2$. Frequently this absorption is accompanied by weaker absorption at 3.9 μm and 2.3 μm; the latter being due to CO (see figure 13.15). Of these features the most interesting is the emission at 11.5 μm, which is probably due to dust near the star.

Detailed studies at lower plate factors have added many important details, of which we shall consider three in particular.

The first is the enhancement of Sr and Ba lines in C stars later than C5

Figure 13.15. Spectrophotometry of carbon stars in the 2–13 μm region. In the third part (Var.): SR, semiregular; M, Mira variable. From Merrill and Stein (1976).

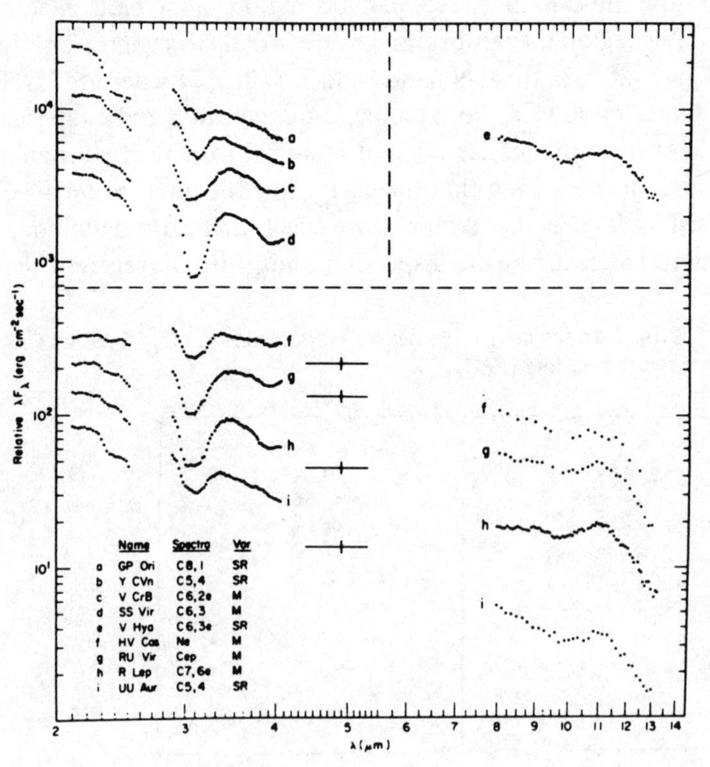

(Yamashita 1967). This was put in a more precise form by Dominy (1984): *s*-type elements (Y, Zr, Mo, Ba, La, Ce, Nd and Sm) are enhanced in stars later than C3. Utsumi (1985) has shown, however, that in cool C stars the *s*-type anomalies disappear again. This is an important fact when looking for the origin of the observed spectral peculiarities. At classification dispersion only the enhancement of Sr and Ba is seen, because the other elements have fainter lines.

The second fact is that in some stars the unstable element Tc is present. We say 'unstable' because the isotope with the longest half life has one of 2.1×10^5 years. Merrill (1955) and Peery (1971) detected Tc in several C stars, whereas in others it is absent. Peery suggested that this is related to the type of variability; irregular variables have it, semiregulars do not.

The third fact is the discovery of carbon isotopes. Sanford (1929) was able to detect bands due to $C^{12}C^{13}$, $C^{13}C^{13}$ and $C^{13}N^{14}$ and this allows the ratio C^{12}/C^{13} to be obtained through a comparison of the band strengths. For such purposes the ratios

$$\frac{\lambda 4744\, C^{12}C^{13}}{\lambda 4735\, C^{12}C^{12}} \quad \text{and} \quad \frac{\lambda 6260\, C^{13}N^{14}}{\lambda 6206\, C^{12}N^{14}}$$

may be used. These effects are thus observable even at classification dispersion. Similar isotopic bands can be observed for molecules involving O, N, Mg and other metals. With regard to C^{12}/C^{13} the situation is summarized in table 13.9, where it can be seen that this ratio is definitely different from the solar value.

Photometry. We have already discussed photometry in section 13.0, so that here we shall comment only on the light variability of C stars. The basic fact is that many C stars are variables. All the variables found in late stars are found among them – Miras, semiregulars and irregulars, with perhaps even isolated

Table 13.9. *The C^{12}/C^{13} ratio.*

Type	Ratio	Type	Ratio
Solar system	90	Early C	4–15
		Late C	40–70
GKM giants	10–30	J	3–10
KM supergiants	5–20	CH	6–60
Ba	8–25	S	10–30

cases of RV Tau (AC Her) and Cepheids (RU Cam). Since the classification of variable stars is mostly based upon photographic and visual light curves, it is not surprising that the distinction between the different groups becomes blurred when they are observed in another wavelength range, like infrared. For instance, Peery (1975) remarks that when observed at λ10 400 Å no clear distinction exists between Ib (irregulars) and SRb (semiregulars), so that both groups can be considered together. Typical amplitudes are then of $\sim 0^m.2$. Miras and SRa variables on the other hand have amplitudes of 1^m–2^m.

Eggen (1972) observed a sample of 35 carbon stars and found all N stars (late C) to be variable and practically all R stars (early C) to be constant. Two-thirds of the variables have amplitudes equal to or less than $0^m.2$ in the R band.

Mikami (1975) gave a statistic based on Yamashita's (1972, 1975) catalogs and showed that the percentage of variables increases from 10% in C0–C3 stars to about 90% in C8–C9 stars. If the C-type stars are considered as a single group he found that the percentages of Miras, semiregulars (SR), irregulars (I) and 'others' are 18%, 44%, 34% and 3% respectively. This implies that the majority of the stars are semiregular or irregular variables, with time scales of the order of 10^2 days.

Absolute magnitude. Luminosity determinations of C stars face two difficulties. The first one (already mentioned) is that many C stars are variables, and thus we have to define an average apparent magnitude based upon extended series of observations, since cycle lengths are typically of the order of several hundred days. The second difficulty is the choice of the wavelength at which to carry out these measures. As we have seen, for $\lambda < 7000$ Å the magnitudes are strongly influenced by the band strength of carbon compounds. Baumert (1974) avoided the difficulty by using a magnitude at 1.04 μm. This region is essentially free from molecular blanketing and the light amplitudes of the (variable) C stars are smaller there than in the blue region. Using statistical parallaxes he then found the values given in table 13.10.

Scalo (1976) in a summarizing discussion notices that possible difficulties reside in interstellar extinction corrections and the conversions from m(V) to m(1.04).

Other approaches can be made through C stars in clusters or in other galaxies. We quote one result by Richer, Oleander and Westerlund (1979) based upon stars in the Large Magellanic Cloud: they find for C3–C7 stars $M(\mathrm{I}) = -5.7$. (I corresponds to $\lambda_0 = 9000$ Å.)

We may thus conclude that early type C stars have absolute magnitudes corresponding to giants, whereas later C stars correspond to luminosity class II.

Binaries. C stars often occur in visual binaries; Gordon (1968) lists a few which usually have A-type companions. With regard to spectroscopic binaries, McClure (1985) mentions that early C stars occur in the same proportion as normal giants.

Membership in stellar systems. C stars are known in some open clusters of great or intermediate age and in globular clusters; usually there is one (or at most a few) object per cluster. For this reason C stars were assigned to old disc population, but they are probably present in both young and old disc populations (Eggen 1972).

C stars are also found in other galaxies. At present 1.1×10^4 are known in the Large Magellanic Cloud and 2.9×10^3 in the Small Cloud and in several other galaxies of the Local Group (Catchpole and Feast 1985).

Frequency. Iwanowska (1966) samples stars down to $V = 12$ and concludes that C stars constitute 1% of all red giants. Since red giants are themselves not very frequent, C stars are thus very rare, and this is confirmed by the fact that in the HD catalog there are only 190 C stars.

When examined in detail, the proportion of C stars varies considerably from place to place. Catchpole and Feast (1985) quote the results given in table 13.11.

Atlases. Carbon stars have been illustrated very often. We quote only a few. Sanford (1950) gives six stars at 10 and 20 Å/mm dispersions in the $\lambda\lambda$3600–8800 region. Johnson and Mendez (1970) cover for seven stars the interval

Table 13.10. *Mean absolute magnitudes of carbon and related stars.*

	Number	$\bar{M}$ at 1.04 μm
C0–C3	54	−0.6
C4–C6	11	−5.3
C7–C9	30	−5.1
Ch	14	−3.9
G–K III		−0.9 to −2.3

1.25–4 μm. Keenan and McNeil (1976) illustrate the C sequence on plates of low dispersion (around 80 Å/mm) in the $\lambda\lambda$3000–7000 region.

For additional reading we recommend Alksne and Ikaunieks' *Carbon stars* (1981) and the Strasbourg Colloquium on *Cool stars with excesses of heavy elements* (1985).

13.1.2 *J stars*

J stars are characterized by unusually strong isotopic bands of carbon; in other words they are carbon stars in which the ratio C^{12}/C^{13} is unusually low.

Stars of this type were known to Sanford (1929) and McKellar (1948), but the group as such was defined by Bouigue (1954). He measured in a number of C stars the intensity ratio between λ6260 and λ6210; the first band is mainly due to $C^{13}N^{14}$, whereas the second one is due to $C^{12}N^{14}$. He calls 'J' stars those in which λ6260 is very conspicuous. Gordon (1971) defines, at plate factors of about 80 Å/mm, a 'J' star as a star exhibiting the isotopic (0.2) band of carbon at λ6168 with a strength equal to at least one half of the normal (1.3) band at λ6122. This definition is preferable to the one used by Bouigue, who had used a larger plate factor ($\sim$ 240 Å/mm). The stars of this group are also sometimes called 'λ6168 stars'.

Besides these bands, Gordon remarks that the $C^{13}C^{13}$ λ4754 is easily visible, and that $C^{12}C^{13}$ λ4744 $\sim$ $C^{12}C^{12}$ λ4737.

Almost all stars with strong C have strong Li I λ6708; the reverse, however, is *not* true. Other characteristics may be weak sodium D lines, weak Sc I λ6259 and La I, but high resolution work on more stars is needed to confirm this.

Yamashita (1972) denotes stars showing J characteristics by adding J after their C classification.

J stars apparently do not stand out from other carbon stars with regard to temperature and luminosity. Roughly, they follow the frequency of the

Table 13.11. *Ratio of carbon to M stars.*

Small Magellanic Cloud center	20
Small Magellanic Cloud periphery	5
Large Magellanic Cloud	2
Solar neighborhood	0.01
Galactic center	$< 10^{-3}$

different C types, although it would seem that they tend to be more frequent at early types.

It seems further that J stars are abnormally red for their carbon type. Although colors are only available for a few stars, this would be in line with a suggestion made by Yamashita (1967) that in J stars there is a correlation between the overabundance of carbon and strong isotopic carbon bands.

Utsumi (1985) finds that none of the eight J stars analysed by him has an overabundance of *s*-type elements.

13.1.3 *Li stars*

Lithium stars are carbon stars with extreme enhancement of the resonance line Li I λ6708. The name seems to have been given by Bidelman (1956). Later, Boesgaard (1970) called them 'super lithium rich'. Li stars occur among the coolest C stars, but are rare: fewer than 20 are known.

13.1.4 *Hydrogen deficient C stars*

A star of this group is characterized by weak or absent hydrogen lines and strong C bands. The group was first described by Bidelman (1953) although the hydrogen weakness of R CrB had been discovered by Ludendorff (1906). The spectrum of this star resembles that of a late F-type supergiant, with weak C_2 present (see figure 13.16). Other stars of the R CrB type have stronger C_2 and CN bands and qualify as C stars.

R CrB is the prototype of a group of variables, and would not be discussed in this book were it not for the fact that there is a small group of non-variable stars which share the characteristics of weak or absent H I, strong C_2 and weak or absent CH bands (see figure 13.17). These stars are labeled Hd (hydrogen deficient). Usually Hd is written after the C type, e.g. C 2, 2 Hd. A further characteristic is the absence of $C^{13}C^{13}$ and $C^{12}C^{13}$ bands. This implies that most of the carbon is in the form of C^{12}. All of the stars are G and K supergiants.

Detailed analysis at higher resolution shows that most elements behave normally, except of course hydrogen and carbon. Some of the features found are illustrated in figure 13.18, taken from Cottrell and Lambert (1982).

Photometrically, the stars behave like other C stars, but do not show an infrared excess as R CrB stars do (Feast and Glass 1973).

These stars are infrequent. About 30 R CrB stars are known (Feast 1975), but only five Hd stars. Perhaps these figures are biased because R CrB stars are easy to discover because of their variability, whereas Hd are only detectable in a spectroscopic survey.

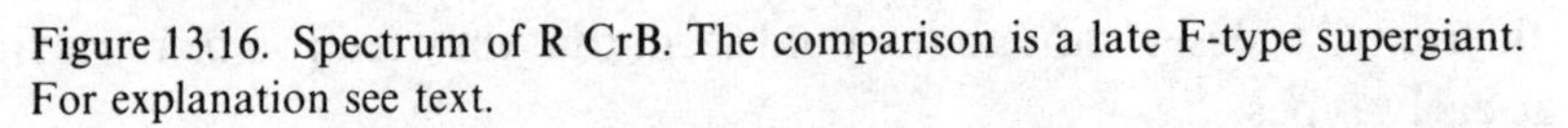

Figure 13.16. Spectrum of R CrB. The comparison is a late F-type supergiant. For explanation see text.

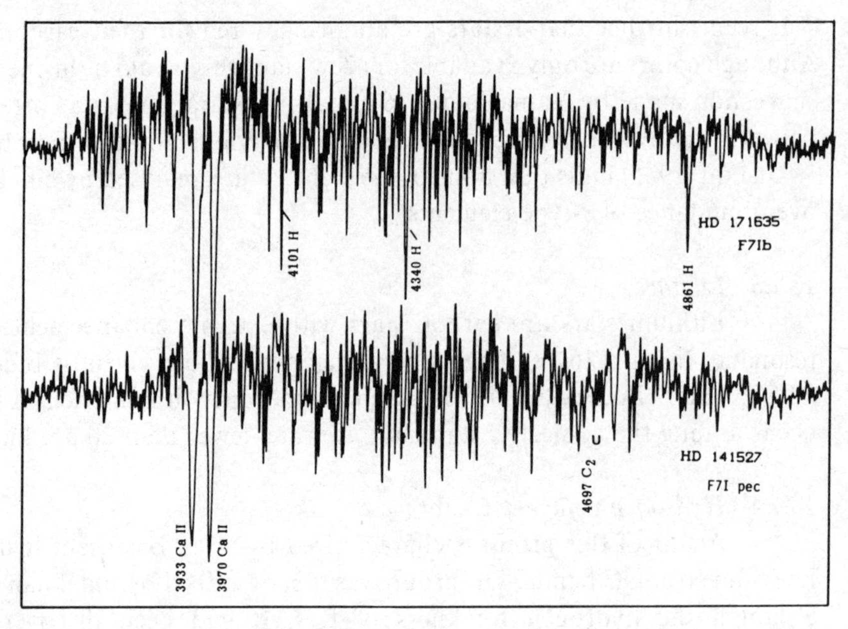

Figure 13.17. Spectrum of the hydrogen deficient carbon star HD 182040. The comparison is HD 223392, a C3, 2 hydrogen normal carbon star.

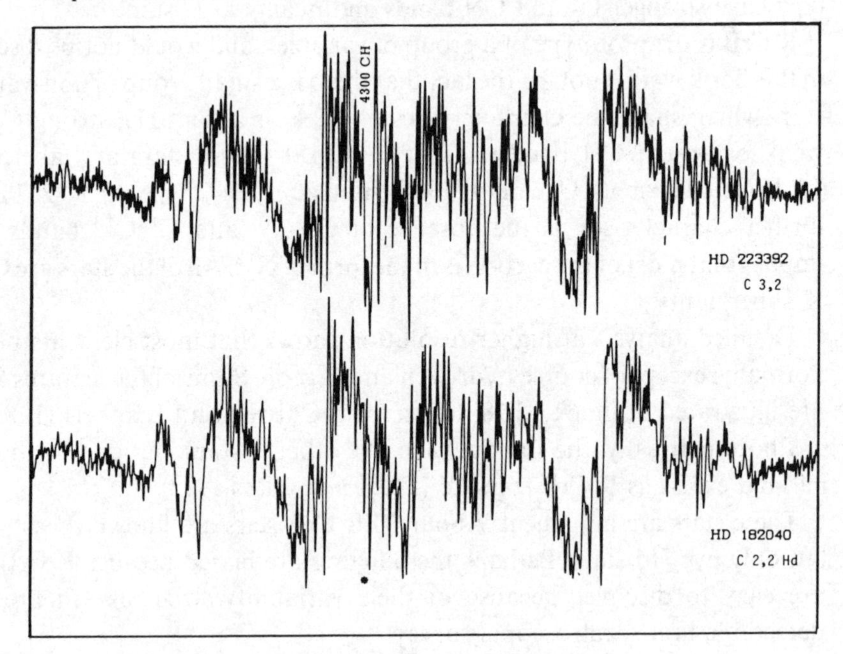

Figure 13.18. Spectral features in three R CrB stars. The comparison, δ CMa, is a normal F8 supergiant. Figures (a), (b), (c) and (d) reproduce four stretches of the spectrum centered on He I λ5876, the N I and O I lines around λ6450 and Hα Li I λ6707.

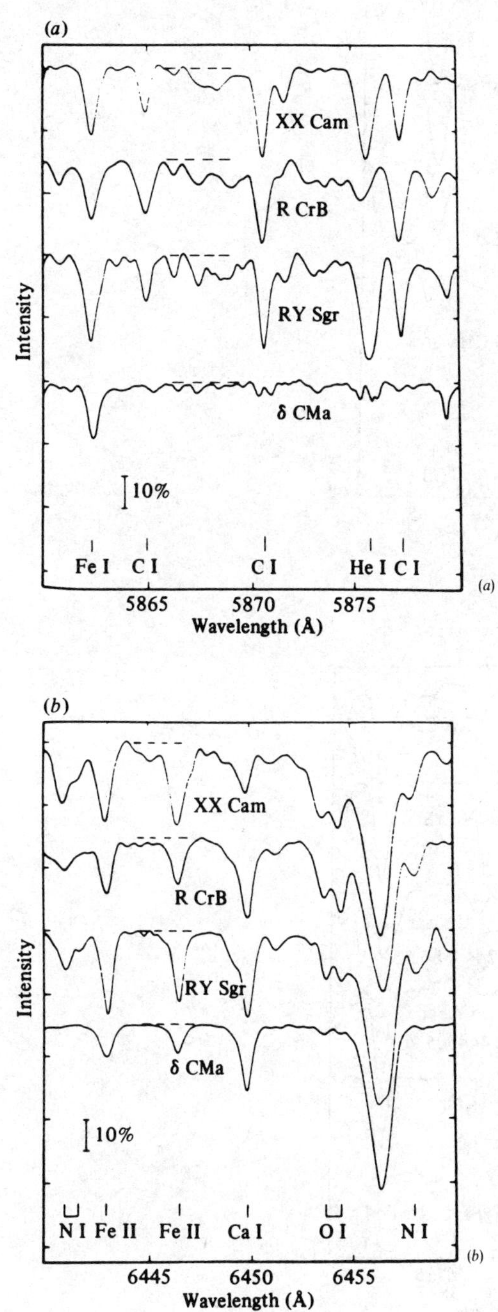

Figure 13.18 *Contd.*

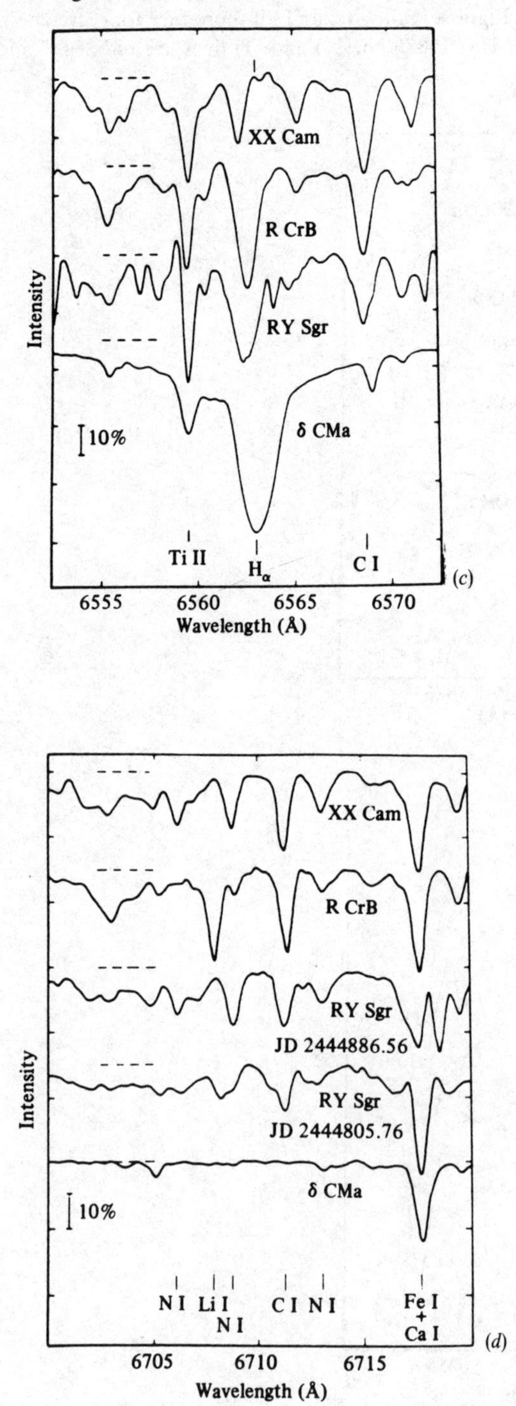

Because Hd stars have weak (or absent) hydrogen lines, they have been linked to the so-called 'early type hydrogen deficient stars' (see section 9.12), which correspond essentially to stars of type O–A. Some authors try to combine all these stars into a single group, but such a procedure seems artificial, firstly because Hd stars are luminous, whereas the 'early type' stars are underluminous. Secondly a gap exists between late A-type stars and the Hd types, which correspond to about G5. Thus it seems best to wait for further clarification.

13.2 Composite spectra stars and related objects

13.2.1 *Composite spectra stars*

A composite spectrum is one which corresponds to the superposition of two spectra. The definition is due to Pickering (1891). This group corresponds to pairs of stars which cannot be observed separately – thus for earthbound telescopes the group includes all doubles whose separation is less than 1–2″. The brightness of both stars, however, must be comparable in the region in which the observations are made. A practical rule is that only stars which differ by less than about one magnitude are observable as composite spectrum. Consequently, the only observable composite spectra are combinations of two spectra lying within a horizontal band one magnitude wide, in the HR diagram. We find, for instance, (a) pairs of dwarfs of similar spectral type, (b) pairs composed of an A-type star and a G–K giant, (c) a B-type star and a K–M bright giant or supergiant. Other cases (two giants, two supergiants) are seldom found (see figure 13.19). Observe, however, that the requirement of 'comparable brightness in the region under observation' has some undesirable consequences: outside the region of comparable brightness, the spectrum corresponds to that of one of the components, whereas the other is invisible. This happens for instance if a composite spectrum, e.g. A3 V + G0 III, is observed in the ultraviolet – only the A3 V spectrum is seen. On the other hand, more composite spectra can be detected by changing the spectral region.

In principle, since both spectra correspond to different stars, their radial velocities should differ. If they do not, the explanation is not valid.

Shajn (1926) and Hynek (1938) were the first to undertake the systematic study of composite spectra. As expected, they found that a large number of composite spectra corresponded to visual or spectroscopic binaries. Hynek added a group of stars, which he labeled 'spectrum binaries', of whose binary nature he was uncertain and for which his material was insufficient to determine

Figure 13.19. Some composite spectra. In the upper part, the composite spectrum HD 190918 (O9.5 I + WN 5.5) is shown with an O9.5 I and WN 6.5 object. In the middle, the composite spectrum HD 215835 (O6 + O6) is shown with an O4 V spectrum. In the lower part, the composite spectrum HD 27639 (K7 III + A) is shown with a K7 III and an A7 V star. Notice that the A-type companion is probably an early A-type.

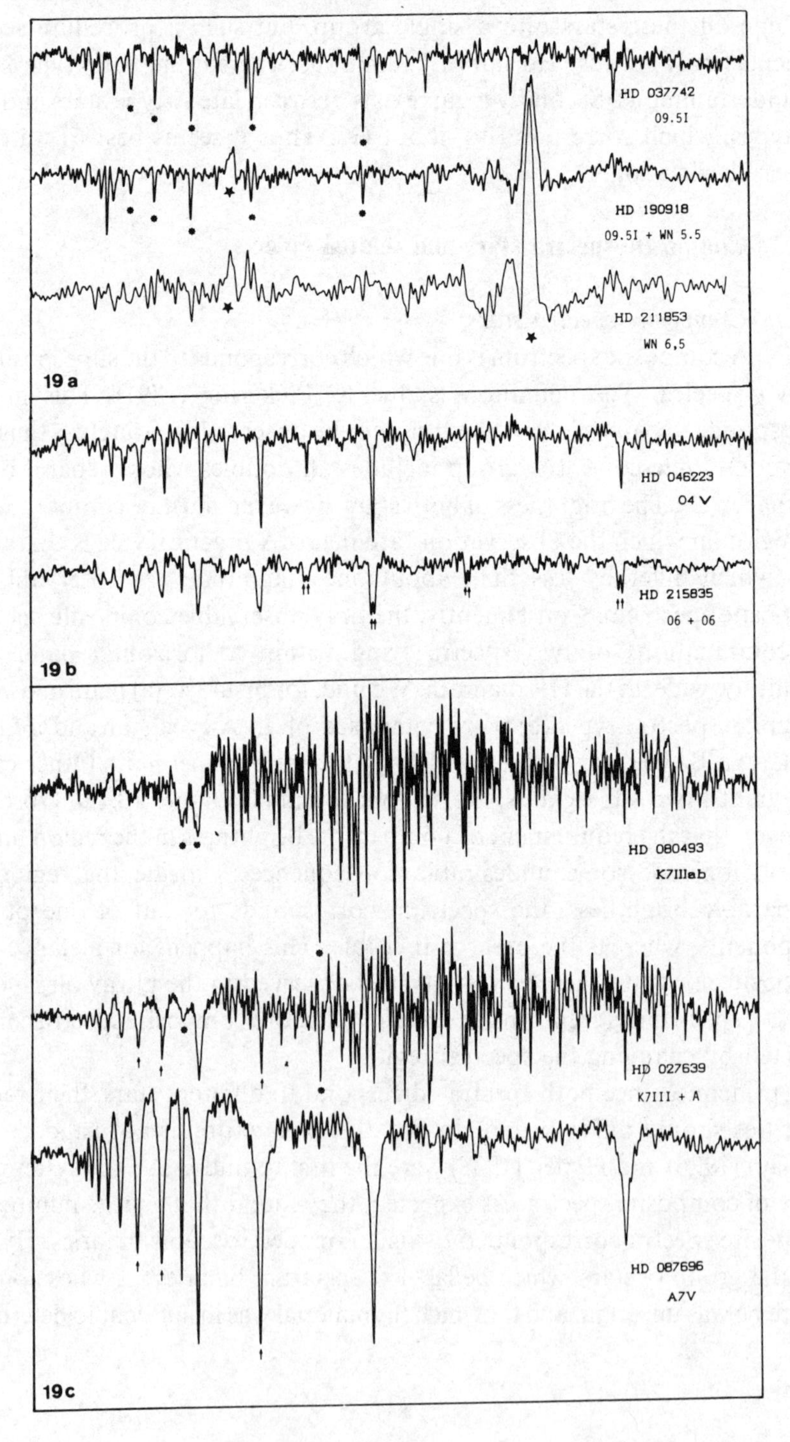

if they had different velocities. He suggested that 'spectrum binaries' corresponded to very close visual binaries (thus escaping visual detection) but sufficiently separated so that their radial velocities differed little.

With better observing techniques and improved classification standards, spectrum binaries dissolved as a class, because all the objects could be shown to be binaries (Ginestet *et al.* 1980).

Composite spectra are classified like any other spectra; the component which is better visible on the plate is written first, as in G0 III + A3 V.

Often the classification of the secondary is difficult, which explains the differences among classifiers. The best classification technique is to compare real composite spectra with pairs obtained artificially by copying together superimposed negatives of a G0 III and an A 3 V star, for example. This technique was first introduced by Hynek (1938).

Photometrically the classification of the two stars is simple: if the spectral type and luminosity of one of the stars are known, its colors are subtracted from the observed colors, giving the spectral type of the secondary and the magnitude difference between them (Δm). The luminosity class of the secondary is fixed by that of the primary and the Δm value. Notice, however, that interstellar reddening can influence the results; it is usually corrected by trial and error until the observations can be reasonably satisfied. Details of the procedure are given by Bahng (1958). As can be expected, the accuracy of the results improves when a large number of colors are used.

Spectrophotometric scans can also be used (Beavers and Cook 1980).

The number of composite spectra is large. Bidelman (1984) lists 175 among the 'bright stars', i.e. at least 1% of all stars. Since the definition of what to include in the class of composite spectra is somewhat arbitrary, the result implies only that they are relatively frequent.

It may be added that before the group of Am stars was defined (section 10.1) some Am star spectra were classified as composite. This is why sometimes there are two HD numbers for one and the same star – as in the case of ζ UMa = HD 78362–3.

13.2.2 *VV Cep stars*

The VV Cephei stars were first distinguished by Bidelman (1954) who defines them as supergiant binary systems whose spectra show emission lines of hydrogen and [Fe II]. The primary is always a K or M supergiant while the secondary is generally an early B-type star.

The emission lines visible in the spectrum are usually of low level of excitation and ionization – singly ionized metals, like [Fe II] for instance.

All VV Cep stars are light variables and some of them are eclipsing binaries. The group as such is clearly a subgroup of 'composite spectrum' stars and has fewer than 20 members. Cowley (1969) provided a description of the group.

13.2.3 ζ *Aur stars*

These objects are spectrum binaries with a late type supergiant component and an early type star, such as ζ Aur, which is classified as K4 Ib + B6 V. No emission lines are seen in the spectrum.

This is a small group of stars, whose main interest lies in the fact that some of its members are eclipsing binaries.

The group was described by Wright (1970, 1973). Some authors combine the ζ Aur stars with the VV Cep stars (de Jager 1980), but the former lack the emission lines of the latter.

13.2.4 *Symbiotic stars*

The term 'symbiotic stars' was first introduced by Merrill (1928) and designates stellar objects whose spectra represent a combination of absorption features characteristic of a low temperature star (generally a giant) with emission lines corresponding to a hot plasma (see figure 13.20).

Figure 13.20. Tracings of the symbiotic star AX Per in 1964 and 1965.

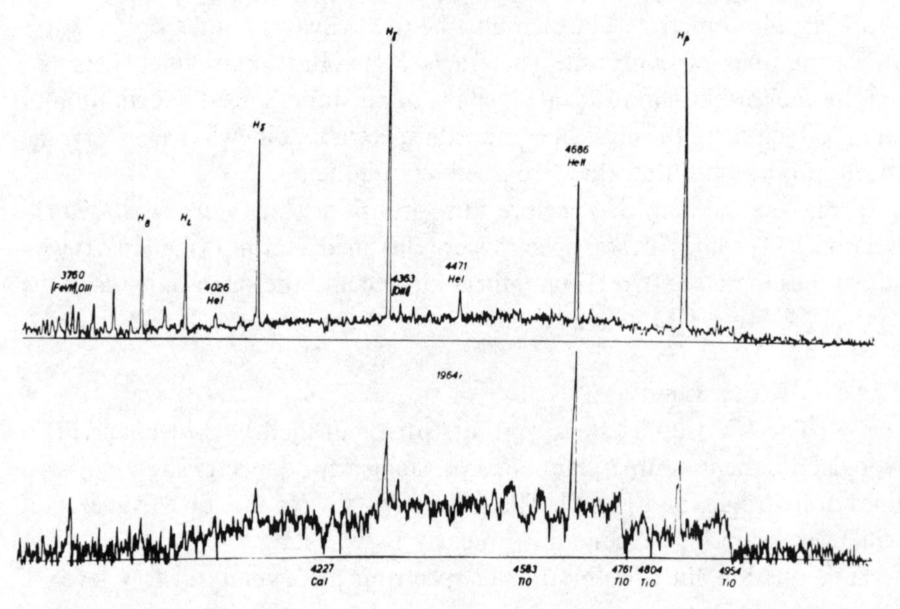

The emission lines present correspond to very high stages of ionization and excitation, as well as forbidden lines characteristic of planetary nebulae. Thus He II and [O III] are regularly seen, and sometimes [Ne III], [Fe V], C III and N III.

All stars of this type are light variables with sometimes very large amplitudes over short periods.

A vast literature exists about these stars. We recommend the various papers in *IAU Coll.* **70** (Friedjung and Viotti 1982) and a paper on the UV spectra by Sahade, Brandi and Fontenla (1984), and the book *The symbiotic stars* by S.J. Kenyon (Cambridge University Press, 1986).

References

Alksne Z.K. and Ikaunieks Ya. (1981) *Carbon stars*, Baumert J. (ed.), Pachart Publ. House

Bahng J.D.R. (1958) *Ap. J.* **128**, 572

Baumert J.H. (1974) *Ap. J.* **190**, 85

Beavers W.I. and Cook D.B. (1980) *Ap. J. Suppl.* **44**, 489

Bidelman W.P. (1953) *Ap. J.* **117**, 25

Bidelman W.P. (1954) *Ap. J. Suppl.* **1**, 179

Bidelman W.P. (1956) *Vistas in Astronomy* **2**, 1428, Beer A. (ed.), Pergamon Press

Bidelman W.P. (1984) *The MK process and stellar classification*, p. 45, Garrison R. (ed.), Toronto

Boesgaard A.M. (1970) *Ap. J.* **161**, 1003

Bouigue R. (1954) *Ann. Aphys.* **17**, 97

Cannon A. (1918) *Annals Harvard Obs.* **91**, 10

Catchpole R.M. and Feast M.W. (1985) *Strasb. Coll, 'Cool stars'*, p. 113

Cottrell P.L. and Lambert D.L. (1982) *Ap. J.* **261**, 595

Cowley A.P. (1969) *PASP* **81**, 297

de Jager C. (1980) *The brightest stars*, D. Reidel Publ. Co.

Dominy J.F. (1984) *Ap. J. Suppl.* **55**, 27

Eggen O. (1972) *Ap. J.* **174**, 45

Feast M.W. (1975) *IAU Symp.* **67**, 129, D. Reidel Publ. Co.

Feast M.W. and Glass I.S. (1973) *MNRAS* **161**, 293

Friedjung M. and Viotti R. (1982) *IAU Coll.* **70**, *'The nature of symbiotic stars'*, D. Reidel Publ. Co.

Ginestet N, Pedoussaut A., Carquillat J.M. and Nadal R. (1980) *AA* **81**, 333

Gordon C.P. (1968) *PASP* **80**, 597

Gordon C.P. (1971) *PASP* **83**, 667

Hansen L. and Kjaergaard P. (1971) *AA* **15**, 123

Harris H.C. and McClure R.D. (1983) *Ap. J.* **265**, L77

Hynek J.A. (1938) *Contr. Perkins Obs.* **10**

Iwanowska W. (1966) *Coll. on late type stars*, Trieste
Jaschek C. (1985) *Strasb. Coll., 'Cool stars'*, p. 3
Johnson H.L. and Mendez M.E. (1970) *A. J.* **75**, 785
Keenan P.C. and Morgan W.W. (1941) *Ap. J.* **94**, 501
Keenan P.C. and McNeil. (1976) *An atlas of the spectra of cooler stars*, Ohio State Univ. Press
Kjaergaard P. (1984) *AA Suppl.* **56**, 313
Lindblad B. (1922) *Ap. J.* **55**, 85
Ludendorff H. (1906) *A.N.* **173**, 1
McClure R. (1985) *Strasb. Coll., 'Cool stars'*, p. 315
McKellar A. (1948) *PASP* **61**, 199
Merrill P.W. (1928) *Ap. J.* **67**, 391
Merrill P.W. (1955) *Ap. J.* **67**, 70
Merrill P.W. and Ridgway S.T. (1979) *Ann Rev. AA* **17**, 9
Merrill P.W. and Stein W.A. (1976) *PASP* **88**, 285
Mikami T. (1975) *PAS Japan* **27**, 447
Mikami T. and Heck A. (1982) *PAS Japan* **34**, 529
Nassau J.J. (1956) *Vistas in Astronomy* **2**, 1361
Peery B.F. (1971) *Ap. J.* **163**, L1
Peery B.F. (1975) *Ap. J.* **199**, 135
Pickering E.C. (1891) *A.N.* **127**, 155
Richer H.B. (1971) *Ap. J.* **167**, 521
Richer H.B., Oleander N. and Westerlund B. (1979) *Ap. J.* **230**, 724
Rufus (1916) *Publ. Michigan Obs.* **2**, 103
Sahade J., Brandi E. and Fontenla J.M. (1984) *AA Suppl.* **56**, 17
Sanford R.F. (1929) *Ap. J.* **111**, 262
Scalo J.M. (1976) *Ap. J.* **206**, 474
Schmidt-Kaler T. (1982) Landolt–Börnstein, group VI, vol. 2b, p. 1
Secchi A. (1868) *Mem. Soc. Ital. Scienze (3)* **2**, 73
Seitter C.W. (1975) *Atlas für Objektiv-Prismen Spektren*, vol. II, F. Dümmler Verlag, Bonn
Shajn G. (1926) *A.N.* **228**, 336
Shane C.D. (1928) *Lick Obs. Bull.* **13**, 123
Spinrad H. and Taylor B.J. (1969) *Ap. J.* **157**, 1279
Spinrad H. and Wing R.F. (1969) *Ann. Rev. AA* **7**, 249
Taylor B.J. (1970) *Ap. J. Suppl.* **22**, 177
Utsumi (1985) *Strasb. Coll., 'Cool stars'*, p. 243
White N.M. and Wing R.F. (1978) *Ap. J.* **222**, 209
Wing R.F. (1971), *Kitt Peak Contr.* **554**, 145
Wing R.F. and White N.M. (1978) *IAU Symp* **80**, 451
Wright K.O. (1970) *Vistas in Astronomy* **12**, 147
Wright K.O. (1973) *IAU Symp.* **51**, 117
Yamashita Y. (1967) *Publ. DAO Victoria* **13**, 67
Yamashita Y. (1972) *Ann. Tokyo Astron. Obs.* **13**, 169
Yamashita Y. (1975) *Ann. Tokyo Astron. Obs.* **15**, 47

Yamashita Y., Nariai K. and Norimoto Y. (1976) *An atlas of representative stellar spectra*, Tokyo Univ. Press

Yamashita Y., Nishimura S., Shimizu M., Noguchi R., Watanabe E. and Okida K. (1977) *PAS Japan* **29**, 731

The complete reference for the Strasbourg Colloquium "Cool stars" is: *Cool stars with excesses of heavy elements*, Strasbourg Colloquium, ed. Jaschek M. and Keenan P.C. (1985)

14

M-type stars

14.0 Normal stars

The spectra of stars of class M are characterized by strong absorption bands of TiO, and by the great number and strength of metallic lines which practically block the spectrum for $\lambda < 4000$ Å.

In the Harvard system four subdivisions of M-type stars were kept: Ma, Mb, Mc and Md. The Mt Wilson observers replaced these types by the decimal subdivisions M0, 1, . . . , 6 in which Ma corresponds to M0–M2, Mb to M3–M5, and Mc to M5 and later. The last group, Md, is characterized by variable emission lines, so that any Md object should be denoted M*x*e, *x* standing for the decimal subtype and e for emission. The Mt Wilson system uses mainly the intensities of the TiO bands for classification and this is also done in the Yerkes system. The reason is that the bands strengthen so rapidly with advancing type that it becomes difficult to find stretches of continuum where atomic lines can be compared reliably, except for the earlier subtypes. The most important TiO bands are listed in table 14.1.

The first bands which appear in the yellow and red are λ5167 and λ7054, and λ4954 in the blue at M0. After M2 many others can be seen; the ones especially useful are marked with an asterisk in table 14.1.

Table 14.1. *TiO bands in M-type stars.*

λ4422	λ5448*	λ6158*
λ4584*	λ5497	λ7054
λ4626	λ5759	λ7589
λ4761*	λ5810	λ7672
λ4954	λ5847*	λ8433*
λ5167	λ5862	

Figure 14.1. Spectra of M-type giants, temperature sequence. For explanation see text.

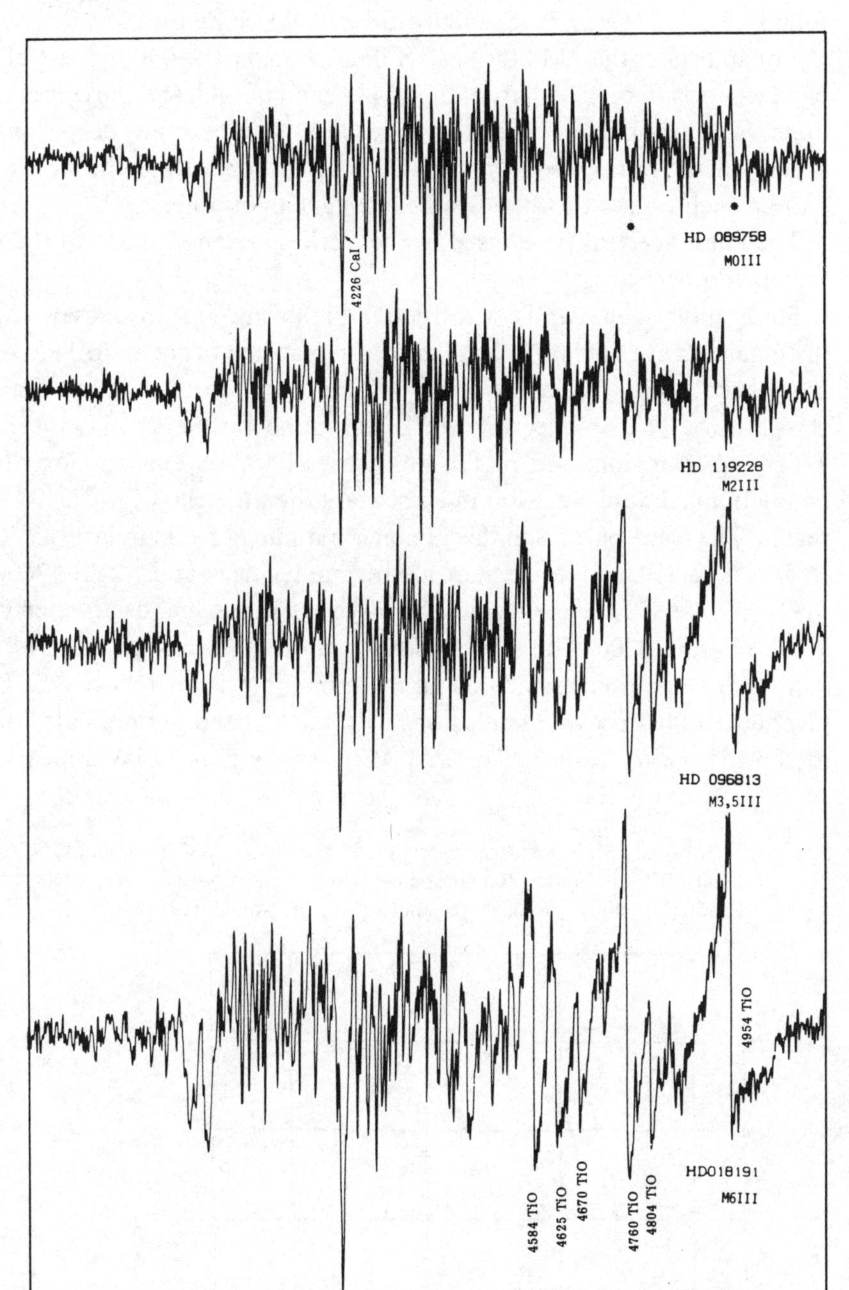

We should, however, keep in mind that the bands saturate with advancing type, so that λ4761 and λ4954 cannot be used after M3, or λ5759 and λ5810 after M5. At M7 the bands of VO become conspicuous; among them are found λ5737, λ7373, λ7865, λ7896 and λ7939 (see figure 14.1).

For stars later than M4 the CaOH band at λλ5500–60 is also useful. This band is superimposed on part of the λ5446 band of TiO, and the combination appears as a box-like absorption, whose red side becomes deeper in later types. There is also additional CaOH absorption within the λ6158 TiO band, which causes this band to be stronger than others.

The latest spectral types used in the Yerkes system are M8 in the giant region and M6 for dwarfs.

Since molecular bands are wide and strong features, they can be distinguished even at large plate factors, so plate factors of up to 3400 Å/mm have been used, for recognition of late type spectra more than for classification. The pioneering work was done by Nassau (1956), in the λλ6800–8800 region. This region was chosen because it comes closer to the region of maximum emission in these stars than the classical λλ3600–4800 region. The main features in this region are summarized in figure 14.2, taken from Nassau (1956). Notice there are several telluric bands; 'a' at λ7600 due to O_2, 'b' at λ6870 due to O_2, and two large and diffuse bands of water-vapor at λ7190 and λ8200. The stellar features are summarized in table 14.2.

M-type stars are classified using these bands. Of the TiO bands, λ7054 must be avoided because confusions with the 'a' band are possible; for this reason it is better to use λ7126. At M2 the group at λ7589 appears as a

Figure 14.2. Main features observable in the λλ6800–8800 region. (A) VO bands; (B) LaO bands; (C) CN bands (also the position of the O_2 λ7600 – 'a' band); (D) telluric bands (H_2O and O_2). From Nassau (1956).

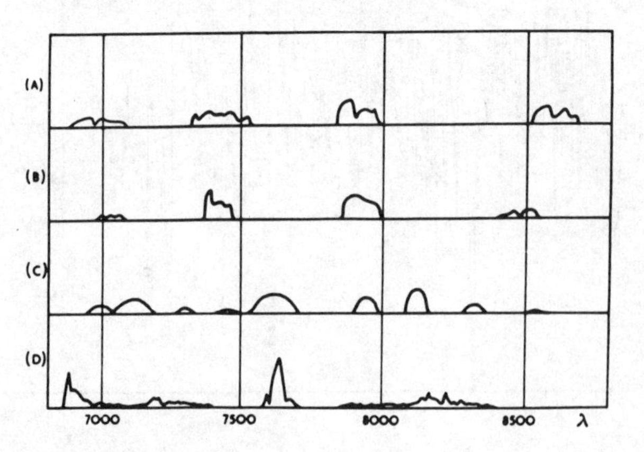

Table 14.2. *Molecular bands in late type stars in the λλ6800–8800 region.*

TiO	CN	LaO	VO
λ7054, λ7088, λ7126, λ7194	λ7945	λ7403	λ7000 weak
λ7589*, λ7667, λ7744, λ7821	λ8125	λ7910	λ7400
group at λ8342	λ8320		λ7900
λ8859			λ8600 weak

*affected by 'a' band.

widening of the 'a' band, which increases toward later classes. The group at λ8342 becomes visible around M5. All features increase toward later types and can be used with plate factors of 3400, 1700 and 850 Å/mm. (See also Seitter (1975) for an illustration at 645 and 1280 Å/mm.)

Carbon stars and S-type stars are segregated by using the bands of CN and LaO. The reader can find more details, as well as reproductions of characteristic spectra, in Nassau (1956). The VO are found as mentioned in the coolest M stars, where they become strong compared to TiO bands, and in long period variables.

The classification of M-type stars runs into two specific difficulties which need to be mentioned briefly. The first is that there are few bright M-type stars, especially later than M4 giants. The question is even more difficult for M-type dwarfs, because the brightest M5 dwarf is of magnitude $V \simeq 9$. This implies that to classify a large number of stars, plate factors higher than is usual for spectral classification must be used.

The second difficulty is that the cooler the star is, the larger is the likelihood of variability in both luminosity and spectral type. It has often been remarked that it is doubtful if any late M-type star has really constant brightness, and this variability implies a variability in spectral type.

The result of these difficulties is that classifications of M stars show a somewhat large scatter. The systematic differences between observers were discussed by Wing and Yorka (1979) and we mention just a few of them, for illustration.

Kuiper (1942) classified M-type stars in the Yerkes scale up to M3, his M5 include M3 to M7 stars and his M8 stars are M6 on the Yerkes scale. Vyssotsky's types (Vyssotsky 1943, 1956; Vyssotsky *et al.* 1946; Vyssotsky and Mateer 1952) show a large scatter – among their M0 and M2 there are K5 to M3.5 dwarfs. Their M5 group corresponds to M3–M4 of the MK scale. Joy and Abt (1974) differ symmetrically in the following way.

Joy and Abt	Yerkes
M0	K5
M1	M0
M3	M2.5
M5	M4.5
M6.5	M6

A large set of consistent classifications was published recently by Keenan and Yorka (1985), and the standards will hopefully homogenize further studies of M-type stars.

Since M-type dwarfs are very rare among bright stars, they were usually observed at larger plate factors than M-type supergiants and giants. This fact has, for a number of years, made it difficult to establish luminosity criteria, until Keenan and his collaborators provided a homogeneous set of standards (see figure 14.3).

Keenan and McNeil (1976) use the following features in early M-type stars: the Ca I $\lambda 4226$ line, which has a pronounced negative luminosity effect, and the Cr I lines $\lambda\lambda 4254$–74–90 which show a similar behavior. These effects are more striking here than in K-type stars. In addition, the ratios

$$\frac{\text{Sr II } \lambda 4077}{\text{Fe I } \lambda 4263} \quad \text{and} \quad \frac{(\text{Y II} + \text{Fe I})\ \lambda 4376}{\text{Fe I } \lambda 4383}$$

have a positive luminosity effect. All these luminosity criteria become difficult to use in stars later than M5.

Several line ratios in the far infrared can be used, although they are less sensitive than the blue lines. At about 50 Å/mm Keenan and Hynek (1945) used

$$\frac{\text{Fe II } \lambda 7712}{\text{Ni I } \lambda 7714} \quad \text{and} \quad \frac{\text{Fe I } \lambda 8689}{\text{Fe I } \lambda 8675}$$

to distinguish dwarfs, giants and supergiants. The K I $\lambda 7699$, Na I $\lambda 8183$ and $\lambda 8195$ have a minimum in giants and are fairly strong in dwarfs and supergiants. An excellent luminosity indicator is also the Ca II triplet $\lambda 8498$, $\lambda 8542$, $\lambda 8662$ (see figure 14.4).

At lower dispersions the intensity ratio of the violet and red sides of $\lambda 4226$ may be used. (In supergiants the left side is almost non-existent.) The

Figure 14.3. Luminosity effects in M-type stars. For explanation see text.

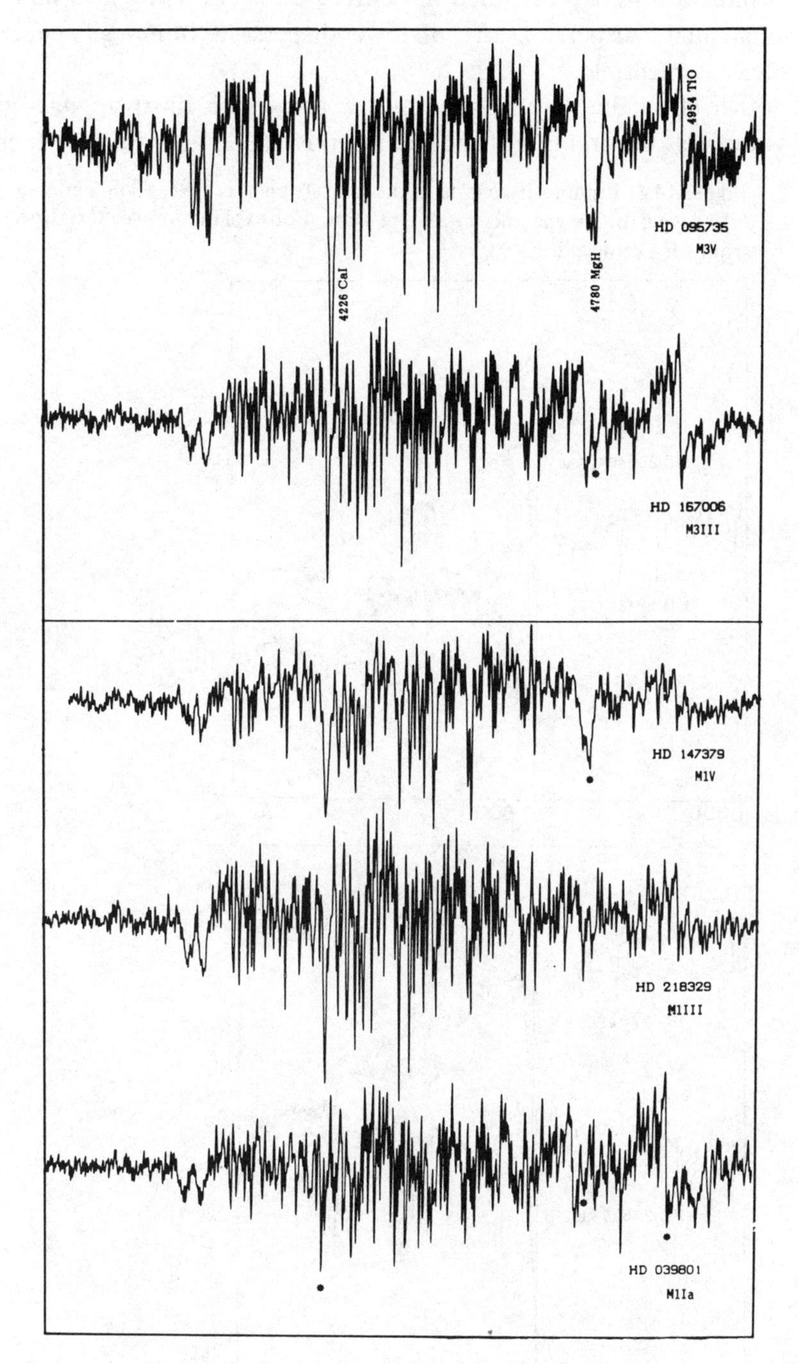

strength of the Na I doublet (λλ5896–90) has a negative luminosity effect. Good illustrations are provided in Seitter's atlas (1975) for 645 and 1280 Å/mm prismatic dispersions. At still lower dispersions luminosity effects are no longer perceptible.

Another effect that can be used at low dispersion to distinguish giants and dwarfs is the strength of the CaH bands λ6382 and λ6389, which are

Figure 14.4. Luminosity effects in the Ca II triplet (λλ8498–8548–8662) in (a) G-type, (b) K-type and (c) M-type stars. For explanation see text. Spectra from M. Dennefeld.

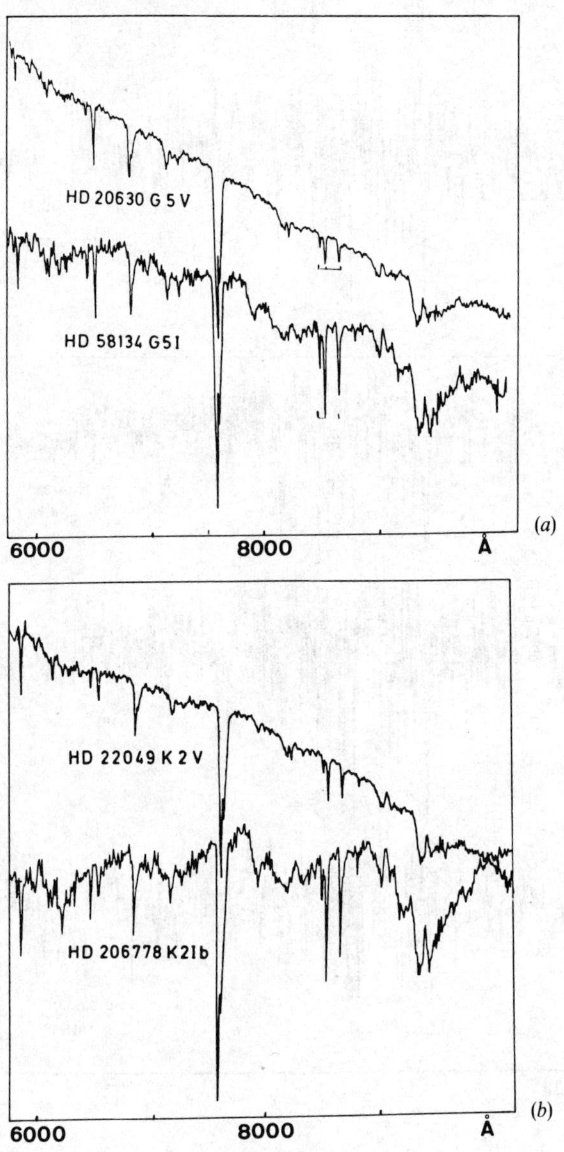

stronger in dwarfs and are probably further enhanced in subdwarfs. The same is true for the MgH band at λ4780 and λ5211 (Spinrad and Wood 1965).

Infrared. Among the important features in the infrared are CO and SiO bands. CO bands at 2.0–2.4 μm are present in virtually all stars later than G and strengthen with type and luminosity (Hyland 1974). SiO bands at 4.0 μm (Rinsland and Wing 1982) are present in all cool oxygen rich stars (M and S), strengthening toward later type. Neither CO nor SiO seem to exhibit a clear luminosity dependence. The far infrared region of giants and supergiants is characterized by broad features, the most prominent of which are seen at 9.7 and 11 μm.

The 9.7 (or 10) μm feature has a width of about 2–3 μm and is attributed to silicates originating in dust clouds surrounding normal M stars. It can be seen in either emission or absorption (see figure 14.5). The 11 μm feature has a width of $\simeq 2$ μm and can be present either in absorption or in emission. Its presence is associated with C-type stars (see figure 14.6).

Emission line stars. The presence of emission lines in late type dwarfs is frequent. According to Joy and Abt (1974) the proportion of stars having emission in the Balmer lines grows with spectral type, as can be seen in table 14.3 taken from their work.

Figure 14.4 (*Contd.*)

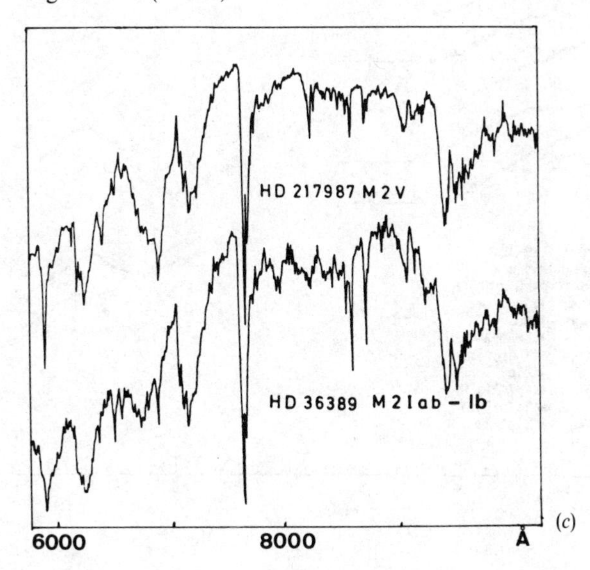

Beside this, a number of stars show emission in the H and K lines of Ca II, which are not accompanied by Balmer lines in emission (see section 11.2). Giant or supergiant stars may also show H and K in emission. Balmer lines in emission are linked to variability, as in the long period variables. Besides the Balmer lines, H and K of Ca II, [Fe II] and AlH lines are also seen in emission. Their behavior varies, however, over the cycle. The presence of such emission lines allows us to conclude directly that the star is variable.

Peculiar stars. The most important groups of peculiar stars occur in the giant (and supergiant) region, where the S stars may be found. These are

Figure 14.5. The infrared region 2–13 μm M-type stars. For explanation see text. From Merrill and Stein (1976).

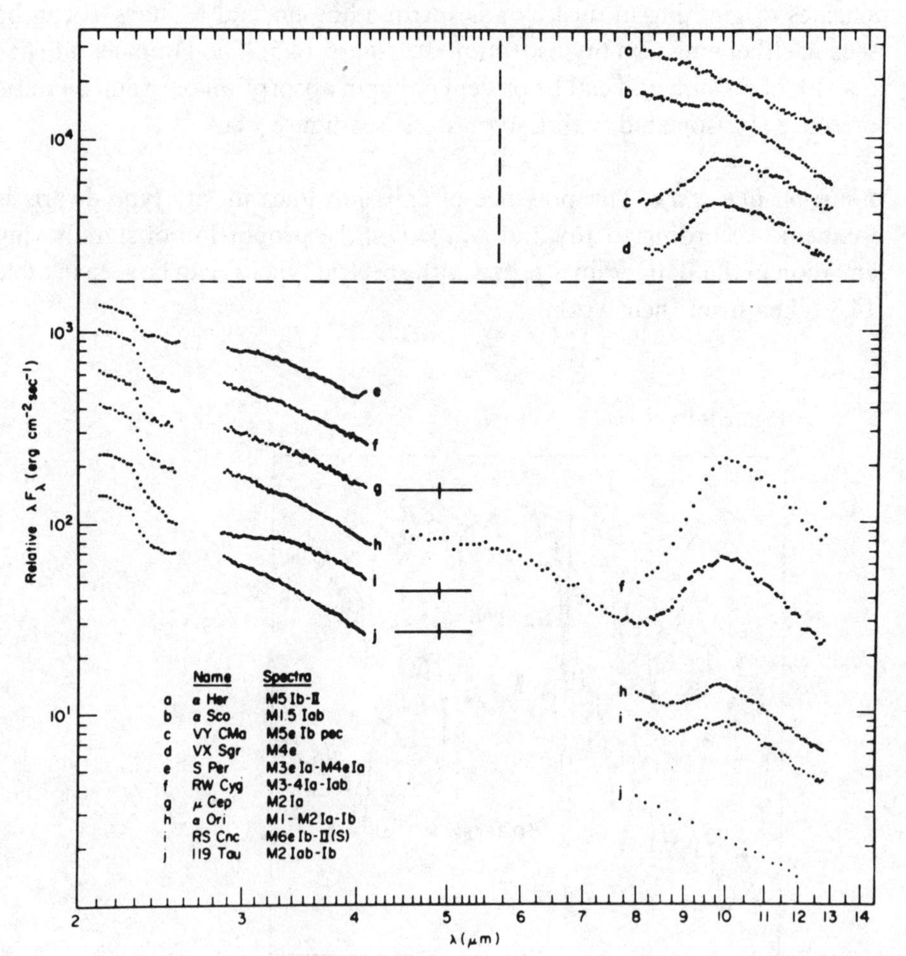

characterized by the presence of ZrO bands and the absence of TiO (see section 14.1).

Between S and M stars, there are transitions called MS and SM stars, which present bands of ZrO and TiO simultaneously.

On the other hand, late C stars are M-type objects (see section 13.1). Here also intermediate objects of the SC- or CS-type exist.

Other late peculiar stars include the flare stars and the BY Dra stars (see section 14.2).

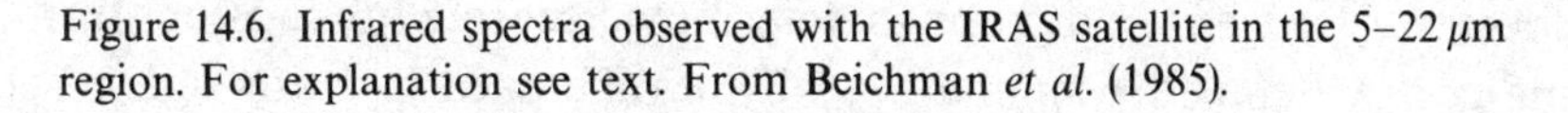

Figure 14.6. Infrared spectra observed with the IRAS satellite in the 5–22 μm region. For explanation see text. From Beichman *et al.* (1985).

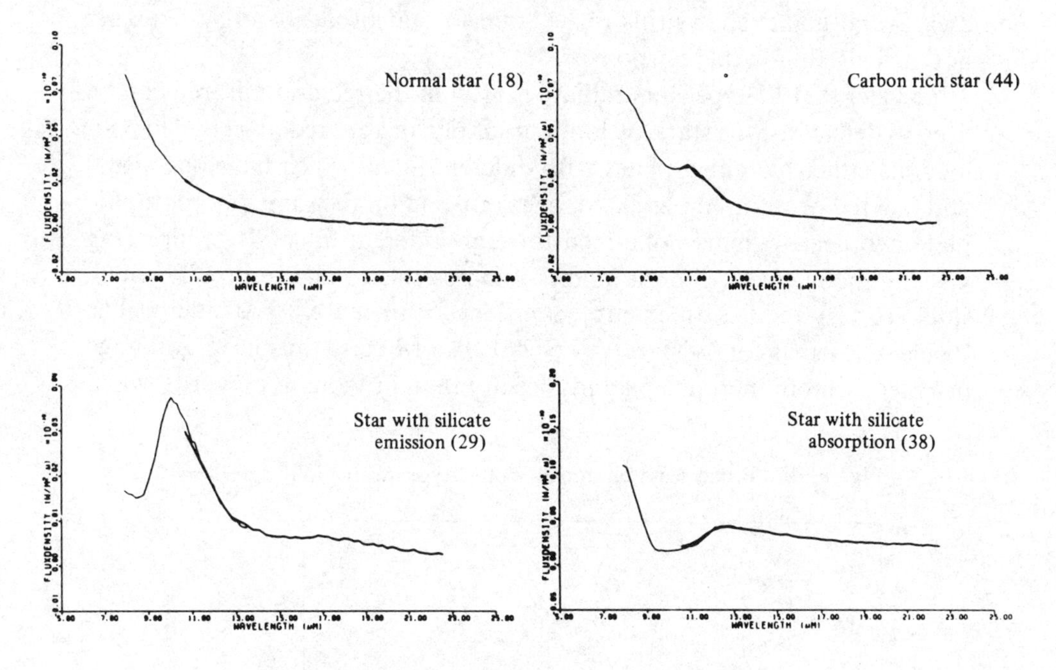

Table 14.3. *Proportion of emission line dwarfs.*

Sp.type	dM	dMe	dMe/dM
M0–0.5	144	5	4%
M1–1.5	88	7	8
M2–2.5	47	6	13
M3–3.5	25	15	60
M4–4.5	25	18	72
M5–later	2	16	800

Photometry. We start as usual with broad band photometry. Figure 14.7 shows the behavior of the UBV...L system as a function of spectral type for dwarfs; it is obvious that spectral type or temperature can be obtained easily from infrared indices. In the visual region, the differences are rather small. On the other hand, it is clear that the indices have very large values, with a considerable slope as a function of λ. Unfortunately, this implies that large differences in color indices may appear if the mean wavelength of the photometric system slides a merest fraction to either side.

It is also clear that in view of the large differences in color indices with spectral type we cannot hope to determine accurate luminosities from these indices. A detailed examination of the difference in color index between dwarfs and giants shows this to be true, so that broad band systems are not very useful for this purpose.

The fact that M-type stars radiate mostly in the red and infrared can be used to detect M-type stars, by looking simply for very red objects. This can be done either by taking plates with different filters which have one visual and one red or infrared passband, or producing on the same photographic plate two nearby images obtained through different filters. By comparing visually the intensity of both images it is possible to decide rapidly which stars are very red. As an example consider for instance an A0 star. It has $V–I \simeq 0$, whereas for M stars $V–I > 2^{m}0$. The M stars thus have an image brighter by more than two magnitudes in I than in V; in other words, when

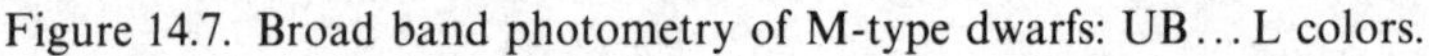

Figure 14.7. Broad band photometry of M-type dwarfs: UB...L colors.

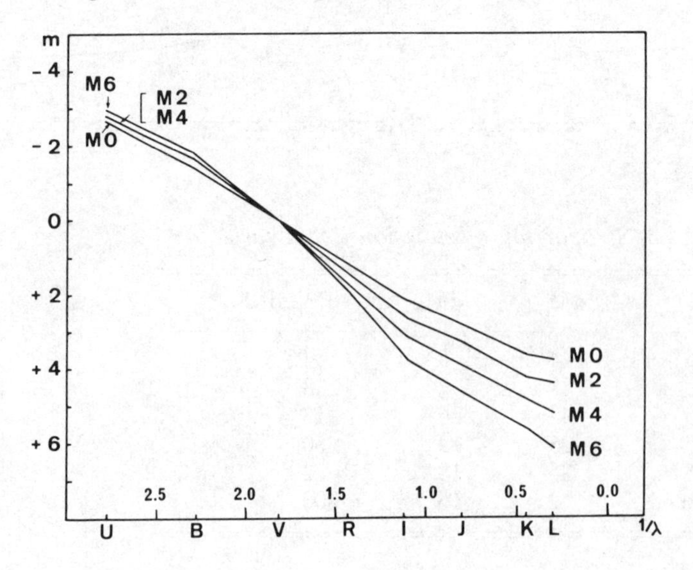

looking at infrared wavelengths, the brightest objects are the infrared ones. The infrared satellite IRAS which photographed the sky at even longer wavelengths (12, 25, 60 and 100 μm) is very useful for a (complete) survey of M-, C- and S-type stars.

Because band filter photometry is little used for subjects other than spectral type, the tendency has been to use narrow band filters placed at strategic points. We have already quoted the Wing system (see section 13.0) which was developed specifically to deal with late type stars and which uses indices centered on TiO (for the spectral types up to M5), VO (for later spectral types) and CN (for the luminosity of stars earlier than M4). Figure 14.8 (Wing 1979) shows the behavior with regard to temperature. The use of the Wing system allows critical examination of the homogeneity and the precision of the different systems used by spectral classifiers; Wing and Yorka (1979) find that the TiO index correlates very well with the Yerkes types, but less well with earlier classifications, as was to be expected.

An important aspect of M-type stars is that most of them are variable. This became obvious with the first photoelectric measurements by Stebbins

Figure 14.8. Eight-color measurements of M-type giants. For explanation see text. The continuous curves represent black bodies with the temperatures indicated above. From Wing (1979).

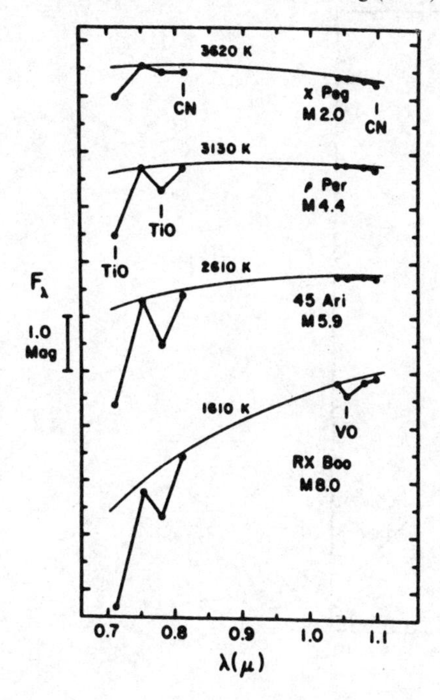

(1930). In general, variability becomes more pronounced toward later types. Long period, semiregular and irregular variables are found among the giants and supergiants, and flare stars and BY Dra variables among the dwarfs. One general fact for the giant variables is that the light amplitude diminishes with increasing wavelength.

Absolute magnitudes. As for K-type stars, the sources of absolute magnitudes are trigonometric parallaxes for dwarfs, cluster and statistical

Table 14.4. *Absolute visual magnitudes of M-type stars.*

	V	III	II	Ia	Ia
M0	+8.8	−0.4	−2.5	−4.5	−7.0
M2	+9.9	−0.6	−2.6	−4.7	−6.9
M4	+11.3	−0.5	−2.6	−4.8	−6.8
M5	+12.3	−0.3		−4.8	−6.8

Figure 14.9. HR diagram of the lower main sequence. All stars are nearby ($r < 20$ pc). On the abscissa the TiO index of Wing's eight-color photometry. From Wing and Dean (1983).

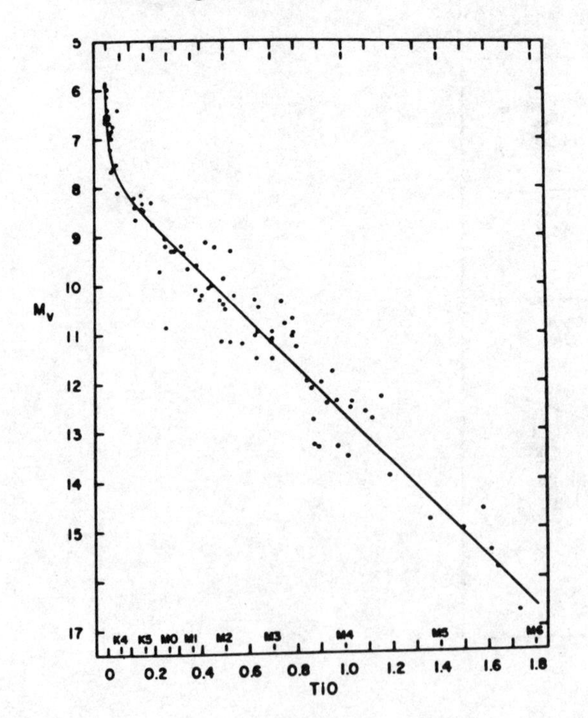

parallaxes for giants and a mixture of various possible procedures for supergiants. Indicative values are given in table 14.4, taken from Schmidt-Kaler (1982).

The tabulated values can be compared with those obtained from nearby dwarfs (Wing and Dean 1983), which are very precise (see figure 14.9). Wing has however not used spectral types as given in the literature, as Schmidt-Kaler did, but spectral types obtained photometrically from the TiO band of his system. This shows up in rather large differences toward later spectral types. But the important fact is that even so a considerable scatter around the average relation does exist, which is of the order of $\pm 0^m5$. This implies that the absolute magnitude of any M-type dwarfs may be uncertain by this amount.

An interesting question is how faint dwarfs can be. Luyten (1968) showed that stars become infrequent after $M_{ph} \sim 19$ and this corresponds to about $M_v \simeq 17$ (Wielen, Jahreiss and Kruger 1983). Thus must be true because only a handful of stars with $M_v > 16$ have been found up to now. The lowest value found corresponds to $M_v = 19 \pm 1$ (Greenstein, Neugebauer and Becklin 1970). Such stars have magnitudes $V \simeq 24$ and are found by proper motion studies of very red stars. It is uncertain as to what spectral type such a luminosity corresponds.

Number and distribution in the galaxy. M-type stars are moderately concentrated toward the galactic plane, with $\beta = 400$ pc for giants and 210 pc for supergiants (Mikami and Ishida 1981). For dwarfs we may guess that $\beta \simeq 2 \times 10^3$ pc.

As for space densities, Mikami and Ishida derive $3 \times 10^{-6}/\text{pc}^3$ for giants and $5 \times 10^{-8}/\text{pc}^3$ for supergiants, which can be compared to $10^{-1}/\text{pc}^3$ for dwarfs (Dawson 1981). The obvious conclusion is that M-type dwarfs are very numerous and M-type giants very rare.

In fact M-type dwarfs constitute 80% of the total number of stars in the solar neighborhood. However, because of the vastly different absolute magnitudes, we find a completely different picture when we examine magnitude limited surveys.

In the bright star catalogue there are 520 M-type stars, i.e. 5% of the total number. All of them are giants or supergiants, because the brightest M-type dwarfs have $V \sim 6$–7.

In the HD, M-type stars constitute 2% of the total number and here again the majority are giants and supergiants, with only a handful of dwarfs.

14.1 S-type and related stars

14.1.1 *S-type stars*

An S-type star is a late type giant (K5–M) showing distinct bands of ZrO in the blue and visual spectral regions.

The class as such was introduced by Merrill (1922) but the result was somewhat heterogeneous because it included a number of objects which, although of late type, were neither normal stars exhibiting TiO bands nor carbon stars; in particular some Ba stars were called 'S'.

The clarification came through the work of Keenan (1954) who accepted as S-type objects only those exhibiting ZrO bands. The spectrum of ZrO is characterized by three band systems, all degrading toward the red from triplet heads. The first (α) system has its strongest head at λ4640, which can be seen in stars only if the overlapping TiO band at λ4626 is weak. In other

Figure 14.10. Spectrum of an S-type star. HD 00 1967 is classified S7e; the comparison is M6 III. For explanation see text.

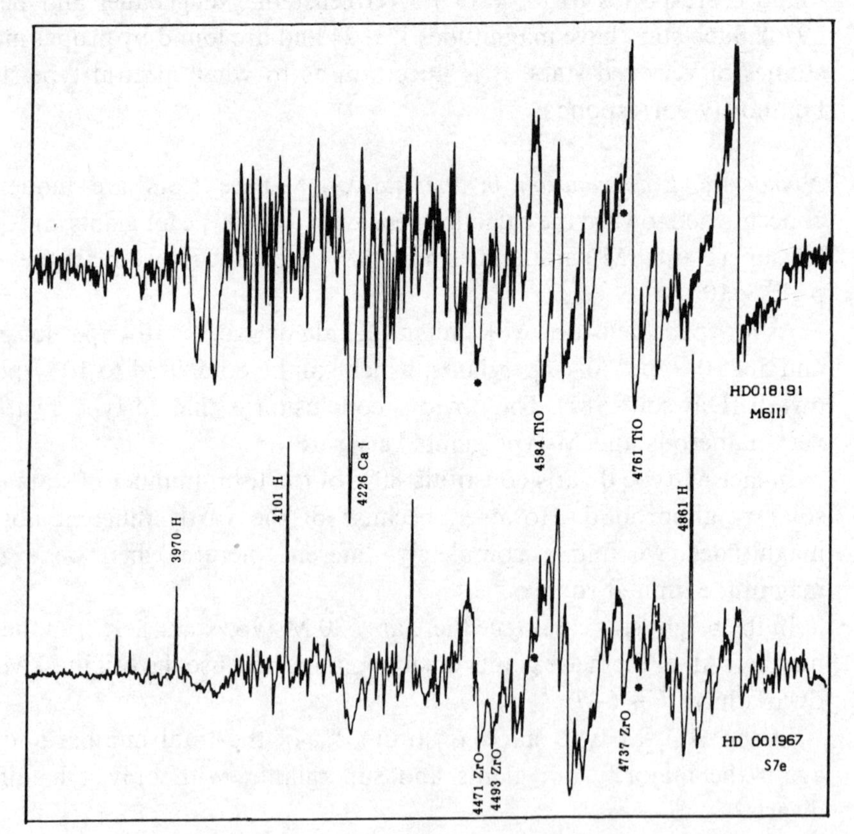

cases the λ4620 band must be used. The β system in the yellow has its strongest head at λ5551. The γ system in the red has a band at λ6474 which is very persistent; however in stars the head of TiO λ6479 may interfere at very low dispersions. These three band systems all arise from the normal electronic level of ZrO and their strongest sequences originate in the lowest vibrational state (see figure 14.10).

In addition to ZrO, the S stars usually also show LaO bands at $\lambda > 7900$ Å. However, not all S stars show them, nor are they always present in a star when the star is a variable. Therefore LaO bands cannot be substituted for ZrO to characterize S-type stars.

Using line intensity estimates of the band intensities of ZrO and TiO on spectra of intermediate and high plate factors (100–200 Å/mm at Hγ), Keenan (1954) defined an S star as an object exhibiting ZrO bands. If these bands are weak or can be seen only at higher dispersion, the star is called MS; a classification like M4 S describes an M4 object with ZrO bands. The S stars themselves were characterized by two indices, one for the temperature and one for the ZrO/TiO ratio.

In later work, Keenan and McNeil (1976) changed to a slightly different procedure using an temperature index plus an estimate of the strength of the ZrO and TiO bands on a scale (0 very weak, 5 very strong).

The main objection against such a procedure is that it uses essentially oxide bands, which are visible only if oxygen is more abundant than carbon. If the converse is true we have a carbon star, and in the classification scheme S and C stars are sharply distinct entities. However, nature produces SC stars, i.e. stars with characteristics intermediate between S and C stars, which a comprehensive classification scheme should also be able to accommodate. This criticism was voiced among others by Ake (1978, 1979), who proposed a

Table 14.5. *Ake's classification scheme of S-type stars.*

Index	Characteristics
1	TiO $\gg$ ZrO and YO
2	TiO > ZrO > 2 × YO
3	2 × YO > ZrO > TiO
4	ZrO > 2 × YO > TiO
5	Zr > 2 × YO, TiO = 0
6	ZrO weak, YO = 0, TiO = 0
7	CS and carbon stars

solution using the strength of the Na I lines. These lines are used in C stars (see section 13.1) so that their use seems natural. Ake also uses two indices to characterize an S-type star, the first being based upon a weighted mean of the intensity estimates of the features of TiO and ZrO, or Na I if TiO is weak or absent. This index is similar to the temperature index of Keenan, but has the advantage that it can also be used for C and SC stars. Ake then adds an abundance index based upon the intensity of ZrO, TiO and YO, which is summarized in table 14.5, taken from Ake (1979).

His 'scale' is probably better than a description, although the word 'index' suggests more than that. Ake thinks that his 'abundance index' is related to the ratio C/O, as suggested by work by Scalo and Ross (1976).

Boeshaar and Keenan (1979) and Keenan and Boeshaar (1980) rediscussed the classification problem. They adopt finally the scheme laid down in table 14.6, taken from the second of the publications quoted.

The temperature index is written before the slanting line, and the C/O index after. Notice that band intensity estimates are no longer necessary, because they are incorporated in the second index.

The scheme has eleven subdivisions instead of Ake's seven and seems rather complex in view of the small number of known S and SC stars. Keenan and Boeshaar (1980) give a list of a hundred objects classified in their scheme, which provides a convenient reference frame.

Despite these improvements, there still remains a problem with the temperature scale. In the classification of carbon stars, when C_2 becomes visible, the star is called C0 and corresponds to a G5 giant. In this scale C5 $\simeq$ M0. In the other groups, MS 0 and S0 were chosen so as to correspond to M0. Obviously when we shift from S to C stars through SC stars there is some artificial shift in the temperature. The temperature types which work well within each isolated sequence are questionable when applied to the total, especially in the case of transition objects.

Emission lines are seen in those S-type stars which are variables (see section 14.0). Most prominent are the Balmer line emissions, usually present in Hα and Hβ but sometimes up to H7; if Balmer lines are seen in emission at some phase, an 'e' is added at the end of the spectral type, as in S6/5e. Feast (1953) has detected other lines in emission, like Si I λ3905 and Fe I λ3852, λ4202 and λ4308.

So far we have referred to classification work done at intermediate plate factors in the 80–100 Å/mm range. However, S stars can be singled out at 250 Å/mm and at even higher plate factors of about 1000 Å/mm in the $\lambda\lambda$5800–6800 region (Nassau and Stephenson 1960) and 3000 Å/mm in the

Table 14.6. *Classification criteria for the sequence MS–D–SC–C (X = temperature type).*

Spectral type	Criteria for C/O	Estimated C/O	Criteria for temperature
MXS	Strongest ZrO bands (λ6473, λ6345, λ5718, λ5551, λ4641) just visible. Also YO λ6132		Same bands and lines as in M-type stars.
SX/1	TiO ≫ ZrO or YO	< 0.95	Same bands and lines as in M-type stars
SX/2	TiO > ZrO	0.95:	Same bands and lines plus total intensity ZrO + TiO
SX/3	TiO = ZrO. YO strong	0.96	TiO + ZrO intensity. Sr I λ4607/Ba II λ4554 from S0 to S5. ZrO(?) λ5305/ZrO λ5551 Infrared LaO becomes strong after S5
SX/4	ZrO > TiO	0.97	TiO + ZrO intensity. Sr I λ4607/Ba II λ4554 from S0 to S5. ZrO(?) λ5305/ZrO λ5551 Infrared LaO becomes strong after S5
SX/5	ZrO ≫ TiO		ZrO intensity. Same ratios as above
SX/6	ZrO strong, No TiO	0.98	ZrO intensity. Same ratios as above
SX/7 = SCX/7	ZrO weaker. D lines strong	0.99	Same ratios as above. λ6456/λ6450
SC X/8	No ZrO or C. D lines very strong	1.00	λ4607/λ4554, λ6456/λ6450
SC X/9	C very weak. D lines very strong	1.02	λ4607/λ4554, λ6456/λ6450
SCX/10 = CX,2	C weak. D lines strong	1.1:	λ4607/λ4554

$\lambda\lambda$6800–8800 region (Nassau, Blanco and Morgan 1954). In the latter case the LaO bands at λ7900 are used (see section 14.0).

Photometry. Eggen (1972) was the first to study the S stars in some detail. He used the UBVRI system, which includes two red filters. Since, however, TiO and ZrO have bands in both of these filters, it seems clear that the UBVRI system will not give a clear separation of M and S stars. This is shown in figure 14.11, in which variable stars are denoted by hatched blocks. It is clear that the majority of the stars are light variables. Out of 107 stars, at least 84, i.e. approximately 80%, are variables. The variables are not a homogeneous group, however, out of 14 stars studied by Eggen (1972), seven are long period variables with very large amplitudes in V and lesser amplitudes in R and R–I ($220^d < P < 420^d$), and nine are small amplitude erratic variables with amplitudes of a few tenths of a magnitude and characteristic time scales less than about 10^2 days.

In view of the difficulties experienced with broad band filters, let us examine a tracing of one M-type and two S-type stars (figure 14.12, from Piccirillo 1976) and see how the filters should be chosen to separate the two types of stars. Piccirillo uses a modification of the Wing system (see section 13.0) with eight colors whose characteristics are given in table 14.7 and whose interpretation can be seen (partially) in figure 14.12. Notice that there are two specific filters placed so as to measure a ZrO index, namely

Figure 14.11. Distribution of (R–I) colors of red peculiar stars. Hatched areas denote variable stars. Ordinate: number of stars. From Eggen (1972).

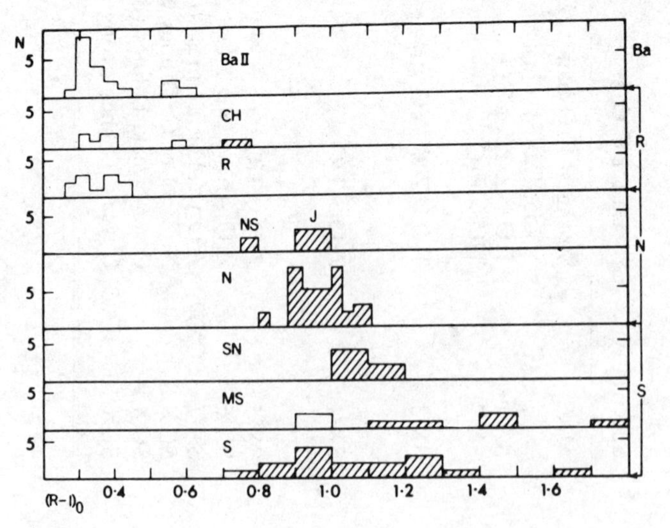

Figure 14.12. Scanner observations of M- and S-type stars in the λλ5000–8000 region. R Cyg and ω Vir are M-type; S UMa is S-type. The positions of various molecular bands are indicated, as well as the place of the filters of the Wing photometric system in this region (arrows). From Piccirillo (1976).

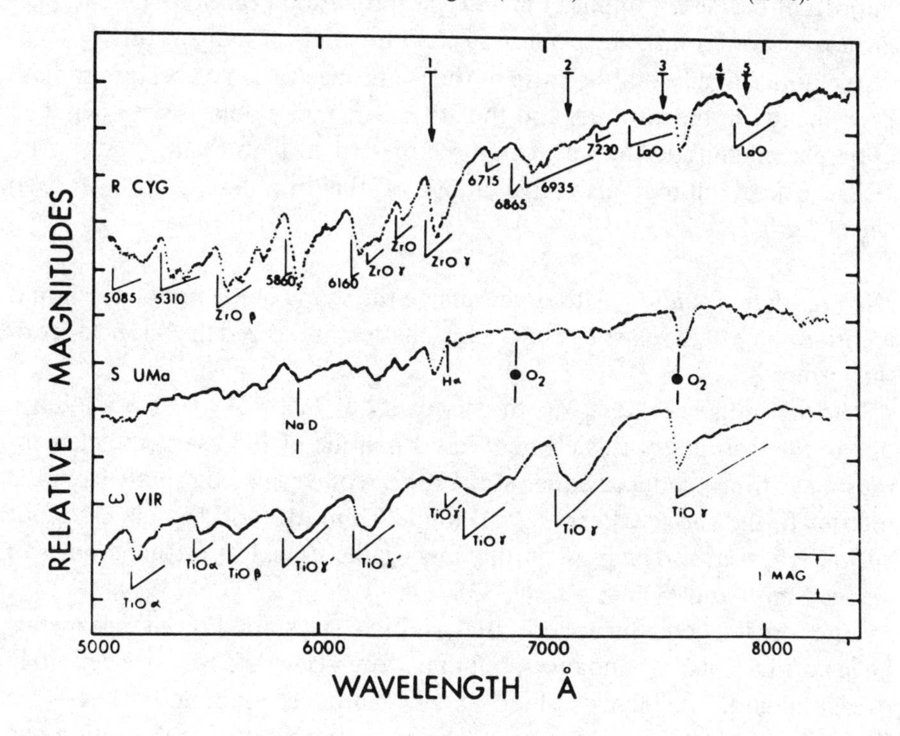

Table 14.7. *Characteristics of the eight filters.*

Filter	λ (Å)	Half band width (Å)	Function	Contaminants
1	6510	40	ZrO	TiO, CN
2	7120	53	TiO	CN, unidentified bands
3	7545	50	Continuum M0–M7, Mild S	Weak CN, VO, LaO
4	7810	42	Continuum G, K, C, Pure S	Weak TiO
5	7945	50	LaO	Weak TiO, VO, CN
6	10 395	55	Continuum I(104)	
7	10 810	74	Continuum for C, S	Weak VO
8	10 975	73	CN	

I(6510)/I(7810), which enables the system to distinguish M and S stars. Piccirillo finds moreover on the basis of his measures that the temperature index of Keenan correlates closely with his ZrO index I(6510)/I(7810). This is important because it implies that S-type stars are on the whole a compact group, without a large dispersion in ZrO for a given temperature.

We notice finally that because of their late spectral types, S stars radiate prominently in the infrared, and thus there are many objects with negative apparent magnitudes at wavelengths larger than 4 μm (Wing and Yorka 1977). These authors also give a list of the brightest object at each wavelength.

High resolution studies. Although outside the scope of the book, we should mention some of the classic results which have emphasized the importance of this group.

The first important fact was the discovery of Tc by Merrill (1952). Since one of the isotopes of this element has a half life of 10^6 years, the element must have been produced either at the surface or circulated rapidly from the interior to the surface. Peery (1971) found Tc in MS and C stars also, but curiously it seems to be present only in variable stars. For further studies of Tc see Smith and Wallerstein (1983).

Later on detailed atmospheric analysis of S-type stars showed that metals have normal (solar) abundances, but that elements with $Z > 38$ are definitely overabundant. Elements which show definitely enhanced lines are Zr, Sr, Y, Ba and La. Such overabundances were attributed to the action of a flux of slow neutrons (the thermonuclear *s*-process) on material of solar composition. Confirmation of this was sought through the presence of isotopes which can only be produced by the *s*-process. The search for such isotopes is feasible because of the molecular bands attributable to the different oxides. This can be done for instance for the Zr isotopes through the search for the bands of ZrO, isotopes 90, 91, 92, 93, 94 and 96 (Peery and Beebe 1970; Zook 1978). Similar studies can be carried out for other elements like Ti, Mg and Cl (Clegg and Lambert 1979) and confirm in general the operation of the *s*-process.

Among the many problems remaining, let us quote the case of R Cam, a star which has a nearly pure S-type spectrum at maximum, whereas increasingly strong TiO bands appear as minimum is approached. This implies that stratification effects probably come into play – or that not all the material has uniform composition.

Absolute magnitude. Before it was recognized that some elements (like Sr) are overabundant in S-type stars, a straight application of luminosity criteria (using line ratios of Sr II and Fe I) resulted in the attribution of supergiant characteristics to these stars. The first to insist on a lower value was Feast (1953) who used the companion of π^1 Gru to show that the absolute magnitude was similar to those of common giants ($M_v \simeq -1^m$). This conclusion was confirmed through later work; Yorka and Wing (1979) obtain a value of $M_v = -1^m$ for non-variables and -2^m for Mira variables at maximum light. Chanturia (1980) obtained $-1^m.8$ and $-3^m.9$ respectively. All these values are rather uncertain and suffer from the fact that V magnitudes are used, which are strongly affected by molecular bands. S-type stars in the globular clusters of the Magellanic Cloud place the M_v of these stars between the absolute magnitude of M giants and that of C stars (Lloyd Evans 1985).

Distribution and population characteristics. Yorka and Wing (1979) analysed the distribution on the sky and concluded that S-type stars are not associated with spiral arm objects. Their concentration toward the galactic plane is not very strong either: 64% of the stars have $|b| \leqslant 10°$, in contrast to supergiants, where this proportion is 96%. A very curious fact is that S-type stars are sparse in the direction to the galactic center; their distribution differs from that of the M-type giants.

The available information suggests that S stars are older population I objects, comparable to the G–M giants. There is, however, some evidence that S stars can also belong to either the very young or the very old (halo) population; this parallels the behavior of C stars and M giants.

Frequency. In general, S stars are rare. They are less abundant than C stars, even, in the ratio of about five to one, if MC and SC stars are excluded (Wing and Yorka 1977). This ratio diminishes if objects in infrared sky surveys are considered; here a ratio of three to one is obtained. Since C stars are already rare when compared with M giants (we quoted 1% earlier), it is easy to see that S stars constitute at most of the order of 0.5% of normal stars. The intermediate groups SC and SM are very infrequent, compared with S-type stars. For field stars, they constitute about 5% of the S-type stars.

Atlases. A valuable atlas of low dispersion spectra was published by Stephenson and Ross (1970). They used 280 and 580 Å/mm at Hγ.

Wyckoff and Clegg (1978) have provided an atlas of R Cyg at 42 Å/mm in the region $\lambda\lambda$4300–8200.

14.1.2 *SC and CS stars*

Stars of these types share the characteristics of both S and C stars – that is, they exhibit clear but not strong bands of ZrO and C_2. In these stars the O/C ratio is probably very close to unity, with most O and C locked up in CO.

Besides these characteristics, the lines of heavy metals (La, Y, Ba and Sr) are enhanced and the bands of CN are prominent. Another important feature is the very strong Na I D lines, for which the equivalent widths range from 13 to 56 Å (Feast 1953).

These stars have also been called semicarbon stars or D-line stars ('D-line' is the Na I doublet at $\lambda\lambda$5889–95 seen at low resolution) (Gordon 1967). Sometimes the notations SC and CS are used in the sense that an SC star is closer to S and a CS star closer to C-type. In such cases the standard stars are R CMi for CS and UY Cen for SC. Stephenson (1973) warns against the use of such distinctions for objective prism surveys, because they are rather uncertain. Keenan and Boeshaar (1980) also avoid the explicit use; they call the stars SC.

Li λ6708 is very strong in SC stars, stronger than in both S- or C-type stars. Its equivalent width may reach up to 10 Å.

14.2 **Flare and related stars**

14.2.1 *Flare stars*

A flare star is an object that undergoes sudden marked changes in brightness, giving rise to a nova-like light curve (amplitude 1^m–6^m) with a time scale of minutes. This definition is purely photometric and we shall only mention some spectroscopic aspects of the group. Flare stars are also called UV Ceti stars, after a prominent member of the group. Those UV Ceti stars associated with nebulosity are denoted UVn.

During the quiescent phase, flare stars exhibit a late spectral type, typically dM, and present emissions in the Ca II and the Balmer lines (see figure 14.13, from Bopp and Moffett 1973).

During the flare-up phase the spectrum changes markedly (Joy and Humason 1949). In the first ('spike') phase of the flare, the spectrum is dominated by a strong continuum radiation with little contribution from emission lines; this continuum is especially strong in the ultraviolet and blue

Figure 14.13. Intensity tracings of the spectrum of UV Cet at different times. Flare 2, strong flare; flare 1, weak flare; quiescent phase at bottom. From Bopp and Moffett (1973).

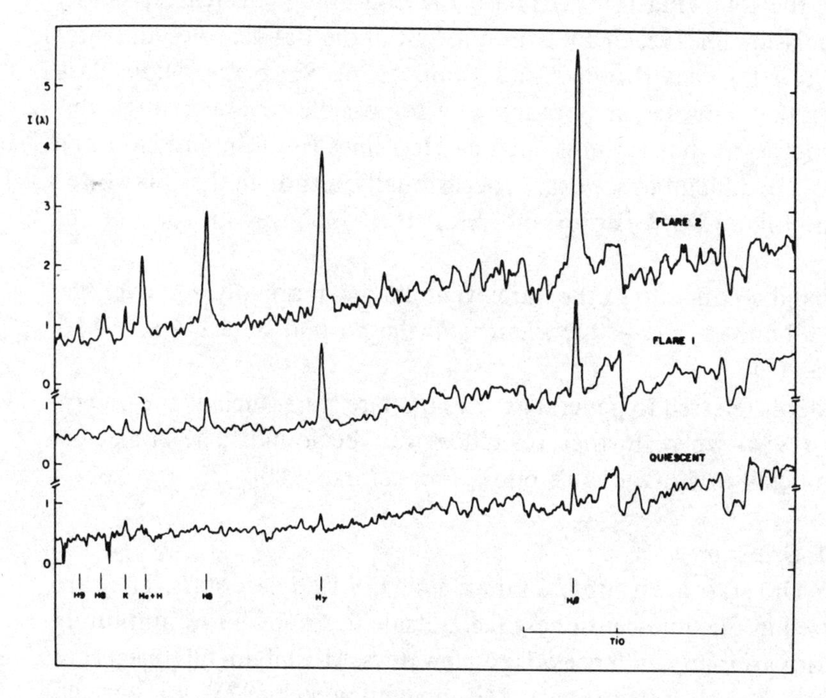

Figure 14.14. Light curve of flare 2 (figure 14.13) of UV Cet. Intensity in U passband, integration time one second. Vertical arrows indicate the times at which the different spectral features were noted. From Bopp and Moffett (1973).

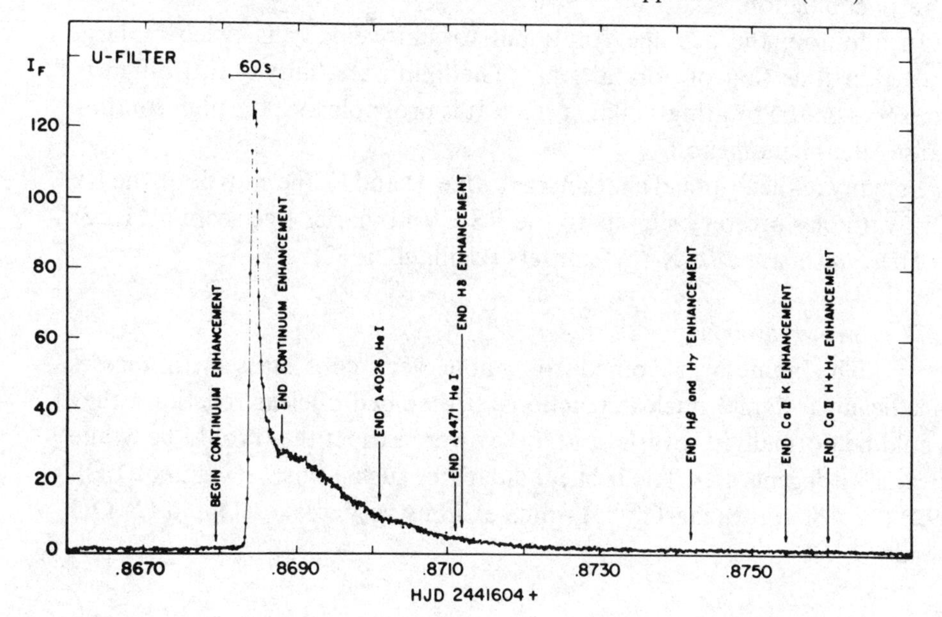

region. In the following ('slow') phase the continuum decreases rapidly whereas emission lines reach a maximum, first in the Balmer lines and later the Ca II (K + H) lines (Moffett and Bopp 1976). Since the whole flare lasts only a few minutes, it is practically impossible to obtain high dispersion spectra, so that features such as He I lines (both in emission and absorption) are difficult to see except occasionally, although they may often be present. Figure 14.14 (Bopp and Moffett 1973) gives an idea of the process.

Broad band colors during the flare event change markedly. On average, $U-B \simeq -1.3$ and $B-V \simeq -0.1$, whereas in the normal state $U-B \simeq +1.1$ and $B-V \simeq +1.6$.

The reader is referred to general articles on flare stars, such as the one by Mirzoyan (1984) where further references can be found. The subject is covered in detail in Gurzadyan's book, *Flare stars* (1980).

14.2.2 *BY Dra stars*

BY Dra stars are a subset of the classic UV Ceti flare stars. They are characterized in the quiescent phase (i.e. outside flares) by a low amplitude photometric variability with periods of a few days. Most if not all objects are double lined spectroscopic binaries (Bopp and Fekel 1977). In thirteen systems these authors found eight SBII, one SBI and four single stars. Thus binarity is apparently a sufficient but not a necessary condition for the BY Dra phenomenon.

In addition, the average rotational to star velocity is twice as large ($\simeq 10$ km/s) as that of normal stars. The light variability is attributed to starspots on the rotating stellar surface. It is probable that the high rotation causes the phenomenon.

In many of their optical characteristics (i.e. H and K line emission), the BY Dra variables are very similar to the RS CVn binaries (see section 11.2.2).

These stars are also X-ray emitters (Caillault 1982).

14.3 **Brown dwarfs**

The name was coined to denote very cool stars with masses insufficient to ignite nuclear reactions. If they had nuclear reactions, they would be normal red dwarfs, and if they were extinct they would be 'white dwarfs' or degenerates. The best candidate for such a type seems to be LHS 2924 (Probst and Liebert 1983) which exhibits very weak TiO and Ca OH bands.

References

Ake T.B. (1978) *IAU Symp.* **80**, 409, D. Reidel Publ. Co.
Ake T.B. (1979) *Ap. J.* **234**, 538
Baumert J.H. (1970) *Contr. Kitt. Peak* **554**, 155
Beichman C.A., Neugebauer G., Habing H.J., Clegg P.E. and Chester T.J. (1985) 'Explanatory supplement', Joint IRAS Science Working Group, Jet Propulsion Laboratory D–1855
Boeshaar P.C. and Keenan P.C. (1979) *IAU Coll.* **47**, 39 = *Spec. Vaticana* **9**, 39
Bopp B.W. and Fekel F. (1977) *A.J.* **82**, 490
Bopp B.W. and Moffett T.J. (1973) *Ap. J.* **185**, 239
Caillault J.P. (1982) *A.J.* **87**, 558
Chanturia S.M. (1980) *Abastumani Bull.* **53**, 95
Clegg R.E.S. and Lambert D.L. (1979) *Ap. J.* **234**, 188
Dawson P.C. (1981) *A.J.* **86**, 1200
Eggen O.J. (1972) *Ap. J.* **177**, 489
Feast M.W. (1953) *Inst. Liege Mem.* **357**, 413
Gordon R.C.P. (1967) Thesis, Univ. Michigan
Greenstein J.L., Neugebauer G. and Becklin E.E. (1970) *Ap. J.* **161**, 519
Gurzadyan G.A. (1980) *Flare stars*, Pergamon Press
Hyland A.R. (1974) *Highl. of Astron.* **3**, 307, D. Reidel Publ. Co.
Joy A.H. and Abt H.A. (1974) *Ap. J. Suppl.* **28**, 1
Joy A.H. and Humason M.L. (1949) *PASP* **61**, 133
Keenan P.C. (1954) *Ap. J.* **120**, 484
Keenan P.C. and Boeshaar P.C. (1980) *Ap. J. Suppl.* **43**. 379
Keenan P.C. and Hynek J.A. (1945) *Ap. J.* **101**, 265
Keenan P.C. and McNeil R.C. (1976) *An atlas of spectra of the cooler stars*, Ohio State Univ. Press.
Keenan P.C. and Yorka S. (1985) *BICDS* **29**, 25
Kuiper G.P. (1942) *Ap. J.* **95**, 201
Lloyd Evans T. (1985) *Strasb. Coll., 'Cool stars'*, p. 163
Luyten W.J. (1968) *MNRAS* **139**, 221
Merrill P.W. (1922) *Ap. J.* **56**, 457
Merrill P. (1952) *Science* **115**, 484
Merrill P. and Stein (1976) *PASP* **88**, 294
Mikami T. and Ishida K. (1981) *PAS Japan* **33**, 135
Mirzoyan L.V. (1984) *Vistas in Astronomy* **27**, 77
Moffett T.J. and Bopp B.W. (1976) *Ap. J. Suppl.* **31**, 61
Nassau J.J. (1956) *Vistas in Astronomy* **2**, 1361
Nassau J.J. and Stephenson C.B. (1960) *Ap. J.* **132**, 130
Nassau J.J., Blanco V.M. and Morgan W.W. (1954) *Ap. J.* **120**, 478
Peery B.F. (1971) *Ap. J.* **163**, L1
Peery B.F. and Beebe R.F. (1970) *Ap. J.* **160**, 619
Piccirillo J. (1976) *PASP* **88**, 680
Probst R.G. and Liebert J. (1983) *Ap. J.* **274**, 245

Rinsland C.P. and Wing R.F. (1982) *Ap. J.* **262**, 201
Scalo J.M. and Ross J.E. (1976) *AA* **48**, 219
Schmidt-Kaler T.H. (1982) Landölt-Bornstein, group VI, vol. 2b, p. 1
Seitter W.C. (1975) *Atlas für Objektiv-Prismen Spektren*, vol. II, F. Dummler, Bonn
Smith V.V. and Wallerstein G. (1983) *Ap. J.* **273**, 742
Spinard H. and Wood D.B. (1965) *Ap. J.* **141**, 109
Stebbins J. (1930) *Publ. Washburn Obs.* **15**, 139
Stephenson C.B. (1973) *Ap. J.* **186**, 589
Stephenson C.B. and Ross H.E. (1970) *A.J.* **75**, 321
Vyssotsky A.N. (1943) *A.J.* **47**, 381
Vyssotsky A.N. (1956) *Ap. J.* **61**, 201
Vyssotsky A.N. and Mateer B.A. (1952) *Ap. J.* **116**, 117
Vyssotsky A.N., Janssen E.M., Miller W.J. and Walther M.E. (1946) *Ap. J.* **104**, 234
Wielen R., Jahreiss H. and Kruger R. (1983) in *IAU Coll.* **76**, 163
Wing R.F. (1973) in *IAU Symp.* **50**, 209
Wing R.F. (1979) *IAU Coll.* **47**, 347 = *Spec. Vaticana* **9**, 347
Wing R.F. and Dean C.A. (1983) *IAU Coll.* **76**, 385
Wing R.F. and Yorka S. (1977) *MNRAS* **178**, 383
Wing R.F. and Yorka S.B. (1979) *IAU Coll.* **47** = *Spec. Vaticana* **9**, 519
Wyckoff S. and Clegg R.E.S. (1978) *MNRAS* **184**, 127
Yorka S. and Wing R. (1979) *A.J.* **84**, 1010
Zook A.C. (1978) *Ap. J.* **221**, L113

15

Degenerates

15.0 White dwarfs

Objects lying five or more magnitudes below the main sequence are called 'white dwarfs'.

The name was attributed to them because the spectrum of the first star analysed, 40 Eri B (Adams 1914), was of type A and the star white, whereas dwarfs were thought to be all of type M and therefore red. So the term calls attention to the inconsistency between the color or spectral type and the luminosity (or absolute magnitude) of the star. Since, however, the stars are neither white nor dwarfs, it would be more logical to call them 'degenerates' (D) because of the equation of state of the matter composing them. We shall use the latter term.

Detection methods. Degenerates, being intrinsically faint, are difficult to single out among the large number of faint stars. (Let us remember that there are about 2×10^7 stars brighter than $B = 15$.) Detection methods are thus of fundamental importance. Essentially three methods have been used to isolate candidates. The first is to search for nearby stars through their proper motion, and retain those which are blue. The second is to search spectroscopically for blue stars where no intrinsically bright stars are expected. The third is to search for faint companions to brighter nearby stars.

The first method was proposed and exploited by Luyten, from 1931 on. From an examination of about 10^8 star images on plates taken at different epochs, Luyten selected those stars whose proper motion is larger than a given amount (for instance 1″). By estimating the color from the difference in magnitude between two images of the star taken through different filters on the same plate (see section 14.0), he eliminated the late type dwarfs. Luyten produced lists of about 3000 white dwarf candidates for further examination. (See for instance Luyten 1971.) Similar work was carried out by Giclas and associates (Giclas 1971). The second technique was used by Humason and Zwicky (1947) (see also section 10.4); this has proven less successful if only degenerates are looked for, because faint blue objects are discovered, not simply faint objects, as in the first method.

Table 15.1, taken from Luyten (1971), gives an idea of the relative frequency of different types of objects to be found at different magnitudes in a region of the sky.

The table shows clearly that degenerates start coming in after 13^{m}. This explains why the stars are studied mainly by photometric rather than spectroscopic methods – unless very large telescopes are used.

The Palomar-Green survey, which covers 10 000 square degrees north of $|b| = 30°$ and down to $B = 16^{m}_{\cdot}2$, is a recent application of this technique. Three thousand stars were selected on the basis of their blue color, and among them about 700 hot degenerates could be detected.

The third method consists in observing faint companions to nearby stars, like Sirius. Nearby stars are selected either by their parallax or by their large proper motion. We recall incidentally that

$$T = 4.74 \mu r$$

where T = transverse velocity (in km/s), μ = proper motion (in ″/year), r = distance (pc).

The third method is of course restricted to detecting degenerates in binary systems, but let us recall that they are numerous even in the nearest solar neighborhood.

These techniques may be biased towards the discovery of D candidates among those having high velocity, i.e. population II objects. Stephenson (1971) has advocated the use of objective prism surveys with a dispersion of about 600 Å/mm. He thinks that up to $m = 14$ there may well remain more

Table 15.1. *Objects found in surveys.*

Magnitude	QS	MS	HB	SD	D	μ
13	—	15	25	60	—	0″6
15	5	5	10	68	12	0″2
17	5		5	65	25	0″06
19	20			50	30	0″02
21	50				50	

Notes: QS, quasars; MS, main sequence; HB, horizontal branch; SD, subdwarfs; D, degenerates; μ, expected proper motion for a D (in ″/year)

undiscovered degenerate stars. In support of this he notices that the third brightest degenerate known in 1971 was discovered as late as 1968 from an objective prism plate; its proper motion is less than 0″1 per year and would have gone undetected in any proper motion survey.

Spectral classification. Spectral classification of degenerates was started by Kuiper (1940) on the basis of the 40 spectra known as that time. The scheme used today was proposed by Luyten (1952), based upon 340 Å/mm spectra. Greenstein (1958) distinguishes the types given in table 15.2. To these classifications the suffixes p = peculiar, wk = weak lined, e = emission and s = sharp are sometimes added.

Greenstein (1960) provides an atlas of different degenerates, from 180 Å/mm plates. Notice that the plate factor is very important since at lower plate factors lines become so shallow as to escape detection.

The most common of these types is DA; almost two-thirds of all degenerates belong to this group. (DO + DB) follow with some 10%; the DC's have a similar frequency. The latter are, however, deceptive because many of them, when examined at better resolution, very often reveal broad lines of other elements, thereby disappearing from the group. Late type objects, including those with molecular bands, are rare (5–10%). The basic fact is that degenerates can be broadly divided into those which show hydrogen lines and those which do not. This dichotomy is rather sharp, since very few stars show hydrogen lines together with some other feature. Out of 600 degenerates, Greenstein (1979) found only three in which helium and

Table 15.2. *Classification of degenerates.*

Type	Description	Example
DC	Continuous spectrum; no lines deeper than 10% of the continuum	W 1516
DB	He I strong; no H lines	L 1573–31
DO	He II lines strong; He I and/or H lines present	HZ 21
DA	H present; no He I lines	40 Eri B
DA, F	H lines sharper and weaker; Ca II line weak	R 627
DF	Ca II lines; no H lines	L 745–46 A
DG	Ca II, Fe I lines but no H lines	Van Maanen Z
λ4135	Broad unidentified line bands	AC + 70°8247
λ4670	Broad bands at $\lambda\lambda$4670, 5165 of C_2; these stars are also called 'C_2 stars'	L 879–14

hydrogen lines coexist, and some in which both H I and Ca II lines appear.

From the range of colors covered by degenerates, we would expect DA's to cover roughly spectral types O to G, after which Balmer lines should no longer be observable. We expect therefore, in analogy to normal stars, to see lines of He, C, N, O, of the lighter elements Mg and Si and of metals like Ti, Cr and Fe. But none of this is found apart from the exceptions mentioned above.

The Greenstein scheme has been enlarged with two additional groups, namely

Figure 15.1. Spectrophotometry of degenerates. Observe that the abscissa is inverse wavelengths. Spectral types are indicated on the right side. Intensity scale is as indicated on the left side. From Oke (1974).

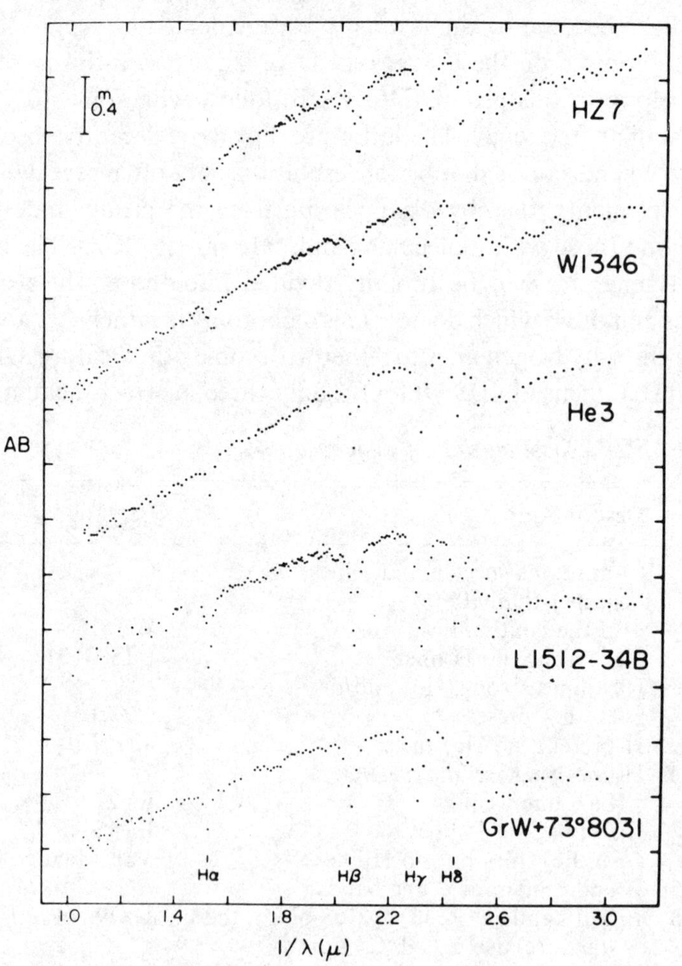

DBA for stars showing He and a trace of H

DAO for stars showing H and weak He II λ4686

This was done in analogy to DAF, where the second letter designates the stronger features and the third the weaker ones. Spectrophotometric tracings of several degenerates are illustrated in figures 15.1, 15.2 and 15.3, taken from Oke (1974).

Recently Sion *et al.* (1983) have proposed a new scheme to classify not only the spectra of the stars but also the degenerate stars themselves. The scheme proposed consists of: (1) an uppercase D for degenerate; (2) an uppercase letter for the primary or dominant spectral type in the optical spectrum; (3) an uppercase letter for secondary spectroscopic features, if present in any part of the electromagnetic spectrum; (4) a temperature index from 0 to 9, defined by $10 \times \Theta_{\text{eff}}$ ($= 50\,400/T$). Additional symbols are provided for

Figure 15.2. Spectrophotometry of degenerates. Observe that the abscissa is inverse wavelengths. Spectral types are indicated on the right side. Intensity scale is as indicated on the left side. From Oke (1974).

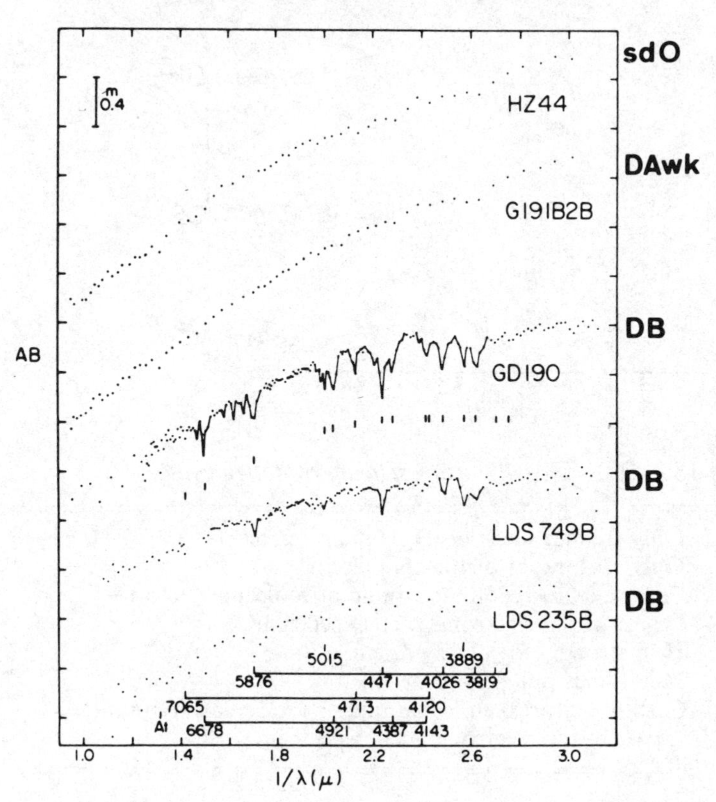

polarized magnetic stars and for peculiar or unclassifiable spectra. It is clear that only (1) and (2) and partially (3) are spectral classifications; if the characteristics observed 'in any part of the electromagnetic spectrum' fall outside the spectroscopic range, they can no longer be used by classifiers. Finally (4) is clearly a theoretical result, based upon photometry. With this in mind let us now consider the spectroscopic part of classification scheme. Sion *et al.* use the symbols given in table 15.3.

Figure 15.3. Spectrophotometry of degenerates. Observe that the abscissa is inverse wavelengths. Spectral types are indicated on the right side. Intensity scale is as indicated on the left side. From Oke (1974).

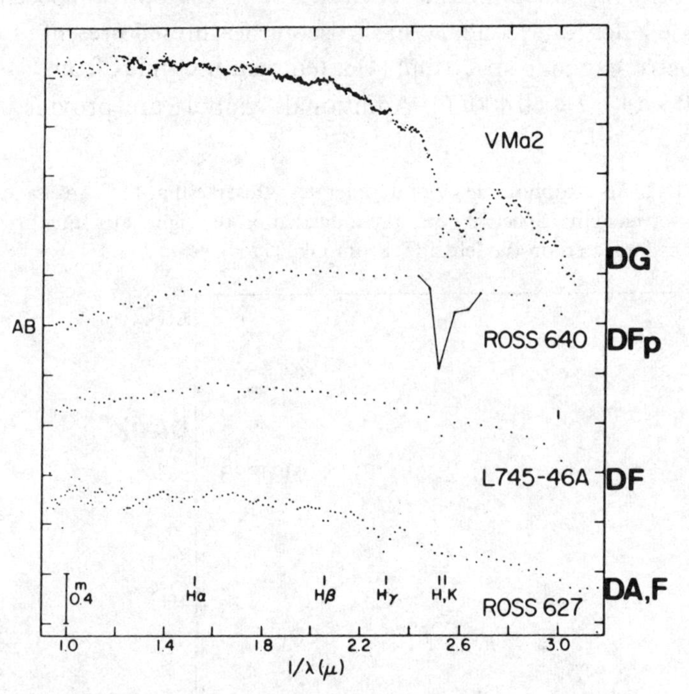

Table 15.3. *New classification scheme of degenerates.*

DA	Only Balmer lines; no H or metals present
DB	Only He I; no H or metals present
DC	Carbon features, either atomic or molecular, in any part of the electromagnetic spectrum
DO	He II strong, He I or H present
DZ	Metal lines only; no H or He
DQ	Carbon features, either atomic or molecular, in any part of the electromagnetic spectrum

In table 15.3, DC, DB, DA and DO essentially retain the meanings of table 15.2. DZ replaces DF, DG and DK (used by the same authors), which puts different objects in the same group, but the total number of DZ objects is rather small (10% of all degenerates). DQ is more difficult since it replaces the C_2 and λ4670 star groups, but also includes objects showing C_2 bands in *any* part of the spectrum.

As mentioned above, the third letter alludes to secondary characteristics if present in any part of the electromagnetic spectrum. The authors of the scheme write: 'a DB star showing Ca II would be classified DBZ, a DB star with weak Balmer lines DBA, a star with Balmer lines dominant but with weak Ca II would become DAZ, a DO star whose ultraviolet spectrum shows N V (λ1240) would become DOZ and so on. With the exception of the DQ stars, non-optical spectra will only be considered in adopting the secondary symbol, not the primary.'

They then propose a temperature index for the fourth place based upon photometric colors and spectrophotometric colors, which supposes that the effective temperature scale for degenerate stars is now the definitive one. (Past experience has shown this to be a very weak assumption.) They proceed to add a few more letters to characterize secondary spectroscopic features: P for polarized magnetic stars, H for magnetic stars showing no detectable polarization, X for peculiar or unclassifiable spectra, and an optional V to denote ZZ Ceti stars or any other variable degenerate star. Table 15.4, taken from their publication, provides examples of the new classification.

As the authors remark, such a scheme represents a departure from the classical approach to spectral classification which is based upon visual inspection of a spectrum. Part of this difficulty disappears if we observe that this is a scheme for classifying stars, not spectra. With the exception of class Q, the first and second letters give the spectroscopic classification, and this is where the pure classification ends.

Spectral lines. We have already remarked that spectral lines are very wide and shallow. Hydrogen line wings often extend 100 and even 150 Å from the line center, but central line depths are less than 0.5 of the continuum intensity for even the strongest lines. These facts render difficult both the tracing of the continuum and the measurement of line profiles. So for instance H_7 and H_8, which are separated by 130 Å, overlap and leave no 'free' continuum in between. Obviously the equivalent widths of both lines are difficult to determine, and the last measurable Balmer line becomes H_5. For lines from

Table 15.4. *Examples of the new classification*

Description	Effective temperature	Example	Old classification	New classification
Only H I	30 000	EG 157	DA	DA1
Only He I	15 000	L 1573–31	DB	DB3
Continuous spectrum	8000	EG 1	DC	DC6
DB with Ca II	14 000	GD 40	DBP	DBZ4
Polarized D with H and He, but He dominant	20 000	Feige 7	DBAP	DBAP3
DA, F	7400	G 74–7	DAF	DAZ7
Peculiar metallic line D with H	8500	Ross 640	DFPec	DZA7
DA with weak He II	50 000	HZ 34	DA	DZO1
Pec D, unidentified composition, no polarization	25 000	GD 229	DXP	DXH3
Cool D, O_2 in the optical and C I in the ultraviolet	8500	L 879–14	C_2	DQ6
DO with N V (λ1240)	70 000		DO	DOZ1

elements other than hydrogen the central line depths are so small that even line recognition poses problems. Table 15.5 furnishes as an example data for the Hγ line, taken from Eggen and Greenstein (1965). These data illustrate very clearly the enormous line widths of degenerates, when compared with dwarfs.

Photometry. As usual we start with the UBV system, in which the largest numbers of degenerates have been observed. Figure 15.4 shows the (U–B, B–V) plot, with different symbols for DA's and non-DA's. Observe that on the whole the colors fall around the black body curve, although in the middle (B–V $\simeq$ O) the DA stars deviate from it.

At both extremes degenerates cannot be separated from other objects by their UBV colors alone: at the bluest colors the degenerates approach the main sequence and at the lower end they join the subdwarfs. We might expect that the deviation in the middle is related to the Balmer line strengths. To examine this we introduce a '(U–B) deficiency', defined as 'observed (U–B)' minus 'black body (U–B)' at a given (B–V); this (U–B) deficiency is then plotted against the Hγ equivalent widths, Figure 15.5, taken from Eggen and Greenstein (1965), shows that the '(U–B) def' increases with W(Hγ), as expected. In the visual classification the descriptive symbols 's' for sharp lines, 'n' for wide lines and 'wk' for weak lines are used; figure 15.5 illustrates that some meaning can be attached statistically to these terms, since DAs lie systematically on top of 'n' and of 'wk' lined stars. However, it must be emphasized that this has only a relative meaning, since even in DA stars the lines are never really sharp – they are just narrower than in most degene-

Table 15.5. *Hγ line data.*

	Degenerates		Dwarfs	
U–V	W	$w(0.5)$	W	$w(0.5)$
– 1.4	8	22	4	4
– 1.2	18	32	5	9
– 1.0	25	40	6	12
– 0.8	30	46	7	14
– 0.6	38	54	8	16
– 0.4	36	42	9	18
– 0.2	8	10	10	

Notes: W, equivalent width (in Å); $w(0.5)$, line width at half central intensity (in Å).

Figure 15.4. UBV photometry of degenerates. For explanation see text.

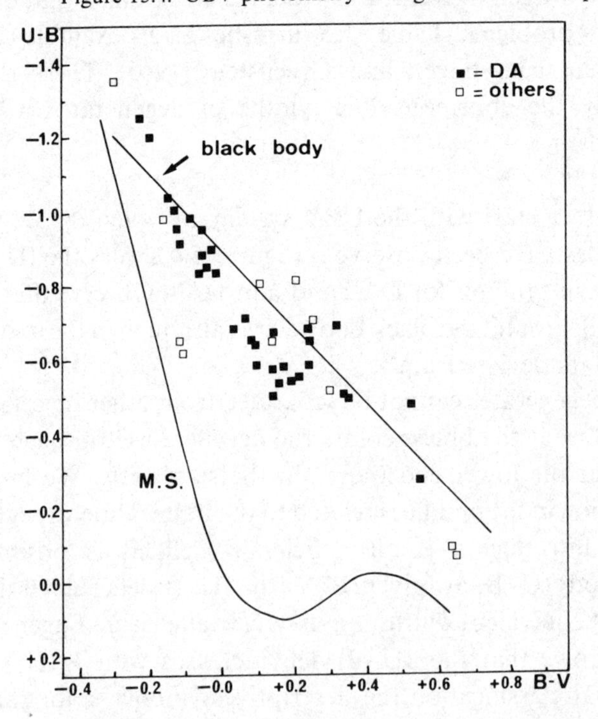

Figure 15.5. U–B deficiency versus Hγ equivalent width. For explanation see text. Crosses represent continuous spectra. The three lines represent averages for sharp (s), wide (n) and weak (wk) lined DA's. From Eggen and Greenstein (1965).

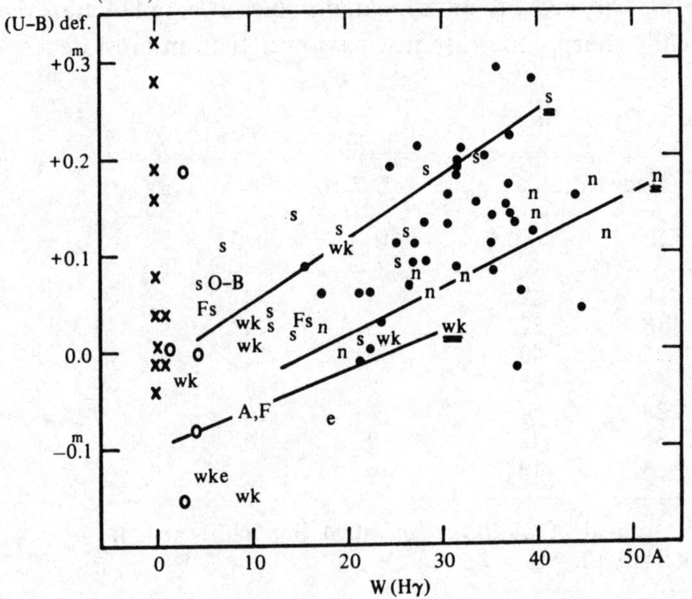

rates. Also the same number of Balmer lines is observed in both DAs and DAn stars – about $n = 8$.

Returning to figure 15.4, it can be seen that there is no separation between DA stars and the other type of objects. A more detailed examination reveals that DO and DB occupy the upper extreme with B–V < – 0.06, whereas DC stars appear around B–V ~ 0. DZ stars appear later on. Among the DA's, the 'n' occupy the bluer region, whereas the 's' occur after B–V ⩾ 0.15. All this is again only true in a statistical sense.

We consider Strömgren photometry next. In this photometry filter b is centered on Hβ, v on Hγ and u after the Balmer discontinuity. Thus if hydrogen lines are present, they should be detectable. This was done by Graham (1972) and we reproduce his figure for DA stars (see figure 15.6). Clearly, the u–b, b–y diagram reproduces the UBV diagram, with a maximum deviation from black body at b–y ~ 0. This deviation should be related to the intensity of Hγ, and this is shown in figure 15.7. On the other hand, stars not showing Balmer lines should not show much deviation from black body, which is illustrated in figure 15.8 taken from Wegner (1979).

Figure 15.6. Strömgren photometry of DA stars. The cross in the upper right represents the average standard error of a star. From Graham (1972).

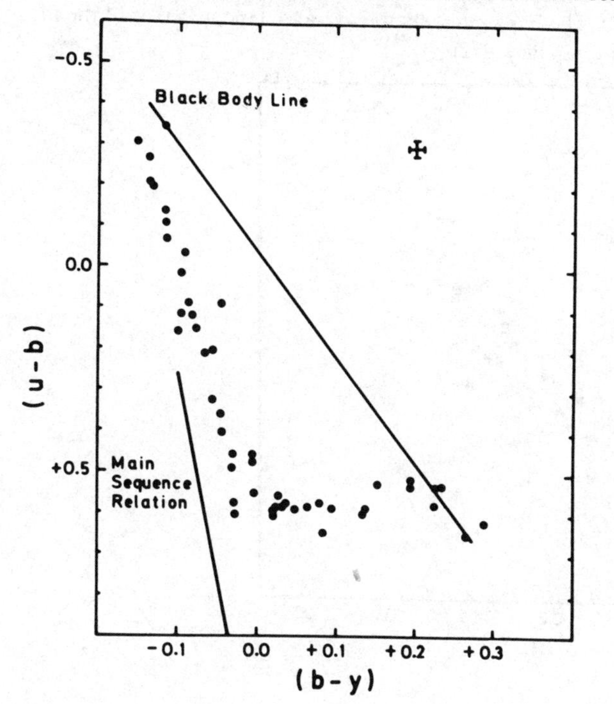

In summary, some, but not all, objects can be recognized with both kinds of photometry. This has led Greenstein in his later papers to prefer a system based on direct scans of the spectra. We have already reproduced some of these tracings in figures 15.1, 15.2 and 15.3, taken from Oke (1974). Greenstein (1984) uses spectrophotometric scans with a resolution of 40 Å below 5700 Å and 80 Å above 5700. He derives from the scans magnitudes at different wavelengths. His wavelengths are given in table 15.6.

These colors can be combined to separate the different types of objects. This can be achieved by an index characterizing the Balmer depression; Greenstein uses [(U–G) + (U–V)]. This is plotted against a temperature index [(G–R) + (V–I)]. The result is given in figure 15.9, which permits the separation of DA's and others, when the Balmer lines are intense. Other stars, except composite systems, fall mostly on or near the same sequence. This is in agreement with what we found in UBV and uvby photometry. Greenstein finds that the cooler non-DA stars show a flux deficiency in band B, the origin of which is unexplained.

Radial velocities. Because of the broad lines, radial velocities are difficult to measure except at high plate factors, and in these cases naturally the

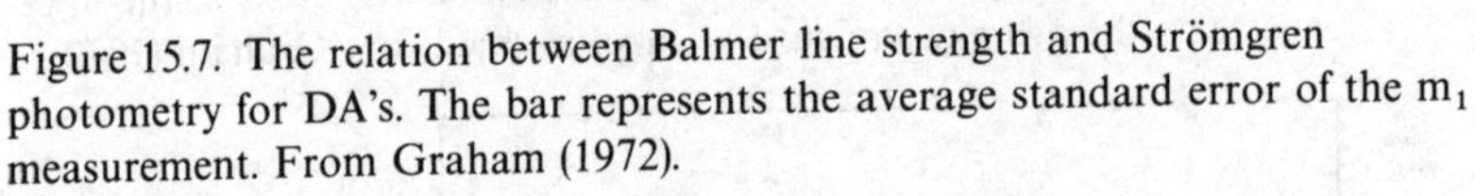

Figure 15.7. The relation between Balmer line strength and Strömgren photometry for DA's. The bar represents the average standard error of the m_1 measurement. From Graham (1972).

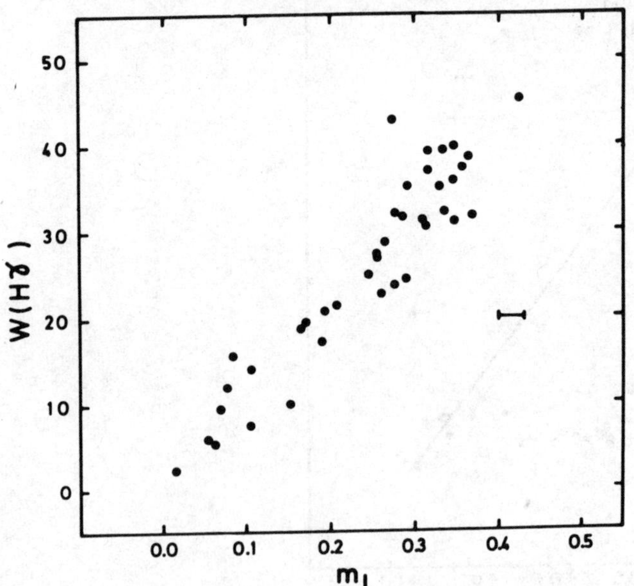

Table 15.6. *Monochromatic magnitudes of the Greenstein system.*

$1/\lambda$	Name	λ	Color	Remarks
1.25	I	8000		
1.44	R	6900		
1.85	V	5400		
2.12	G	4700	G–R	Affected by C_2 in λ4670 stars
2.35	B	4250	B–V	Affected by Hγ in DA stars
2.80	U	3550	U–B	Balmer jump

Figure 15.8. Strömgren photometry for non-DA stars. The cross shows the average external error of measurements. The straight line represents a black body. From Wegner (1979).

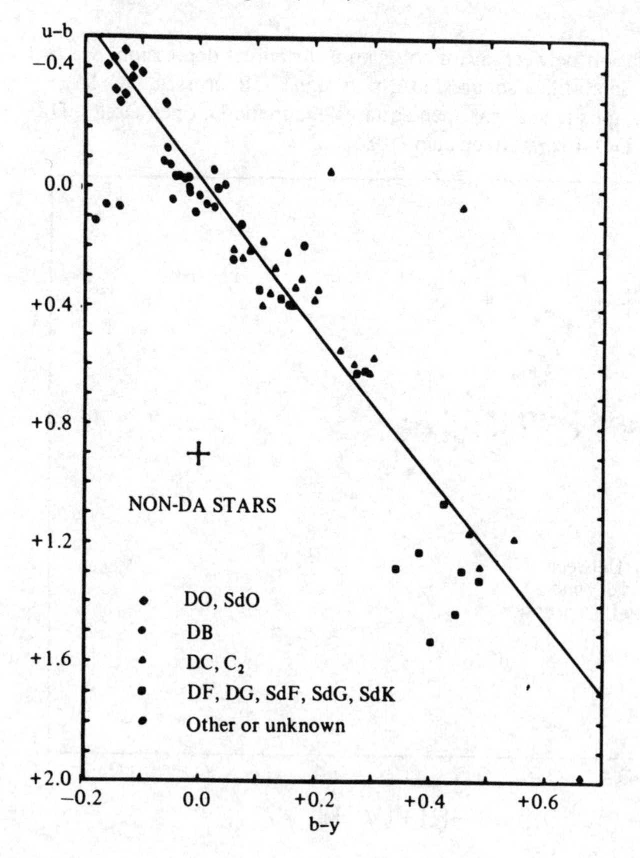

uncertainties are also large. A somewhat surprising fact is that when DA stars are observed at very low plate factors (which is, however, very difficult because of the exposure times involved) sharp Balmer line cores are seen (Greenstein and Peterson 1973; Greenstein *et al.* 1977) which can be used for accurate measurement. Altogether, fewer than 200 radial velocities are available (Trimble and Greenstein 1972).

The analysis of known radial velocities shows that the average absolute radial velocity $|\rho|$ is 27 km/s, with a range between 0 and 70 km/s, if some very large velocities are omitted. This implies that degenerates do not belong to an old population. Another important result is that in the analysis of radial velocities a new term has to be introduced. Usually the observed radial velocity (ρ_{obs}) is decomposed into a term reflecting the solar motion component ($V_0 \cos \lambda$) directed toward the apex and a peculiar component (ρ_p) thus:

$$\rho_{obs} = \rho_p + V_0 \cos \lambda$$

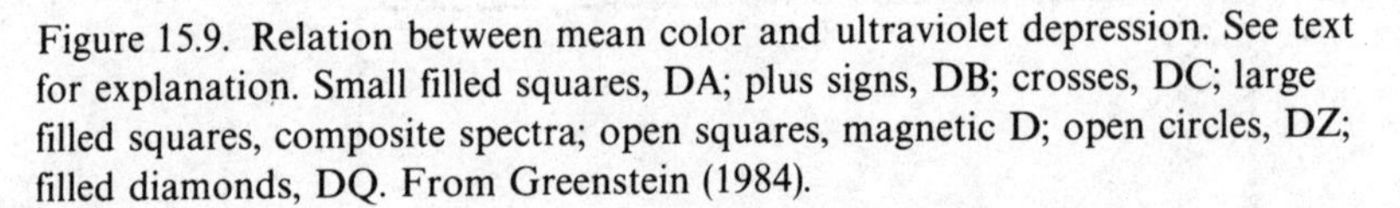

Figure 15.9. Relation between mean color and ultraviolet depression. See text for explanation. Small filled squares, DA; plus signs, DB; crosses, DC; large filled squares, composite spectra; open squares, magnetic D; open circles, DZ; filled diamonds, DQ. From Greenstein (1984).

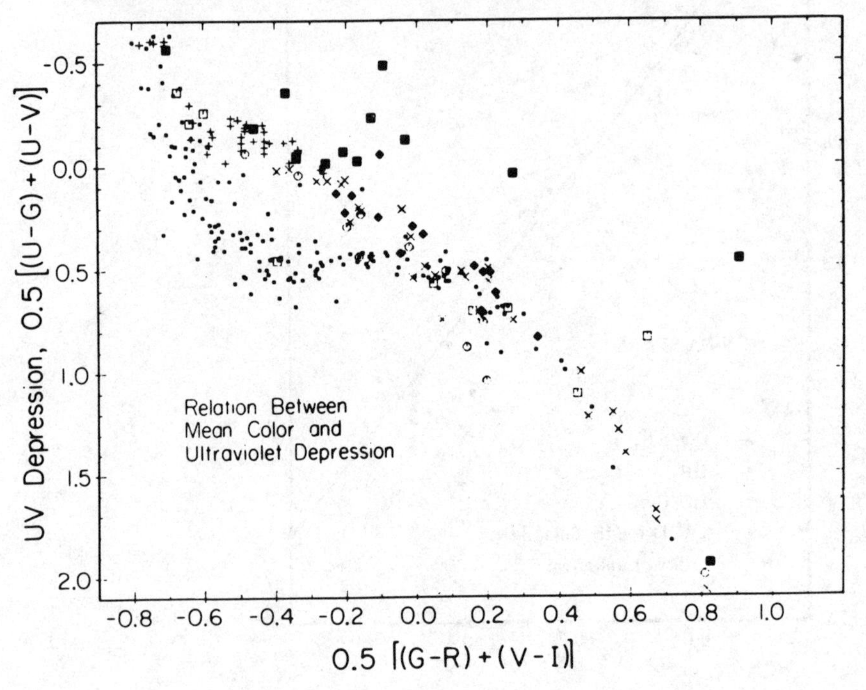

where ρ_p has a Maxwellian distribution. This formula applies to all kinds of stars except degenerates, for which a constant term 'K' has to be introduced:

$$\rho_{obs} = \rho_p + K + V_0 \cos \lambda$$

Greenstein *et al.* (1977) derive $K = 45$ km/s. Such a term is expected on the basis of special relativity and should have the value

$$K = 0.635\ M/R$$

M and R being the mass and radius of the degenerate, expressed in solar units. This relation introduces a direct link between radii and masses of degenerates.

Magnetic fields. Very large magnetic fields are observable in some degenerates. If the approximate formula (see section 2.3.3) is used,

$$\Delta\lambda \sim 10^{-12}\lambda^2 H$$

with $H \simeq 5 \times 10^6$G and $\lambda = 4867$ (Hβ), we obtain $\Delta\lambda \sim 120$ Å. With such a field we should thus observe Hβ split into components separated by about 120 Å. Such splitting can be observed in some stars, like GD 90 (Angel *et al.* 1974), for which λ4793, λ4857 and λ4918 are seen. Figure 15.10 reproduces a portion of the spectrum. Up to the present, magnetic fields have been observed in two dozen degenerates, with field strengths of up to 10^8 G. The majority of the degenerates however do not exhibit multiple Balmer lines, which implies that large magnetic fields are not common.

Binaries. Wide binaries can be detected by searching for nearby stars with similar proper motions, or by searching for faint companions of bright stars. The first technique was applied by Luyten (1969, 1979) who provides a long list of proper motion pairs. He remarks that binaries consisting of two degenerates are rare and that in the case of a dwarf–degenerate pair, the dwarf is usually the brighter component.

As for the fainter degenerate components of bright star systems, they seem to be present in many nearby systems – within 5 pc from the sun there are at least five, namely α CMa B, α CMi B, 40 Eri B, Stein and L 145–141.

Closer binaries are more difficult to detect. The most promising technique seems to be to search for stars with abnormal colors (Greenstein and Sargent 1974). The most favorable cases for detection are those in which a late (red) object combines with an (early) degenerate. Up to now eight spectroscopic binaries of this type are known (Lanning 1982) including some eclipsing binaries. Some more candidates were added by Probst (1983) who made infrared observations of degenerates to discover M-type companions. Besides these well behaved binaries there are also some strongly interacting close binaries, in which the degenerate component plays a key role, like the cataclysmic binaries in which the degenerate may or may not have a (strong) magnetic field. Many of these systems are X-ray emitters. If a strong magnetic field is present, the system is called an AM Her type object or polar.

Absolute magnitude. Although by definition degenerates are stars lying below the main sequence, the definition was hard to apply in practice because of the small number of stars with good parallaxes. There were 19 good parallaxes in 1965, about 50 in 1976 and about 80 in 1984. From these data (cluster stars and reliable trigononometric parallaxes) Greenstein (1984) derives

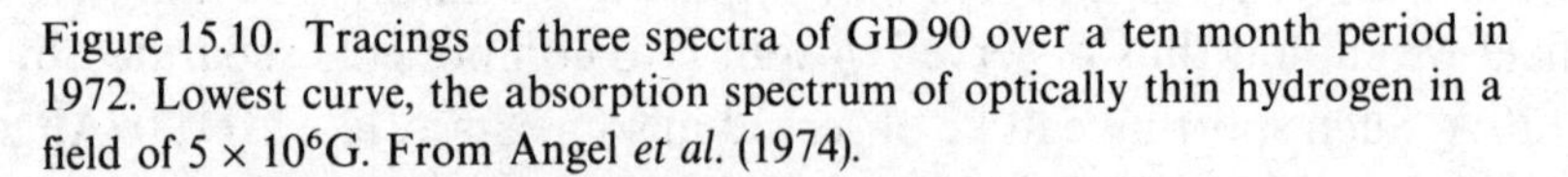

Figure 15.10. Tracings of three spectra of GD 90 over a ten month period in 1972. Lowest curve, the absorption spectrum of optically thin hydrogen in a field of 5×10^6G. From Angel *et al.* (1974).

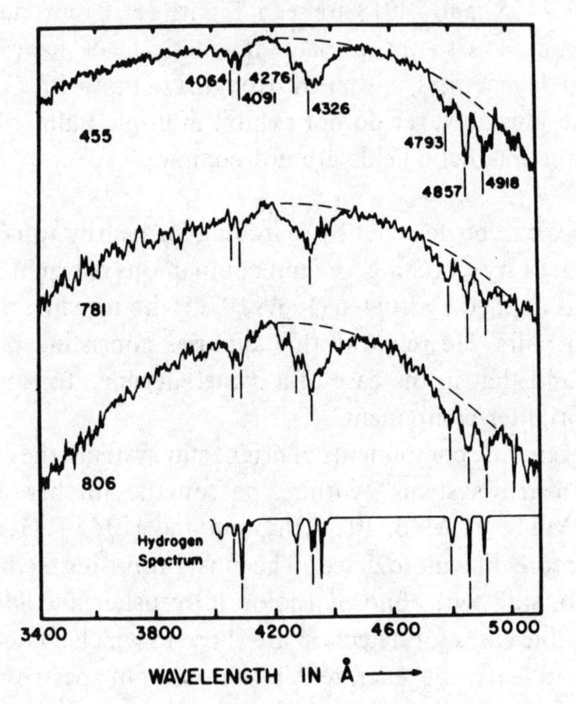

$$M_v = 13.03 + 3.11(\mathrm{G-R}) - 0.80(\mathrm{G-R})^2$$

This is approximately equivalent to

$$M_v = +6.77(\mathrm{B-V}) - 5.22(\mathrm{B-V})^2$$

and places M_v in the range $+11^m$ to $+16^m$; i.e. $8^m - 12^m$ below the main sequence. For the moment it seems that all degenerates fit the same relation, within $\pm 0\overset{m}{.}5$.

Degenerates in clusters. Degenerates are found in open clusters; for example, about twelve are known in the Hyades. Fewer degenerates have been found in about half a dozen other clusters and there may be some in a globular cluster (NGC 6752; Richer 1978).

Number of objects. The number of spectroscopically confirmed degenerates has been increasing regularly since their discovery. Three in 1917, 18 in 1939, 267 in 1953, and in 1985 there were about 1500 spectroscopically recognized degenerates. These are listed in a catalog by McCook and Sion (1977); a new edition has been announced. However, only ten degenerates are brighter than $V = 12$. Due to the different discovery techniques used, present surveys are sure to be incomplete, although more complete surveys of certain sky regions enable us to predict the total number of degenerates. The relative proportion of different subtypes is also biased. Sion (1984) shows for instance that for bright degenerates, the DA's are six times more numerous than non-DA's, whereas for fainter stars the ratio diminishes to about two.

Further reading. The review papers by Liebert (1980) and Angel (1978) are recommended, as well as various papers in the *IAU coll.* **53**, *White dwarfs and variable degenerate stars* (van Horn and Weidemann 1979).

15.1 ZZ Ceti stars

ZZ Ceti stars are pulsating degenerates of type DA which are rapidly variable and multiperiodic. Usually if the amplitude is small ($< 0\overset{m}{.}05$) the light curve is sinusoidal; if it is larger ($0\overset{m}{.}05$ to $0\overset{m}{.}3$) the oscillation is multiperiodic and non-sinusoidal. The periods fall in the range 1×12^2–1.2×10^3 seconds (McGraw 1979; Robinson 1979).

It seems that most if not all DA's in the region $-0.50 > \mathrm{U-B} > -0.68$ and $0.21 > \mathrm{B-V} > 0.15$ are variables (Fontaine *et al.* 1982). About 20 objects of this type are known.

References

Adams W.S. (1914) *PASP* **26**, 198

Angel J.R.P., Carswell R.F., Strittmatter P.A., Beavér E.A. and Harms R. (1974) *Ap. J.* **194**, L48

Angel J.R.P. (1978) *Ann. Rev. AA* **16**, 487

Eggen O.J. and Greenstein J.L. (1965) *Ap. J.* **141**, 83

Fontaine G., McGraw J.T., Dearborn D.S.P., Gustafson J., and Lacombe P. (1982) *Ap. J.* **258**, 651

Giclas H.L. (1971) *IAU Symp.* **42**, 24

Graham J.A. (1972) *A. J.* **77**, 144

Greenstein J.L. (1958) *Handbuch der Physik* **50**, 161, Springer Verlag

Greenstein J.L. (1960) in *Stars and stellar systems*, vol VI, Kuiper G.P. and Middlehurst B.M. (ed.), Univ. of Chicago Press.

Greenstein J.L. (1979) *Ap. J.* **233**, 239

Greenstein J.L. (1984) *Ap. J.* **276**, 611

Greenstein J.L. and Peterson D.M. (1973) *AA* **25**, 29

Greenstein J.L. and Sargent A.I. (1974) *Ap. J. Suppl.* **28**, 157

Greenstein J.L., Boksenberg A., Carswell R. and Shortridge K. (1977) *Ap. J.* **212**, 186

Humason M.L. and Zwicky F. (1947) *Ap. J.* **105**, 85

Kuiper G.P. (1940) *PASP* **53**, 248

Lanning H.H. (1982) *Ap. J.* **253**, 752

Liebert J. (1980) *Ann. Rev. AA* **18**, 363

Luyten W.J. (1952) *Ap. J.* **116**, 283

Luyten W.J. (1969) *Proper motion survey with the 48-inch Schmidt telescope*, No. XVIII

Luyten W.J. (1971) *IAU Symp.* **42**, 1

Luyten W.J. (1979) *Univ. Minnesota L.* II

McCook G.P. and Sion E.M. (1977) *Villanova Contr.* **2**

McGraw J.T. (1979) *Ap. J.* **229**, 203

Oke J.B. (1974) *Ap. J. Suppl.* **27**, 21

Probst R.G. (1983) *Ap. J. Suppl.* **53**, 335

Richer H.B. (1978). *Ap. J.* **224**, L. 9

Robinson E.L. (1979) *IAU Coll.* **53**, 343

Sion E.M. (1984) *Ap. J.* **282**, 612

Sion E.M., Greenstein J.L., Landstreet J.D., Liebert J., Shipman H.L. and Wegner G.A. (1983) *Ap. J.* **269**, 253

Stephenson C.B. (1971) *IAU Symp.* **42**, 61

Trimble V. and Greenstein J.L. (1972) *Ap. J.* **177**, 441

van Horn H.M. and Weidemann V. (eds.) (1979) *White dwarfs and variable degenerate stars, IAU coll.* **53**, Univ. of Rochester Press

Wegner G. (1979) *A.J.* **84**, 1384

16

Further developments

We shall examine in this last chapter some issues which are relevant to further progress in the field of classification. We shall group the issues into three sections. The first concerns the incorporation of additional information into the 'classical' scheme. The second is about groupings of superior hierarchical order. In the third section, we consider the future of classification.

16.1 Incorporation of new information

A question we have considered briefly in various chapters is the incorporation of new data into the classical scheme. To give an example, suppose that a large number of spectra covering the region $\lambda\lambda$1200–3000 became available and that we are interested in a particular group of objects, for instance HB stars. In the classical region ($\lambda\lambda$3600–4800) this is a homogeneous group (see section 10.4), but in the UV region we discover that half of the stars observed exhibit a feature at λ3040 not present in the other stars. A similar situation arises if some DC stars (i.e. degenerates having a continuous spectrum with no lines) are discovered in the UV to display carbon features. We could imagine these stars being studied in yet another region of the spectrum, for instance the 10–100 μm region, and finding there that an HB or a DC star has an infrared excess, indicating the presence of a circumstellar dust cloud.

The incorporation of such data into the 'classical' classification scheme is now urgent, because for the first time large amounts of information from other wavelength regions are available. In spite of this, the incorporation of new data is not a new problem. In fact, we already have information derived from ground-based optical observations which could be incorporated into a broader classification scheme. Up to the present, the problem has been given only occasional attention.

Of the several kinds of existing information, the most abundant is kinematic information. The fact that a star has low, intermediate or high velocity is of fundamental importance for the adjudication of an object to a given stellar population (see section 16.2).

Kinematic information is, however, not the only sort which should eventually be incorporated; we also have rotational velocities and magnetic fields. In addition, information about light variability or binarity characteristics is relevant to our knowledge of a star. We can imagine that each object can be described by a large (but finite) number of parameters, so that progress in astronomy is characterized by the speed with which our knowledge of all these different parameters for all stars is completed.

Before indulging in high flying speculation, let us examine how many data are available now. If we return to table 1.1 of chapter 1, we see that kinematic data exist for some 10^5 stars, whereas other data exist for only 10^3 stars. To this we can add figure 16.1, taken from Ochsenbein (1980). The figure shows rather convincingly that even basic information is rather scanty.

Since we are so far from having complete data for a large set of stars, the attention of astronomers has focused instead on two smaller samples of stars for which we can expect eventually to have a complete set of data.

The first sample is that of the 'bright stars', with $V \leqslant 6^{m}5$ and the second

Figure 16.1. Percentage of stars with simultaneously known data. μ, proper motion; HD, unidimensional spectral type; MK, two-dimensional spectral type; RV, radial velocity; UBV, UBV photometric data. Percentages with respect to $N = 187\,620$. Data from Ochsenbein (1980).

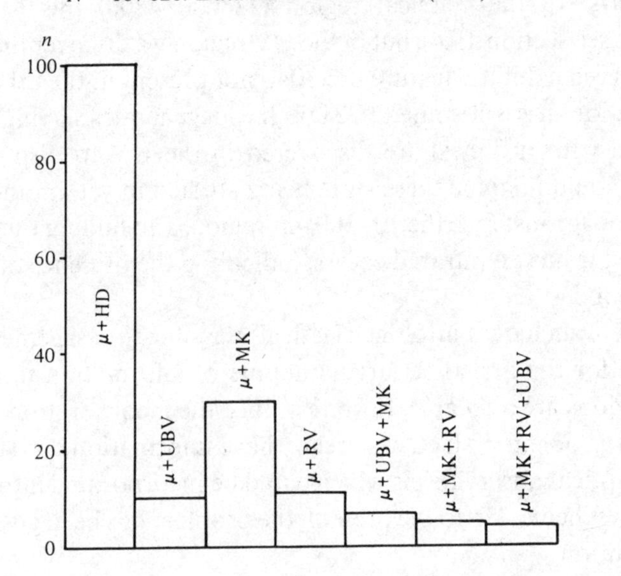

that of 'nearby stars', with distances $r \leqslant 20$ pc. Data on the first sample were collected by Hoffleit and Jaschek (1982) and on the second sample by Gliese (1969) and Gliese and Jahreiss (1979). However, even in these two restricted samples, the data are still incomplete. The incompleteness of the first sample was examined by Hoffleit and Jaschek (see their preface) and for the nearby stars this was done by Jahreiss and Gliese (1985) and Hauck and Mermilliod (1984).

On the other hand, a perusal of either catalog shows that the data concerning each object can hardly to be expected to fit into a short additional classification index. The solution to such a problem seems simple – namely to store all available information in a data center, where it can be easily accessed. The databank SIMBAD (Set of Identifications, Measurements and Bibliography for Astronomical Data) of the Centre de Données Stellaires at Strasbourg is one such solution. SIMBAD contains at present information about 6×10^6 stars; the information and the bibliographic references (to papers in which the objects were studied) are regularly updated. For a description of SIMBAD see Egret (1983) and the various issues of the *Bulletin d'information du Centre de Donnée Stellaires.* Efforts are being made to make SIMBAD available to an increasing number of astronomers.

Let us now return to the problem of incorporating additional data into the classification. We have seen that additional data, such as kinematics, photometry, rotation, magnetic field, binary nature, variability, etc. can be stored in databanks. What is left is the problem of how to incorporate into spectral classification significant data from other wavelength ranges. As we have seen, the amount of such data is not yet large, but it is growing fast.

The best solution for the moment seems to be 'notes' appended to the spectral classification. Notes should always refer to observed characteristics and not to their interpretation, because the latter tends to vary with time. So a C IV strong absorption line at $\lambda 1549$ is an observed fact which may be important, whereas its interpretation ('indicator of superionization') is much less certain. Notes should be clear, specific and, if they are from other sources, they should carry complete bibliographic references. Notes like 'peculiar spectrum in the UV' are worse than useless, because it is unclear what we mean by 'UV' or 'peculiar'; similarly with a note 'discovered by Bidelman' without any further reference.

Notes present no problems provided they refer to small numbers of objects – say fewer than 10% of the stars of a catalog. They become cumbersome when they affect more than 10% of the stars. Time is then ripe to create a new set of (additional) parameters for classification.

Actual classification should *not* include elements coming from other wavelength ranges. A B0 III star classified as such in the classical region (which in the ultraviolet range $\lambda < 3000$ Å seems to be a B0 supergiant) *must* be quoted as B0 III if the catalog refers to classification in the Yerkes system.

We need to be strict on such matters in order to prevent total confusion over what is observed; the supergiant appearance in the ultraviolet must find its place in a note. Systems which disobey these rules, like the classification system for degenerates (see section 15.0), generate confusion.

For photometric classification, the Geneva–Lausanne group has a project (at a well advanced stage) to establish a catalog which collects all references to all photometric systems in which a given star has been observed. This catalog will thus provide access to all known photometric information (Mermilliod 1984).

16.2 Groupings of superior order

We described in chapter 1 the classification methods in biology, with its grouping into species and genera. Naturally we can ask if further groupings exist above these two. In biology the answer is clear – we have the kingdom, the class, the order, the genus and the species.

We can ask similarly if the groups we introduced in stellar classification can be further synthetized. The answer is 'yes': such syntheses have been introduced. To facilitate their description, let us consider two approaches to the problem, the 'partial' and the 'general'.

We call 'partial' groupings those in which a few groups lying in related spectral types are grouped together. 'General' groupings on the other hand are those in which large groups are synthetized into a new one.

Partial groupings. One partial grouping which has already been discussed in section 9.5 is that of 'chemically peculiar stars'. Preston (1974) groups under this title all early type main sequence stars identified by the presence of anomalously strong (or weak) absorption lines of certain elements in their spectra. His scheme is shown in table 16.1, taken from this paper.

The scheme is a synthesis of previously disconnected groups, the common trait is that they satisfy the spectroscopic definition given above.

A similar group of 'peculiar late type giants' could be defined in a similar way: 'late type giants identified by the presence of anomalously strong (or weak) absorption lines of certain elements in their spectra'. A list of the

Table 16.1. *Classification scheme for chemically peculiar (CP) stars.*

CP group	Classical name	Discovery criteria	Other properties			Temperature domain
			Rotation	Binary frequency	Binary Periods	
1	Metallic line (Am)	Weak Ca II and/or Sc II; enhanced heavy metals	Slow	High	Abnormal	7000–10 000 K
2	Magnetic Ap	Enhanced Si, Cr, Sr, Eu, etc.	Slow	Low	Abnormal	8000–15 000 K
3	Hg–Mn	Enhanced Hg II λ3984, Mn II	Very slow	Normal?	Abnormal?	10 000–15 000 K
4	He weak	Q(Sp) > Q(UBV)	Slow?	?	?	13 000–20 000 K

subgroups has been given in section 13.0 and illustrated in figure 13.3, so we shall not provide further details.

A common trait of both previous groups is that they unite several small subsets of normal stars – 'small' here implies $\leqslant 10\%$.

To the two groupings we considered we might add the 'natural groups' discussed in section 3.6, which represent the coalescence of various nearby groups into a larger one when a larger plate factor, i.e. lower resolution, is used. But this is really a 'change in optics' rather than a grouping of the same type as the peculiar red giants.

General groupings. The most general groupings are related to 'stellar populations'. Since this term meant different things to different scientists, we must retrace a little of the history of the term.

The concept of 'stellar populations' was introduced by Baade (1944). He states in the conclusion of his paper that 'in dealing with galaxies we have to distinguish two types of stellar populations, one which is represented by the ordinary HR diagram (type I), the other by the HR diagram of the globular clusters (type II). Characteristic of the first type are highly luminous O- and B-type stars and open clusters; of the second, globular cluster and short period Cepheids'.

The term 'population' is used as an abbreviation for the membership of stars to a given type of HR diagram. Since some stars may appear at the same location in the two-dimensional HR diagram but belong to either population I or population II – for instance K-type dwarfs – it is clear that the attribution of a star to a population has a statistical meaning, and cannot be carried out unambiguously for each object, if only absolute magnitude and a color index are used.

Through subsequent theoretical work, the different types of HR diagrams were found to correspond to stars of different age and composition. Ten years later, at the Vatican meeting, great advances had been made and Schwarzschild (1958) commented that 'all stars we now observe differ from another in three basic characteristics, their mass, their initial composition and their age'. He asks then that if all these stars are classified into stellar populations, which of the three basic characteristics is the one in which stars from one population differ from those of another. In his opinion the main parameter is age, although age and initial composition are closely linked. Because of this we should expect to find a continuous sequence of stellar populations. Schwarzschild describes such a scheme, which is reproduced in table 16.2.

Table 16.2. *Age sequence of stellar populations.*

Population	Typical members	Age (10^9 years)	Heavy element abundance	Velocity dispersion (km/s)	Subsystem
Young population I	Young galactic clusters	0–1	0.04	10	Flat
Intermediate population I	'Strong line' stars	1–3	0.03	20	
Old population I	'Weak line' stars	3–5.5	0.02	30	Intermediate
Mild population II	Oort's 'high velocity stars'	5.5–6	0.01	50	
Extreme population II	Globular clusters	6–6.5	0.003	130	Spherical

At the same conference Oort (1958) proposed a slightly different scheme given in table 16.3.

The examples which are quoted in table 16.3 ('typical members') are mainly selected by spectroscopic criteria (bright red giants, A-type stars, strong line stars etc) or by light curve characteristics (RR Lyr, nova, etc.) although 'high velocity stars' are selected by a third, kinematic criterion. This latter group, as we have seen in section 11.5, is however not very homogeneous.

The two preceding schemes are based upon the assumption of a strong coupling between composition (i.e. abundance of metals with respect to hydrogen) and age. It is not the place here to recount the history of the subject, which can be followed in the review paper by King (1971). Let us summarize here the post-Vatican situation by saying that the assumption of strict coupling was shown to be incorrect in the next decade. So for instance globular cluster stars, which according to tables 16.2 and 16.3 should have low metallicity, were found to have a whole variety of metallicities, which contradicts the 'strong coupling' hypothesis. The interpretations of these facts are complicated because the position of an object in the HR diagram is fixed by its internal evolution, whereas what we observe is the composition of the outer atmosphere, which may or may not be linked strongly to the former. Let us also notice in passing that the derivation of composition effects ('metallicity') for a group of stars is made in successive steps. First the composition of a few 'typical' objects is established by means of a stellar atmosphere analysis. Then the metallicity effects are measured in a convenient photometric system. The third step consists of calibrating these indices on the compositions established in the first step, and the last step applies the result to the group members. It is clear that the definition of the group and the selection of 'typical' members is a subject for stellar classification, which in most cases is carried out by spectral classification.

More recent reviews of the subject (for instance Mould 1982) use an even more general definition of stellar populations. He writes: 'A stellar population is a collection of stars of similar age, composition and kinematics.' Although to a certain extent this was inherent in the schemes of tables 16.2 and 16.3, it is here spelled out more clearly. Notice that the definition is difficult to apply in practice, because only kinematics can be obtained more or less directly from observations, whereas age and composition are dependent upon a chain of steps involving complex interpretations. Since we have already described the essential steps in deriving 'composition', let us do the same for 'age'. Dating an object starts by relating a position in the HR diagram to an age. Such a relation is, strictly speaking, not unambiguous,

Table 16.3. *Stellar populations.*

Population	Halo population II	Intermediate population II	Disc population		Older population I	Extreme population I
Some typical members	Subdwarfs; globular cluster RR Lyr with $P > 0.3$	High-velocity F–M; long-period variables	Planetary nebulae bright red giants; novae	Weak line stars	Strong line stars; Me dwarfs	Gas supergiants
$\lvert z \rvert$ (parsecs)	2000	700	450	300	160	120
$\lvert Z \rvert$ (km/s)	75	25	18	15	10	8
Axial ratio	2	5	25	?	?	100
Concentration toward center	Strong	Strong	Strong	?	Little	Little
Distribution	Smooth	Smooth	Smooth	?	Patchy, spiral arms	Extremely patchy, spiral arms
Metal abundance (Schwarzschild)	0.0003	0.01	—	0.02	0.03	0.04
Age (10^9 years)	6	6.0 to 5.0	5	1.5 to 5	0.1 to 1.5	<0.1
Total mass (10^{10})	16	47	47	47	5	2

Note: z, height above galactic plane; Z, velocity component perpendicular to the galactic plane; masses are expressed in solar units.

since an object may cross the same point of the HR diagram more than once during its evolution. Besides this difficulty of principle, there are several practical difficulties in applying the above calibration. The first is that the calibration itself depends upon composition; the second is that the luminosity has to be known quite precisely – large uncertainties in M produce large incertitudes in age. We have stressed several times in this book the uncertainties in present-day calibrations. Locations in the HR diagram are thus uncertain, except if the objects lie in clusters, and ages share the same fate. Field stars are therefore labelled with the ages and compositions derived for a few well studied members of their groups. The selection of the group member is done by classification methods, either spectroscopic or photometric or by a combination of both.

On the top of these difficulties we have still to decide which one of the three properties used to define populations (age, composition, kinematics) is to be used in practice with the greatest weight. Let us take the case of those O-type stars with normal spectra and high velocity. If the composition is normal and they are of type O, they are probably young and of solar composition. They would thus be population I, were it not for the velocity. From the strict point of definition, we can only conclude that they do not belong to the same population, but we do not know how to assign them to a group within the scheme.

In recent years it seems that the emphasis is being put mainly on composition, so that the populations are differentiated by their composition, although it is difficult to find precise definitions. A consensus seems to be established on three populations given in table 16.4.

These definitions are not used the same way by all authors. Other authors prefer to place the limit of halo stars at [Fe/H] = – 1 and of normal stars at – 0.4; one has to find out in each case what definition an author used.

A question of considerable practical importance is how to position a group of stars in this scheme, if the group does not belong to the few ones

Table 16.4. *Populations.*

Population	Other names	[Fe/H]
Young	Young disc	> – 0.2
Intermediate	Old disc	– 0.2 to – 0.6
Old	Halo	– 0.6 to – 2.0
III	Extreme halo	– 2.0 to – 4.0

listed as examples. The answer is implicit in the definitions and in what we said earlier. The kinematic properties, composition and age have to be determined. To do so we need a knowledge of proper motions, radial velocities and distances (individual or statistical), photometric parameters (for the determination of absolute magnitude and of composition) and spectral classification. Since such a number of data are hard to obtain even for small groups of stars, authors are most often content to study these parameters in a small number of objects (10 or 20) and to apply the result to the whole group defined either spectroscopically or photometrically or by a combination of both methods.

Users of comprehensive schemes, like those set forth in studies of populations, should be aware of the fact that many lofty speculations are still based largely on the results of visual inspection of spectrograms.

16.3 The future of classification methods

As the example of other natural sciences shows, classification methods share the general growth of scientific output. New times bring new problems or shed new light on old problems, and despite assertions that classification problems belong to the past they will be still with us for some time. The reason for such optimism lies in the simple fact that the number of stars in our own galaxy is of the order of 10^{11}, a number which is too large even to dream of studying them one by one; and this completely ignores the stars in other galaxies.

With present-day techniques, spectra can be obtained for an object with $V = 15$, but even if all the existing telescopes worked full time throughout the year they would be capable of observing only an insignificant portion of the 4×10^8 stars of $V \leqslant 15$.

Assuming thus the survival of classification, let us ask what will probably be the great problems of the near future.

The easiest prediction is that more effort will go into automated classification, since this is the only way of handling very large quantities of data. The classification by Cannon of the 2.2×10^5 stars contained in the Henry Draper Catalogue, and its reclassification by Houk, probably represents the order of magnitude of what a single astronomer can do – a project of classifying 10^6 spectra seems clearly out of question for any individual.

We mentioned automated methods in section 2.6, and add here recent work done along these lines. Classification methods in the statistical sense were discussed at length in the meeting on *Statistical methods in astronomy*

(1983), with emphasis on applications. Other papers can be found in a meeting on *Automated classification methods* (1982). A serious attempt to define spectral groups without knowledge of existing classifications was carried out by Heck *et al.* (1986). They used ultraviolet spectra, taken by the IUE satellite in the range $\lambda\lambda$1200–2500 and having a resolution of 2 Å. So far the mathematical procedures used have been successful in that they have been able to reproduce a scheme set up through visual examination of the spectra.

It is clear that such attempts were mostly done to test a particular methodology; what is still needed is to compare their output with that of visual classifiers, comparing several thousand spectra. Such a comparison would help to define accurately the limitations of both approaches.

A second prediction is that in the near future every astronomer will be linked to a (regional) data center. This will mean that we can use all the known information on the star we are interested in, whether it be photometric, spectroscopic or kinematic. Observe that such information is rarely consulted if it involves using manually a series of catalogs in which any object may be designated with half a dozen different names. Instant availability of the necessary data will permit checks of the data, a judgement of their compatibility, a critical evaluation of their precision and perhaps recalibrations. This represents a deep revolution in data handling, comparable only to the introduction of printed books, or the advent of computers.

A third prediction is that with the opening of new wavelength regions through satellite astronomy, we shall have to reexamine the existing classification system critically to determine whether it can be used to describe correctly the information coming from other wavelength regions. Consequently spectral classification in the future will deal predominantly with data from these 'new regions', with the clear mandate to produce new 'standard stars' for these regions.

It seems probable that much more emphasis shall be put on statistical techniques for data description to analyse the information provided by classification techniques. Curiously, classification methods in astronomy have largely ignored a number of well-known statistico-mathematical techniques readily available. Their use could have a large impact on photometric classification methods which adapt easier to numerical treatment than the discontinuous spectral classification techniques. It seems reasonable to think that rapid progress can be achieved in this field.

We have left aside up to now what can be expected from the traditional spectral classification methods – i.e. visual examination of spectra covering the classical region. Here also it is clear that we are facing a set of new problems. First of all, thanks to modern technologies, we can observe the brightest stars in the nearby galaxies and we are confronted with the crucial problem whether these stars are similar – and to what extent – to the intrinsically brightest stars in our galaxy. If they are similar we can use the Yerkes system to classify them – if not, what kinds of modifications are to be introduced? The same thing is also true for stars in globular clusters and/or in open clusters – here again recent years have provided the first large amount of spectra of cluster objects. Are they identical to the nearby field stars? Do we have to set up a new classification system for population II?

Another series of problems derives from the availability of lower plate factors for fainter stars. If we were able to reclassify all brighter stars at 20 Å/mm, would the reclassification be compatible with the Yerkes system, or would we discover finer subdivisions in luminosity, or other composition indicators? The importance of lower plate factors is pointed out by Keenan (1984) who said: 'If I could start my life over again, I would undertake spectral classification at resolution no worse than 1 Å, and as much better than that as possible.'

The series of new problems seems endless and the most exciting developments seem to be 'just around the corner'.

References

Baade W. (1944) *Ap. J.* **100**, 137

Eggen O. (1983) *Ap. J. Suppl.* **51**, 183

Egret D. (1983) *BICDS* **24**, 109

Gliese W. (1969) *Veroff Heidelberg N.* **22**

Gliese W. and Jahreiss H. (1979) *AA Suppl.* **38**, 423

Hauck B. and Mermilliod J.C. (1984) *BICDS* **26**, 17

Heck A., Egret D., Nobelis P. and Turlot J.C. (1986) *Astroph. Sp. Sc.* **120**, 223

Hoffleit D. and Jaschek C. (1982) *The bright star catalogue*, 4th edn., Yale Univ.

Jahreiss H. and Glies W. (1985) *BICDS* **28**, 19

Keenan P.C. (1984) *The MK process and stellar classification* (ed. Garrison R.), p. 29

King I. (1971) *PASP* **83**, 377

Mermilliod J.C. (1984) *BICDS* **26**, 3

Mould J.R. (1982) *Ann. Rev. AA* **20**, 91

Ochsenbein F. (1980) Ph. D. thesis, Strasbourg

Oort J.H. (1958) *Ric. Astr.* **5**, 415

Preston G. (1974) *Ann. Rev. AA* **12**, 257

Schwarzschild M. (1958) *Ric. Astr.* **5**, 204

Automated classification methods (1982) *BICDS* **23**

Statistical methods in astronomy (1983) *Strasbourg Coll.* ESA SP. 201

Index

Detailed subdivisions of each group are given for the main groups (O, B, A, . . .) and some of the larger groups (Be, C, WR). For all other groups no subdivision according to subjects is provided.